Graduate Texts in Mathematics **34**

Editorial Board

F. W. Gehring

P. R. Halmos
Managing Editor

C. C. Moore

Frank Spitzer

Principles of Random Walk

Second Edition

Springer-Verlag
New York   Heidelberg   Berlin
1976

Frank Spitzer
Cornell University
Department of Mathematics
Ithaca, New York 14850

Editorial Board

P. R. Halmos
Managing Editor
Indiana University
Department of Mathematics
Swain Hall East
Bloomington, Indiana 47401

F. W. Gehring
University of Michigan
Department of Mathematics
Ann Arbor, Michigan 48104

C. C. Moore
University of California
at Berkeley
Department of Mathematics
Berkeley, California 94720

AMS Subject Classification
60J15

Library of Congress Cataloging in Publication Data

Spitzer, Frank Ludvig, 1926–
  Principles of random walk.

  (Graduate texts in mathematics; 34)
  First edition published by D. Van Nostrand, Princeton, N.J.,
in series: The University series in higher mathematics,
edited by M. A. Stone, L. Nirenberg, and S. S. Chern.
  Bibliography: p. 395
  Includes index.
  1. Random walks (Mathematics) I. Title.
II. Series.
QA274.73.S65 1975   519.2'82   75-26883

All rights reserved.

No part of this book may be translated or reproduced
in any form without written permission from Springer-Verlag.

© 1964, by Frank Spitzer

Printed in the United States of America.

ISBN 0-387-90150-7 Springer-Verlag New York

ISBN 3-540-90150-7 Springer-Verlag Berlin Heidelberg

PREFACE TO THE SECOND EDITION

In this edition a large number of errors have been corrected, an occasional proof has been streamlined, and a number of references are made to recent progress. These references are to a supplementary bibliography, whose items are referred to as [S1] through [S26].

A thorough revision was not attempted. The development of the subject in the last decade would have required a treatment in a much more general context. It is true that a number of interesting questions remain open in the concrete setting of random walk on the integers. (See [S19] for a recent survey). On the other hand, much of the material of this book (foundations, fluctuation theory, renewal theorems) is now available in standard texts, e.g. Feller [S9], Breiman [S1], Chung [S4] in the more general setting of random walk on the real line. But the major new development since the first edition occurred in 1969, when D. Ornstein [S22] and C. J. Stone [S26] succeeded in extending the recurrent potential theory in Chapters II and VII from the integers to the reals. By now there is an extensive and nearly complete potential theory of recurrent random walk on locally compact groups, Abelian ([S20], [S25]) as well as non-Abelian ([S17], [S2]). Finally, for the non-specialist there exists now an unsurpassed brief introduction to probabilistic potential theory, in the context of simple random walk and Brownian motion, by Dynkin and Yushkevich [S8].

In view of the above mentioned developments it might seem that the intuitive ideas of the subject have been left far behind and perhaps lost their vitality. Fortunately this is false. New types of random walk problems are now in the stage of pioneering work, which were unheard of when the first edition appeared. This came about because the simple model of a single particle, performing a random walk with given transition probabilities, may be regarded as a crude approximation to more elaborate random walk models. In one of these a single particle moves in a random environment, i.e. the transition probabilities are themselves random variables. In other models one considers the simultaneous random walk of a finite or even infinite system of particles, with certain types of interaction between the particles. But this is an entirely different story.

PREFACE TO THE FIRST EDITION

This book is devoted exclusively to a very special class of random processes, namely to *random walk* on the lattice points of ordinary Euclidean space. I considered this high degree of specialization worth while, because the theory of such random walks is far more complete than that of any larger class of Markov chains. Random walk occupies such a privileged position primarily because of a delicate interplay between methods from *harmonic analysis* on one hand, and from *potential theory* on the other. The relevance of harmonic analysis to random walk of course stems from the invariance of the transition probabilities under translation in the additive group which forms the state space. It is precisely for this reason that, until recently, the subject was dominated by the analysis of characteristic functions (Fourier transforms of the transition probabilities). But if harmonic analysis were the central theme of this book, then the restriction to random walk on the integers (rather than on the reals, or on other Abelian groups) would be quite unforgivable. Indeed it was the need for a self-contained elementary exposition of the connection of harmonic analysis with the much more recent developments in potential theory that dictated the simplest possible setting.

The potential theory associated with Markov processes is currently being explored in the research literature, but often on such a high plane of sophistication, and in such a general context that it is hard for the novice to see what is going on. Potential theory is basically concerned with the probability laws governing the time and position of a Markov process when it first visits a specified subset of its state space. These probabilities satisfy equations entirely analogous to those in classical potential theory, and there is one Markov process, namely *Brownian motion*, whose potential theory is exactly the classical one. Whereas even for Brownian motion the study of absorption probabilities involves delicate measure theory and topology, these difficulties evaporate in the case of random walk. For arbitrary subsets of the space of lattice points the time and place of absorption are automatically measurable, and the differential equations encountered in the study of Brownian motion reduce to difference equations for random walk. In this sense the study of random walk leads one to potential theory in a very simple setting.

One might ask whether the emphasis on potential theory is a natural step in the development of probability theory. I shall try to give a brief but affirmative answer on two different grounds.

(a) After studying the probability laws governing a Markov process at fixed (non-random) times, one usually introduces the notion of a stopping time. (We do so in definition 3 of section 3, i.e., in D3.3) A stopping time $\mathbf{T}$ is a random variable such that the event $\mathbf{T} > t$ depends only on the past of the process up to time t. From that point on we are concerned with a new process, which is precisely the original process, stopped at time $\mathbf{T}$. But unfortunately one cannot say much about this stopped process unless it happens to be Markovian, with transition probabilities invariant under translation in time. Hence one is led to ask: *For what stopping times is the stopped process of such a simple type?* This question leads directly to potential theory, for it is easy to see that *the stopped process is Markovian with stationary transition probabilities if and only the stopping time is of the type:* $\mathbf{T} = $ *time of the first visit of the process to a specified subset of its state space.*

(b) Classical Newtonian potential theory centers around the Green function (potential kernel) $G(x,y) = |x-y|^{-1}$, and in logarithmic potential theory (in the plane) this kernel is replaced by $A(x,y) = \ln|x-y|$. As we shall see, both these kernels have a counterpart in the theory of random walk. For transient random walk $G(x,y)$ becomes the expected number of visits to y, starting at x. For recurrent random walk there is a kernel $A(x,y)$ such that $A(x,0) + A(0,y) - A(x,y)$ represents the expected number of visits to y, starting at x, before the first visit to 0. It is hardly an oversimplification, as we shall see, to describe the potential theory of random walk as the study of existence, uniqueness, and other basic properties such as asymptotic behavior, of these kernels. That raises a natural question: *How much can one learn about a random walk from its potential kernel?* The answer is: In principle, everything. Just as the characteristic function (Fourier transform of the transition function) uniquely determines the transition function, and hence the random walk, so we shall find (see problems 13 in Chapter VI and 8 in Chapter VII) that *a random walk is completely determined by its potential kernel.*

I am uncertain about the "prerequisites" for this book, but assume that it will present no technical difficulties to readers with some solid experience and interest in analysis, say, in two or three of the following areas: probability theory, real variables and measure, analytic functions, Fourier analysis, differential and integral operators. I am painfully

PREFACE TO THE FIRST EDITION

aware, however, of a less tangible prerequisite, namely the endurance required by the regrettable length of the book. In view of the recent vintage of most of the material many examples and extensions of the general theory seemed sufficiently full of vitality to merit inclusion. Thus there are almost 100 pages of examples* and problems, set apart in small print. While many are designed to clarify and illustrate, others owe their inclusion to a vague feeling that one should be able to go farther, or deeper, in a certain direction. An interdependence guide (following the table of contents) is designed to suggest paths of least resistance to some of the most concrete and intuitive parts of the theory—such as simple random walk in the plane (section 15), one-sided absorption problems, often called fluctuation theory (Chapter IV), two-sided absorption problems (Chapter V), and simple random walk in three-space (section 26).

Since most of my work as a mathematician seems to have found its way into this book, in one form or another, I have every reason to thank those of my teachers, Donald Darling, William Feller, and Samuel Karlin, who introduced me to the theory of stochastic processes. I also owe thanks to J. L. Doob for his suggestion that I plan a book in this area, and to the National Science Foundation for enabling me to spend the year 1960-61 in Princeton, where an outline began to take form under the stimulating influence of W. Feller, G. Hunt, and D. Ray. In the fall of 1961 much of the material was presented in a seminar at Cornell, and I owe a great debt to the ensuing discussions with my colleagues. It was particularly fortunate that H. Kesten's interest was aroused by some open problems; it was even possible to incorporate part of his subsequent work to give the book a "happy ending" in the form of T32.1 in the last section. (As explained in the last few pages of Chapter VII this is not really the end. Remarkably enough one can go further, with the aid of one of the most profound inventions of P. Lévy, namely the theory of the dispersion function in Lévy's Théorie de l'addition des variables aléatoires (1937), which was the first modern book on the subject of random walk.)

Finally it is a pleasure to thank all those who were kind enough to comment on the manuscript. In particular, J. L. Doob, H. Kesten, P. Schmidt, J. Mineka, and W. Whitman spotted a vast number of serious errors, and my wife helped me in checking and proofreading.

---

*Throughout the book theorems are labeled T, propositions P, definitions D, and examples E. Theorem 2 in section 24 will be referred to as T2 in section 24, and as T24.2 when it is mentioned elsewhere in the text.

TABLE OF CONTENTS

CHAPTER I. THE CLASSIFICATION OF RANDOM WALK   PAGE

1. Introduction   1
2. Periodicity and recurrence behavior   14
3. Some measure theory   24
4. The range of a random walk   35
5. The strong ratio theorem   40
   Problems   51

CHAPTER II. HARMONIC ANALYSIS

6. Characteristic functions and moments   54
7. Periodicity   64
8. Recurrence criteria and examples   82
9. The renewal theorem   95
   Problems   101

CHAPTER III. TWO-DIMENSIONAL RECURRENT RANDOM WALK

10. Generalities   105
11. The hitting probabilities of a finite set   113
12. The potential kernel $A(x,y)$ 121
13. Some potential theory   128
14. The Green function of a finite set   140
15. Simple random walk in the plane   148
16. The time dependent behavior   157
    Problems   171

CHAPTER IV. RANDOM WALK ON A HALF-LINE

17. The hitting probability of the right half-line   174
18. Random walk with finite mean   190
19. The Green function and the gambler's ruin problem   205
20. Fluctuations and the arc-sine law   218
    Problems   231

CHAPTER V. RANDOM WALK ON A INTERVAL

		PAGE
21.	Simple random walk	237
22.	The absorption problem with mean zero, finite variance	244
23.	The Green function for the absorption problem	258
	Problems	270

CHAPTER VI. TRANSIENT RANDOM WALK

24.	The Green function $G(x,y)$	274
25.	Hitting probabilities	290
26.	Random walk in three-space with mean zero and finite second moments	307
27.	Applications to analysis	322
	Problems	339

CHAPTER VII. RECURRENT RANDOM WALK

28.	The existence of the one-dimensional potential kernel	343
29.	The asymptotic behavior of the potential kernel	352
30.	Hitting probabilities and the Green function	359
31.	The uniqueness of the recurrent potential kernel	368
32.	The hitting time of a single point	377
	Problems	392
	BIBLIOGRAPHY	395
	SUPPLEMENTARY BIBLIOGRAPHY	401
	INDEX	403

INTERDEPENDENCE GUIDE

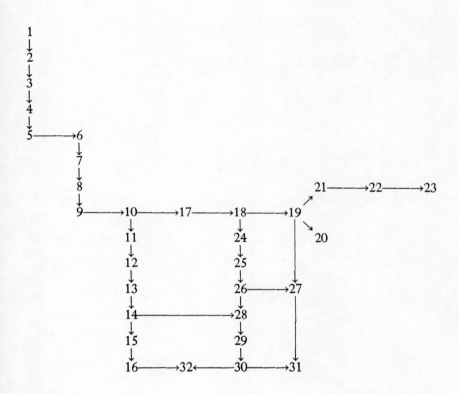

Chapter I

THE CLASSIFICATION OF RANDOM WALK

1. INTRODUCTION

The simplest definition of random walk is an analytical one. It has nothing to do with probability theory, except insofar as probabilistic ideas motivate the definition. In other words, probability theory will "lurk in the background" from the very beginning. Nevertheless there is a certain challenge in seeing how far one can go without introducing the formal (and formidable) apparatus of measure theory which constitutes the mathematical language of probability theory. Thus we shall introduce measure theory (in section 3) only when confronted by problems sufficiently complicated that they would sound contrived if expressed as purely analytic problems, i.e., as problems concerning the transition function which we are about to define.

Throughout the book R *will denote the space of d-dimensional integers*. In other words R is the set of ordered d-tuples (lattice points)

$$x = (x^1, x^2, \ldots, x^d), \qquad x^i = \text{integer for } i = 1, 2, \ldots, d.$$

As soon as we have defined what is meant by a random walk, it will be natural to call R the *state space* of the random walk.

For each pair x and y in R we define a real number $P(x,y)$, and the function $P(x,y)$ will be called the *transition function* of the random walk. It is required to have the properties

(1) $$0 \leq P(x,y) = P(0, y-x), \qquad \sum_{x \in R} P(0,x) = 1.$$

The most restrictive of these properties perhaps is the *spatial homogeneity* expressed by $P(x,y) = P(0, y-x)$, where, of course, $y - x$ is the point in R with coordinates $y^i - x^i$, $i = 1, 2, \ldots, d$. It shows that the transition function is really determined by a single function $p(x) = P(0,x)$ on R with the properties

$$0 \le p(x), \qquad \sum_{x \in R} p(x) = 1.$$

In other words, specifying a transition function is equivalent to specifying a probability measure on R (a non-negative function $p(x)$ whose sum over R is one).

Now we are finished—not in the sense that there is no need for further definitions, for there is, but in the sense that all further definitions will be given in terms of $P(x,y)$. We may even say, informally, that *the transition function $P(x,y)$ is a random walk*. Quite formally, we *define a random walk as a function $P(x,y)$ possessing property* (1) *defined for all pairs x,y in a space of lattice points R*, and a random walk is said to be *d-dimensional* if the dimension of R is d.[1]

E1 The so-called *simple random walks* constitute a particularly important class of random walks. If R is d-dimensional, let

$$|x| = \left[\sum_{i=1}^{d} (x^i)^2 \right]^{1/2}$$

denote the Euclidean distance of the point x from the origin. Then $P(0,x)$ defines *d-dimensional simple random walk* if

$$P(0,x) = \frac{1}{2d} \quad \text{when } |x| = 1,$$
$$ = 0 \quad \text{when } |x| \ne 1.$$

When $d = 1$, a somewhat wider class than simple random walk is of considerable interest. When

$$P(0,1) = p, \quad P(0,-1) = q, \quad p \ge 0, \quad q \ge 0, \quad p + q = 1,$$

we shall call $P(x,y)$ the transition function of *Bernoulli random walk*. Since $P(0,x)$ corresponds to our intuitive notion of the probability of a "one-step" transition from 0 to x, it is tempting to denote by $P_n(0,x)$ the probability of an "n-step" transition from 0 to x (the probability that a "particle," starting at 0, finds itself at x after n transitions governed by $P(x,y)$). Suppose that n and x are both even or both odd and that $|x| \le n$ (otherwise $P_n(0,x)$ will be zero). Then $P_n(0,x)$ should be the probability of $\frac{1}{2}(x + n)$ successes in n Bernoulli (independent) trials, where the probability of success is p (and of failure q). These considerations suggest that

---

[1] This definition will serve us in the first two sections. In D3.2 of section 3 it will be superseded by a more sophisticated version.

we should define $P_n(0,x)$ for arbitrary random walk in such a way that we get for Bernoulli random walk

(1) $$P_n(0,x) = p^{(n+x)/2} q^{(n-x)/2} \binom{n}{(n+x)/2}$$

when the sum $n + x$ is even and $|x| \leq n$, and $P_n(0,x) = 0$ otherwise. It is easy to check that one gets just this result from the definition in D1 below, according to which

(2) $$P_n(0,x) = \sum_{x_1 \in R} \sum_{x_2 \in R} \cdots \sum_{x_{n-1} \in R} P(0,x_1) P(x_1,x_2) \cdots P(x_{n-1},x);$$

for if we define the generating function

$$f(z) = \sum_{x \in R} P(0,x) z^x, \quad z \text{ complex and not zero,}$$

for Bernoulli random walk, then

$$f(z) = pz + \frac{q}{z}, \quad z \neq 0.$$

But equation (2) implies that $P_n(0,x)$ is the coefficient of z^x in the (Laurent) series for $[f(z)]^n$, and it is easy to check, using the Binomial Theorem, that this coefficient is given by (1). Note that this calculation also suggests P1 (below) to the effect that the coefficient of $[f(z)]^{n+m}$ is given by the convolution of the coefficients of $[f(z)]^n$ and $[f(z)]^m$.

D1 *For all x,y in R, $P_0(x,y) = \delta(x,y) = 1$ if $x = y$, 0 otherwise, and $P_1(x,y) = P(x,y)$, $P_n(x,y) =$*

$$\sum_{x_i \in R,\, i=1,\ldots,n-1} P(x,x_1) P(x_1,x_2) \ldots P(x_{n-1},y), \quad n \geq 2.$$

Here the sum extends, as briefly indicated, over all $(n-1)$ tuples $x_1, x_2, \ldots, x_{n-1}$ of points in R.

Our first result, based on D1, is

P1 *For all x,y in R,*

$$P_{n+m}(x,y) = \sum_{t \in R} P_n(x,t) P_m(t,y) \text{ for } n \geq 0,\, m \geq 0,$$

$$\sum_{y \in R} P_n(x,y) = 1, \quad P_n(x,y) = P_n(0, y-x) \text{ for } n \geq 0.$$

Proof: The most natural proof results from the interpretation of definition D1 as the definition of matrix multiplication. To be sure, $P(x,y)$ is an infinite matrix, but that makes no difference at all as the sum in D1 converges absolutely. The first result in P1 is easily obtained if one writes out the definition of $P_{m+n}(x,y)$ according to D1. The resulting sum is then over $x_1, x_2, \ldots, x_{m+n-1}$. Now (using absolute convergence) one may perform the summation over all the variables except x_m. If one then applies D1 to the result, one obtains precisely the first equation in P1, with x_m taking the place of the variable of summation t. We leave it to the reader to check the case when $m = 0$ or $n = 0$, where the preceding argument is not strictly correct.

The last two statements of P1 (which incidentally amount to the assertion that $P_n(x,y)$ is again a transition function of a random walk) are also easy to prove. One simply performs the indicated summation (over n) or translation (by $-x$) in D1.

The probability interpretation of $P_n(x,y)$ is evident. It represents the probability that a "particle," executing the random walk, and starting at the point x at time 0, will be at the point y at time n. The next definition concerns a function of the same type: the probability, again starting at the point x at time 0, that the *first* visit to the point y should occur at time n. This function (unlike $P_n(x,y)$ it is not a transition function) will be called $F_n(x,y)$. In D2 we shall write $\{y\}$ to denote the subset of R consisting of the element y, and $R - \{y\}$ will denote the state space R with the point y excluded.

D2 $F_0(x,y) = 0, \qquad F_1(x,y) = P(x,y),$

$$F_n(x,y) = \sum_{\substack{x_i \in R - \{y\} \\ i=1,2,\ldots,n-1}} P(x,x_1)P(x_1,x_2)\ldots P(x_{n-1},y), \qquad n \geq 2,$$

for all x,y in R.

The most important properties of $F_n(x,y)$ are

P2 (a) $\qquad F_n(x,y) = F_n(0, y - x),$

(b) $\qquad \sum_{k=1}^{n} F_k(x,y) \leq 1,$

(c) $\qquad P_n(x,y) = \sum_{k=1}^{n} F_k(x,y) P_{n-k}(y,y),$

for $n \geq 1$ and arbitrary x,y in R.

Proof: The truth of part (a) is immediate from D2, using only the spatial homogeneity of $P(x,y)$. The proof of part (b) is considerably more delicate. The statement (b) is "probabilistically obvious" as the sum in part (b) represents a probability—the probability that the first visit to y, starting at x, occurs before or at time n. Fortunately it is not hard to base an elementary proof on this idea. (The skeptical reader will observe that we are indeed about to introduce the notion of measure—but measure of a very simple sort, the total number of elementary events being countably infinite.)

Let us define as the set of "elementary events" the set of sequences ω of the form $\omega = \{x_0, x_1, x_2, \ldots, x_{n-1}, x_n\}$ where $x_0 = x$, and where $x_1, x_2, \ldots, x_n$ may assume any value in R. Since R is countable, this set of sequences, which we denote Ω_n, is also countable. With each ω in Ω_n we associate the measure

$$p(\omega) = P(x,x_1)P(x_1,x_2)\ldots P(x_{n-1},x_n).$$

It follows from D1 and P1 that

$$\sum_{[\omega \mid \omega \in \Omega_n, x_n = y]} p(\omega) = P_n(x,y), \qquad \sum_{\omega \in \Omega_n} p(\omega) = \sum_{y \in R} P_n(x,y) = 1.$$

On the other hand, if

$$A_k = [\omega \mid \omega \in \Omega_n; x_1 \neq y, x_2 \neq y, \ldots, x_{k-1} \neq y, x_k = y],$$
$$1 \leq k \leq n,$$

then the sets A_k are disjoint subsets of Ω_n, and it is obvious from D2 that

$$F_k(x,y) = \sum_{\omega \in A_k} p(\omega), \qquad 1 \leq k \leq n.$$

The A_k being disjoint, one obtains

$$\sum_{k=1}^{n} F_k(x,y) \leq \sum_{\omega \in \Omega_n} p(\omega) = 1.$$

Part (c) can be proved in a similar fashion, but we shall use mathematical induction instead. Suppose that (c) holds when $n = j$. Then one can write, using P1 and the induction hypothesis,

$$P_{j+1}(x,y) = \sum_{t \in R} P(x,t)P_j(t,y) = \sum_{t \in R} P(x,t) \sum_{k=1}^{j} F_k(t,y)P_{j-k}(y,y).$$

However, D2 shows that

$$\sum_{t \in R} P(x,t)F_k(t,y) = \sum_{t \in R - \{y\}} P(x,t)F_k(t,y) + P(x,y)F_k(y,y)$$
$$= F_{k+1}(x,y) + P(x,y)F_k(y,y).$$

It follows, using the induction hypothesis again, that

$$P_{j+1}(x,y) = \sum_{k=1}^{j} F_{k+1}(x,y) P_{j-k}(y,y) + \sum_{k=1}^{j} P(x,y) F_k(y,y) P_{j-k}(y,y)$$

$$= \sum_{m=2}^{j+1} F_m(x,y) P_{j+1-m}(y,y) + F_1(x,y) P_j(y,y)$$

$$= \sum_{m=1}^{j+1} F_m(x,y) P_{j+1-m}(y,y).$$

That completes the induction, and also the proof, since (c) is evidently correct when $n = 1$.

Next we define, in D3 below, the function $G_n(x,y)$ to correspond to the expected number of visits of the random walk, starting at x, to the point y within time n. (As soon as we develop the simplest probabilistic apparatus, this function will of course be an expectation, being defined as a sum of probabilities.) Then we prove, in P3, a result comparing $G_n(x,y)$ to the expected number of visits to the starting point of the random walk.

D3

$$G_n(x,y) = \sum_{k=0}^{n} P_k(x,y), \qquad n = 0, 1, \ldots, x,y \in R.$$

P3

$$G_n(x,y) \leq G_n(0,0) \text{ for } n \geq 0 \text{ and all } x,y \text{ in } R.$$

Proof: As $G_n(x,y) = G_n(x - y, 0)$ in view of P1, it suffices to prove P3 in the case $y = 0$ and $x \neq 0$. Using part (c) of P2 we have

$$G_n(x,0) = \sum_{k=1}^{n} P_k(x,0) = \sum_{k=1}^{n} \sum_{j=0}^{k} F_{k-j}(x,0) P_j(0,0)$$

and a simple interchange of the order of summation gives

$$G_n(x,0) = \sum_{j=0}^{n} P_j(0,0) \sum_{i=0}^{n-j} F_i(x,0).$$

Using part (b) of P2,

$$G_n(x,0) \leq \sum_{j=0}^{n} P_j(0,0) = G_n(0,0).$$

The stage is now set for the most important classification of random walks, according to whether they are *recurrent*[2] or *transient* (nonrecurrent). The basic idea is that the sum $\sum_{k=1}^{n} F_k(0,0)$ represents the probability of a return to the starting point before or at time n. The sequence of the sums $\sum_{k=1}^{n} F_k(0,0)$ is nondecreasing as n increases, and by P2 they are bounded by one. Hence they have a limit, which we shall call F, and $F \leq 1$. Therefore it is reasonable to call the random walk recurrent if $F = 1$ and transient if $F < 1$.

Actually it turns out that there is another, equivalent, classification, based on the number G, the limit of the monotone sequence $G_n(0,0)$. G may be finite or infinite (in which case we write $G = +\infty$) and it will be shown (in P4) that $G < \infty$ when $F < 1$ and $G = +\infty$ when $F = 1$. But first we make two more definitions designed mainly to simplify the notation.

D4

$$G(x,y) = \sum_{n=0}^{\infty} P_n(x,y) \leq \infty; \qquad F(x,y) = \sum_{n=1}^{\infty} F_n(x,y) \leq 1,$$
$$G_n(0,0) = G_n, \quad G(0,0) = G; \quad F_n(0,0) = F_n, \quad F(0,0) = F.$$

D5 *The random walk defined by the transition function P is said to be recurrent if $F = 1$ and transient if $F < 1$.*

P4 $G = \dfrac{1}{1-F}$, *with the interpretation that $G = +\infty$ when $F = 1$ and $F = 1$ when $G = +\infty$.*

Proof: It would perhaps be natural to use the method of generating functions, applied to the convolution equation

(1) $$P_n(0,0) = \sum_{k=0}^{n} F_k P_{n-k}(0,0), \qquad n \geq 1,$$

which is a direct consequence of P2 (part (c)) and the notation introduced in D4. But P4 can also be obtained directly, as follows.

---

[2] In the general theory of Markov chains it is possible that the probability of return to the starting point is one for some, but not all, points of the state space. Such points are then called *recurrent* or *persistent*, and the points with return probability less than one are *transient*, cf. [31], Vol. 1, p. 353, and [9], p. 19. As every random walk has the property that either all states are recurrent or all are transient, we shall apply these adjectives directly to the random walk rather than to its states.

Summing the convolution equation (1) over $n = 1, 2, \ldots, m$, and adding $P_0(0,0) = 1$ on each side, gives

$$(2) \qquad G_m = \sum_{k=0}^{m} F_k G_{m-k} + 1, \qquad m \geq 1.$$

Letting $m \to \infty$,

$$G = 1 + \lim_{m \to \infty} \sum_{k=0}^{m} F_k G_{m-k} \geq 1 + G \sum_{k=0}^{N} F_k,$$

for every integer N, and therefore

$$G \geq 1 + GF.$$

This proves, by the way, that $G = +\infty$ when $F = 1$, since the inequality $G \geq 1 + G$ has no finite solutions.

On the other hand, equation (2) gives

$$(3) \quad 1 = G_m - \sum_{k=0}^{m} G_k F_{m-k} \geq G_m - G_m \sum_{k=0}^{m} F_{m-k} \geq G_m(1 - F),$$

so that $1 \geq G(1 - F)$, which shows that $G < \infty$ when $F < 1$. That completes the proof of the identity $G(1 - F) = 1$, and hence of P4.

E2 Consider *Bernoulli random walk* with $P(0,1) = p$, $P(0,-1) = q$. An easy calculation (see E1) gives $P_n(0,0) = 0$ when n is an odd integer, and

$$(1) \qquad P_{2n}(0,0) = (pq)^n \binom{2n}{n} = (-1)^n (4pq)^n \binom{-1/2}{n}.$$

Since p and q are not arbitrary, but $0 \leq p = 1 - q$, it follows that $4pq \leq 1$. Thus the Binomial Theorem yields the power series (generating function)

$$(2) \qquad \sum_{n=0}^{\infty} t^n P_{2n}(0,0) = (1 - 4pqt)^{-1/2}$$

valid for all complex t in the unit disc $|t| < 1$. Letting t approach one through the real numbers less than one (we shall habitually write "$t \nearrow 1$" for this type of limit), it is clear that

$$(3) \qquad \lim_{t \nearrow 1} \sum_{n=0}^{\infty} t^n P_{2n}(0,0) = \sum_{n=0}^{\infty} P_{2n}(0,0)$$

$$= \sum_{n=0}^{\infty} P_n(0,0) = G \leq \infty.$$

It follows from (2), compared to (3), that

(4) $$G = \begin{cases} (1 - 4pq)^{-1/2} < \infty & \text{when } p \neq q \\ +\infty & \text{when } p = q = 1/2. \end{cases}$$

In view of P4, we have shown that *Bernoulli random walk in one-dimension is recurrent if and only if $p = q = 1/2$*, i.e., simple random walk is the only recurrent Bernoulli random walk.

For the sake of completeness, let us repeat the above argument, working with F instead of with G. Setting $x = y = 0$ in part (c) of P2,

$$P_n(0,0) = \sum_{k=0}^{n} P_{n-k}(0,0) F_k(0,0) \text{ for } n \geq 1,$$

or

$$P_n(0,0) = \sum_{k=0}^{n} P_{n-k}(0,0) F_k(0,0) + \delta(n,0) \text{ for } n \geq 0.$$

That gives

$$\sum_{n=0}^{\infty} t^n P_n(0,0) = \sum_{n=0}^{\infty} t^n P_n(0,0) \sum_{n=1}^{\infty} t^n F_n(0,0) + 1, \quad 0 \leq t < 1.$$

Replacing t by $\sqrt{t}$, one concludes from equation (2) that

(5) $$\sum_{n=0}^{\infty} t^n F_{2n}(0,0) = 1 - \sqrt{1 - 4pqt}, \quad 0 \leq t < 1.$$

Again one arrives at the conclusion that

$$F = \lim_{t \nearrow 1} \sum_{n=1}^{\infty} t^n F_{2n}(0,0) = 1 - \sqrt{1 - 4pq} = 1$$

if and only if $4pq = 1$, which happens when $p = q = 1/2$.

In the unsymmetric case (when $p \neq q$) we know from P3 that $G(0,x) \leq G(0,0) = G < \infty$ for all x in R. For convenience we shall assume that $p > q$, and proceed to calculate $G(0,x)$, by deriving, and then solving, a *difference equation* satisfied by $G(0,x)$. From P1 one obtains

$$P_{n+1}(0,x) = \sum_{y \in R} P(0,y) P_n(y,x) = p P_n(1,x) + q P_n(-1,x)$$
$$= p P_n(0, x-1) + q P_n(0, x+1),$$

for all $n \geq 0$ and all x in R. Summation over $n \geq 0$ yields

(6) $$G(0,x) - \delta(x,0) = p G(0, x-1) + q G(0, x+1), \quad x \in R.$$

It is not difficult to solve (6). The associated homogeneous difference equation

$$f(x) = p f(x-1) + q f(x+1)$$

has the solutions

(7) $$f(x) = A r_1^x + B r_2^x$$

where $r_1 = p/q$, $r_2 = 1$ are the zeros of the polynomial $qt^2 - t + p$. Now we need a "particular" solution $\varphi(x)$ of the nonhomogeneous equation

(8) $$\varphi(x) - \delta(x,0) = p\varphi(x-1) + q\varphi(x+1).$$

Let us choose $\varphi(0) = \varphi(1) = 0$. Then the function $\varphi(x)$ has a unique extension to R which satisfies (8), and it is simple to calculate, recursively, that

(9) $$\varphi(x) = 0 \text{ for } x \geq 0, \qquad \varphi(x) = (p-q)^{-1}\left[\left(\frac{p}{q}\right)^x - 1\right] \text{ for } x \leq 0.$$

It follows from the elementary theory of difference equations that $G(0,x)$ must be obtainable by superposition of functions in (7) and (9), i.e.,

(10) $$G(0,x) = \varphi(x) + A\left(\frac{p}{q}\right)^x + B.$$

Observe now that the function $\varphi(x)$ in (9) is bounded (since we are assuming that $p > q$). According to P3 we have $G(0,x) \leq G < \infty$ which implies that $A = 0$. Thus it remains only to evaluate the constant B, using equation (4), to the effect that

$$G(0,0) = (1 - 4pq)^{-1/2} = \frac{1}{p-q}.$$

From (9) and (10) one therefore obtains the result that $B = (p-q)^{-1}$, and we have proved that *for Bernoulli random walk with $p > q$,*

(11) $$G(0,x) = \begin{cases} (p-q)^{-1} & \text{for } x \geq 0 \\ (p-q)^{-1}\left(\frac{p}{q}\right)^x & \text{for } x \leq 0. \end{cases}$$

One last result, easily within reach of the elementary methods of this section, is the "weak" *ratio ergodic theorem* (a "strong" version of which is proved in T5.1 (Theorem 1 of section 5)).

P5 *For every random walk*

$$\lim_{n \to \infty} \frac{G_n(x,y)}{G_n(0,0)} = F(x,y) \text{ whenever } x \neq y.$$

Remark: Although the statement of P5 makes no distinction between recurrent and transient random walk, such a distinction will nevertheless arise in the proof. The result of P5 is correct in both cases, but for entirely different reasons! The proof will show further that in the transient case P5 is false, as it stands, when $x = y$, whereas it is obviously correct for recurrent random walk, even when $x = y$.

1. INTRODUCTION

Proof: First we observe that it suffices to prove P5 in the special case when $y = 0$. Thus we may use part (c) of P2 in the form

$$G_n(x,0) = \delta(x,0) + \sum_{k=1}^{n} P_k(x,0)$$

$$= \delta(x,0) + \sum_{j=0}^{n} P_j(0,0) \sum_{k=1}^{n-j} F_k(x,0),$$

so that

$$\frac{G_n(x,0)}{G_n(0,0)} = \frac{\delta(x,0)}{\sum_{j=0}^{n} P_j(0,0)} + \frac{\sum_{j=0}^{n} P_j(0,0) \sum_{k=0}^{n-j} F_k(x,0)}{\sum_{j=0}^{n} P_j(0,0)}, \quad x \in R.$$

In the transient case the denominators have a finite limit, so that one obtains

$$\lim_{n \to \infty} \frac{G_n(x,0)}{G_n(0,0)} = \frac{\delta(x,0)}{G} + F(x,0)$$

for all x in R, and in particular the limit $F(x,0)$ when $x \neq 0$.

To obtain a proof when the random walk is recurrent, let

$$a_n = \sum_{k=0}^{n} F_k(x,0), \qquad b_n = P_n(0,0), \qquad n \geq 0.$$

The problem is then to show that

$$\lim_{n \to \infty} \frac{\sum_{j=0}^{n} b_j a_{n-j}}{\sum_{k=0}^{n} b_k} = \lim_{n \to \infty} a_n = F(x,0).$$

For every positive integer N one can decompose

$$\frac{\sum_{j=0}^{n} b_j a_{n-j}}{\sum_{k=0}^{n} b_k} - \alpha$$

$$= \frac{\sum_{j=0}^{n-N} b_j(a_{n-j} - \alpha)}{\sum_{k=0}^{n} b_k} + \frac{\sum_{j=n-N+1}^{n} b_j a_{n-j}}{\sum_{k=0}^{n} b_k} - \alpha \frac{\sum_{j=n-N+1}^{n} b_j}{\sum_{k=0}^{n} b_k}.$$

This decomposition is valid for all $n > N$ and for every real α, but we shall of course set $\alpha = F(x,0)$. Since b_n is a bounded sequence such that the series $\sum b_n$ diverges, it is clear that the last two terms tend to zero as $n \to \infty$, for each fixed N. We can now choose $N = N(\epsilon)$ so that $|a_n - \alpha| < \epsilon$ when $n > N$. With this choice of N one obtains

$$\varlimsup_{n \to \infty} \left| \frac{\sum_{j=0}^{n} b_j a_{n-j}}{\sum_{k=0}^{n} b_k} - F(x,0) \right| \leq \epsilon$$

and as ϵ is an arbitrary positive number, the proof of P5 is complete.

E3 Let us now apply P5 to *Bernoulli random walk*. When $p > q$, P5 tells us that $F(0,x) = 1$ for every $x > 0$, since

$$F(0,x) = \frac{G(0,x)}{G(0,0)}$$

and $G(0,x) = (p - q)^{-1}$, for every $x \geq 0$, according to E2. Inasmuch as $F(0,x)$ represents the probability that the first visit to x occurs at some finite time (i.e., that x is visited at all), it was of course to be expected that $F(0,x) = 1$ for all positive x when $p > q$. In this case one would also expect that $F(0,x) < 1$ when $x < 1$, and E2 together with P5 actually shows that $F(0,x)$ goes to zero geometrically as $x \to -\infty$. One obtains

$$F(0,x) = \left(\frac{p}{q}\right)^x \quad \text{when } x < 0.$$

Finally, consider the *simple random walk*, with $p = q = 1/2$. According to P5,

$$\lim_{n \to \infty} \frac{G_n(0,x)}{G_n(0,0)} = F(0,x)$$

for every x, but it still remains to evaluate the limit function $F(0,x)$. We know only that $F(0,0) = 1$, but it would indeed be surprising if there were a point x_0 such that $F(0,x_0) < 1$.

In fact, one could argue that $F(0,x) \equiv 1$ on the following intuitive grounds. Return to 0 is certain (since $F(0,0) = 1$). However, a visit to x_0 before the first return to 0 is certainly possible, in fact it has a probability $p(x_0) \geq 2^{-|x_0|}$. But once the random walk has reached x_0, the probability that it will ever return to 0 is $F(x_0,0) = F(0,-x_0)$, and by symmetry $F(0,-x_0) = F(0,x_0)$. Hence we seem to have shown that

$$1 = F(0,0) \leq 1 - p(x_0) + p(x_0)F(0,x_0) = 1 - p(x_0)[1 - F(0,x_0)]$$

which implies that $F(0,x_0) = 1$.

The only trouble is that *we used certain intuitively plausible facts about properties of the probability measures induced by the transition function $P(x,y)$, before having described the method by which probability measures are to be assigned.* In particular we argued that if $p(x_0)$ is the probability of visiting x_0 before the first return to 0, then the probability of visiting x_0 before 0, and then 0, is $p(x_0)F(x_0,0)$. This relation of independence is indeed a property of any reasonable assignment of probability measure under which $P_n(x,y)$ retains the obvious interpretation as n-step transition probabilities, but the proof must wait until section 3.

It is, however, of some interest to see if one can show analytically that $F(0,x) \equiv 1$ when $p = q$, and we offer the following three alternative methods of proof.

(a) From Stirling's formula

$$n! \sim \sqrt{2\pi n}\, e^{-n} n^n$$

(here and in the sequel $a_n \sim b_n$ means that the ratio $a_n/b_n \to 1$ as $n \to \infty$), one obtains

$$P_{2n}(0,0) = (-1)^n \binom{-1/2}{n} \sim (\pi n)^{-1/2}.$$

It is equally easy to show that

$$P_{2n}(0,x) \sim (\pi n)^{-1/2} \quad \text{when } x \text{ is an even integer,}$$
$$P_{2n+1}(0,x) \sim (\pi n)^{-1/2} \quad \text{when } x \text{ is odd.}$$

Finally, summing on n yields, for every x,

$$G_n(0,x) \sim \tfrac{1}{2} \sum_{k=1}^{n} \left(\frac{\pi k}{2}\right)^{-1/2} \sim \tfrac{1}{2}\left(\frac{n}{2\pi}\right)^{1/2},$$

so that

$$F(0,x) = \lim_{n \to \infty} \frac{G_n(0,x)}{G_n(0,0)} = 1.$$

(b) A more careful study of the transition functions $P_n(x,y)$ (easy for instance by the methods of Fourier analysis in Chapter II) shows that one can dispense with Stirling's formula and in fact prove a much stronger result, namely

$$\sum_{n=0}^{\infty} [P_n(0,0) - P_n(0,x)] = \lim_{n \to \infty} [G_n(0,0) - G_n(0,x)] = |x|.$$

Clearly it follows that

$$\lim_{n \to \infty} \frac{G_n(0,x)}{G_n(0,0)} = F(0,x) = 1, \qquad x \in R.$$

(c) In view of P3

$$0 \leq \frac{G_n(0,x)}{G_n(0,0)} \leq 1.$$

Furthermore P1 and D3 give

$$\sum_{t \in R} P(x,t) \frac{G_n(t,y)}{G_n(0,0)} = \frac{1}{G_n(0,0)} [G_n(x,y) + P_{n+1}(x,y) - \delta(x,y)].$$

Now we let n tend to infinity. The left-hand side of the above equation then tends to

$$\sum_{t \in R} P(x,t) F(t,y).$$

The right-hand side goes to $F(x,y)$ since $G_n(0,0) \to \infty$ and $|P_{n+1}(x,y)| \le 1$. Hence

$$\sum_{t \in R} P(x,t) F(t,y) = F(x,y),$$

and specializing by setting $y = 0$, we find that

$$\sum_{t \in R} P(x,t) F(t,0) = \tfrac{1}{2} F(x+1,0) + \tfrac{1}{2} F(x-1,0) = F(x,0).$$

But this equation has solutions only of the form $F(x,0) = ax + b$. Since $0 \le F(0,x) \le 1$ and $F(0,0) = 1$, we conclude again that *simple random walk in one dimension has the property that $F(0,x) = 1$ for all x*.

2. PERIODICITY AND RECURRENCE BEHAVIOR

We saw in section 1 that recurrence of a random walk manifests itself in the limiting behavior of $G_n(0,0)$ as well as of $F_n(0,0)$. But so far it is not clear exactly how $G_n(0,x)$, the expected number of visits of the random walk to x in time n, depends on the recurrence of the random walk. The same question of course applies to $F_n(0,x)$ and to $F(0,x)$ which is the probability of a visit (at some finite time) to the point x.

Certain results are obvious. For example, since $G(0,x) \le G(0,0) \le \infty$ for every x, it is clear that $G(0,x) < \infty$ for all x if the random walk is transient. But suppose that the random walk is recurrent, so that $G(0,0) = \infty$. Does it follow that $G(0,x) = \infty$ for every x in R? The answer is no; it is indeed possible to find a recurrent random walk such that $G(0,x_0) < \infty$ for some $x_0 \in R$. The most trivial example of this type is the random walk which "stands still," i.e., with $P(0,0) = 1$. According to our definitions it is recurrent, as $G_n(0,0) = n + 1 \to \infty$ as $n \to \infty$. But clearly $G(0,x) = 0$ for all $x \ne 0$. Note that we have not even bothered to specify the dimension of the state space R. It is immaterial as the random walk would not "know the difference." It takes place on the state space consisting of the origin alone.

Given a state space R of dimension $d \geq 1$ and a point x_0 in R, we can define a recurrent random walk on R such that $G(0,x_0) = 0$. For example, let

$$P(0,2x_0) = P(0,-2x_0) = 1/2,$$
$$P(0,x) = 0 \text{ for all other } x \text{ in } R.$$

This random walk is nothing but simple symmetric one-dimensional random walk, no matter what the dimension of R. Therefore it is recurrent, but again it is clear that $G(0,x_0) = 0$. The reason is of course again that the random walk takes place on a subset (a subgroup, as it happens) of R which does not contain the point x_0.

There is no need to look for other types of examples where $G(0,0) = \infty$ but $G(0,x) < \infty$ for some x, because there are none. Instead we shall begin by formalizing the ideas suggested by the preceding examples, and then proceed to explain the crucial role of that subset of the state space R which is actually visited by the random walk. Given a random walk, i.e., a transition function $P(x,y)$ defined for x, y in R, we define three subsets of R, called Σ, R^+, and $\bar{R}$

D1
$$\Sigma = [x \mid P(0,x) > 0],$$
$$R^+ = [x \mid P_n(0,x) > 0 \text{ for some } n \geq 0],$$
$$\bar{R} = [x \mid x = y - z, \text{ for some } y \in R^+ \text{ and } z \in R^+].$$

P1 *R^+ is the set of all finite sums from Σ including the origin 0 (the empty sum). It is also the smallest additive semigroup containing Σ. $\bar{R}$ on the other hand is the smallest additive subgroup of R which contains R^+.*

Proof: There is not much to verify apart from the group (semigroup) axioms. (A semigroup is supposed to be closed under addition like a group, but not under subtraction, and although it has an identity (the origin), no other elements in it need have an inverse.)

The origin is in R^+ by definition. If x is in R^+, and $x \neq 0$, then it follows from $P_n(0,x) > 0$ that there is a finite sequence $x_1, x_2, \ldots, x_{n-1}$ in R such that

$$P_n(0,x) \geq P(0,x_1)P(x_1,x_2)\ldots P(x_{n-1},x) > 0.$$

But then $P(0,x_1) > 0$, $P(0,x_2 - x_1) > 0$, and so on, so that $x_1, x_2 - x_1$, etc. are in Σ. Therefore $x = x_1 + (x_2 - x_1) + \cdots + (x - x_{n-1})$, is a representation of x as a finite sum of elements of Σ.

Conversely, if $x = y_1 + y_2 + \cdots + y_n$, with $y_i \in \Sigma$ for $i = 1, \ldots, n$, then $P(0,y_k) = P(y_1 + \cdots + y_{k-1}, y_1 + \cdots + y_k) > 0$, $k = 1, 2, \ldots, n$, so that

$$P_n(0,x) = P_n(0, y_1 + \cdots + y_n)$$
$$\geq P(0,y_1) P(y_1, y_1 + y_2) \ldots P(y_1 + \cdots + y_{n-1}, x) > 0.$$

This proves that R^+ is really the set of all finite sums of elements of Σ.

If we use either the characterization of R^+ just obtained or the one in D1, it is clear that R^+ is closed under addition. Hence R^+ is a semigroup. There can be no smaller semigroups containing Σ, since any such semigroup must contain all finite sums from Σ, and R^+ is just the set of such sums.

$\bar{R}$ is a group since it is closed under differences, by D1. It contains R^+ by definition, and there can obviously be no smaller group than $\bar{R}$ with these two properties.

The ideas that led to the definition of the sets R^+ and $\bar{R}$ enable us to understand the difference between transient and recurrent random walk a little better than in section 1. First consider transient random walk, which offers very little difficulty.

P2 *If $P(x,y)$ is the transition function of a transient random walk with state space R, then*

$$G(0,x) < \infty \quad \text{on} \quad R$$

and

$$G(0,x) = F(0,x) = 0 \quad \text{on} \quad R - R^+.$$

Proof: By P1.3,

$$G(0,x) = \lim_{n \to \infty} G_n(0,x) \leq \lim_{n \to \infty} G_n(0,0) = G < \infty,$$

which proves the first statement. To prove the second one, assume that the set $R - R^+$ is nonempty (otherwise the second statement of P2 is vacuous). In that case, $P_n(0,x) = 0$ for $x \in R - R^+$ for every $n \geq 0$. But then

$$G(0,x) = \sum_{n=0}^{\infty} P_n(0,x) = 0, \qquad x \in R - R^+,$$

$$F(0,x) = \sum_{n=1}^{\infty} F_n(0,x) \leq G(0,x) = 0, \qquad x \in R - R^+.$$

The last inequality came from D1.2 which implies that

$$F_n(0,x) \leq P_n(0,x), \qquad x \in R, \quad n \geq 0.$$

In the recurrent case we first prove a few lemmas (P3, P4, P5) of independent interest, which will later be combined with P2 to give a complete description (T1) of the basic differences between recurrent and transient random walk.

P3 *If a random walk is recurrent and $x \in R^+$, then*
$$G(0,x) = \infty.$$

Proof: If $x \in R^+$, we may assume that $x \neq 0$ and choose $m \geq 1$ so that $P_m(0,x) > 0$. Then, using P1.1,
$$P_m(0,x)P_n(x,x) \leq \sum_{t \in R} P_m(0,t)P_n(t,x) = P_{m+n}(0,x).$$

Summing on n from zero to k, one obtains
$$P_m(0,x)G_k(x,x) = P_m(0,x)G_k(0,0) \leq \sum_{n=0}^{k} P_{m+n}(0,x)$$
$$= \sum_{n=0}^{m+k} P_n(0,x) - \sum_{n=0}^{m-1} P_n(0,x)$$
$$= G_{m+k}(0,x) - G_{m-1}(0,x).$$

Letting $k \to +\infty$, one finds that
$$P_m(0,x) \cdot G \leq \lim_{k \to \infty} G_{m+k}(0,x) - G_{m-1}(0,x).$$

But by P1.4, we know that $G = +\infty$, which proves that
$$\lim_{k \to \infty} G_{m+k}(0,x) = G(0,x) = +\infty.$$

The next step, still for recurrent random walk, is the investigation of $F(0,x)$. The obvious probability interpretation of $F(0,x)$ as the probability of a visit to x, starting at 0, in a finite time is very helpful. It suggests that $F(0,x) = 1$ for all x in R^+. Actually considerable work is required to translate this suggestion into mathematical certainty, and here too the probability interpretation is useful in suggesting what to do. The probabilistic arguments that follow can be made rigorous by careful use of certain measure theoretic facts to be introduced in section 3.

We assume that $x \in R^+$, and also (for the moment) that $-x \in R - R^+$. Then the random walk can go from 0 to x with positive probability. But once at x it can never return to zero, since a transition from x to 0 is impossible, being equivalent to one from 0 to $-x$. Hence it is possible to leave 0 and never to return. This contradicts the hypothesis that the random walk is recurrent. Therefore we have "demonstrated" the impossibility of having $x \in R^+$ while $-x \in R - R^+$. In other words, recurrent random walk has the property that $x \in R^+$ implies $-x \in R^+$. But R^+ is a semigroup, and a semigroup which contains the inverse (negative) of each element is a group. Hence $R^+ = \bar{R}$. Now it is easy to go further, and to conclude that $F(0,x) = 1$ not only when $x \in R^+$, but for all $x \in \bar{R}$.

We shall choose to disregard this interesting argument, which is due to Feller [12],* 1951, to obtain the same result by the more elementary methods of this and the last section.

P4 *For arbitrary random walk, $x, y \in R$ and $n \geq 0$*

(a) $$\sum_{t \in R} P(x,t) G_n(t,y) = G_{n+1}(x,y) - \delta(x,y),$$

and for recurrent random walk

(b) $$\sum_{t \in R} P(x,t) F(t,y) = F(x,y).$$

Proof: Part (a) follows by computation from P1.1 and D1.3 since

$$\sum_{t \in R} P(x,t) G_n(t,y) = \sum_{k=0}^{n} \sum_{t \in R} P(x,t) P_k(t,y)$$
$$= \sum_{k=1}^{n+1} P_k(x,y) = G_{n+1}(x,y) - \delta(x,y).$$

Dividing equation (a) by $G_n(0,0)$ (which is positive), one finds

(1) $$\sum_{t \in R} P(x,t) \frac{G_n(t,y)}{G_n(0,0)} = \frac{G_n(x,y)}{G_n(0,0)} + \frac{P_{n+1}(x,y)}{G_n(0,0)} - \frac{\delta(x,y)}{G_n(0,0)}.$$

Now let $n \to +\infty$ in (1). Since $G_n(0,0) \to G = \infty$, the last two terms on the right-hand side in (1) tend to zero. Next we observe that

$$0 \leq \frac{G_n(x,y)}{G_n(0,0)} \leq 1, \qquad \lim_{n \to \infty} \frac{G_n(x,y)}{G_n(0,0)} = F(x,y),$$

the inequality being due to P1.3, and the limit due to P1.5.

The boundedness of the ratios $G_n(x,y)/G_n(0,0)$ implies, by a dominated convergence argument, that

$$\lim_{n \to \infty} \sum_{t \in R} P(x,t) \frac{G_n(t,y)}{G_n(0,0)} = \sum_{t \in R} P(x,t) \lim_{n \to \infty} \frac{G_n(t,y)}{G_n(0,0)},$$

* Numerals in brackets refer to the Bibliography at the end of the book.

so that one may conclude, by letting n tend to infinity in (1), that part (b) of P4 is true.

Now we shall see that P4 very quickly leads to

P5 *If a random walk is recurrent and $x \in R^+$, then also $-x \in R^+$. Further $R^+ = \bar{R}$ and*
$$F(0,x) = 1 \text{ for all } x \in \bar{R},$$
$$F(0,x) = 0 \text{ for all } x \in R - \bar{R}.$$

Proof: We shall work with the special case
$$\sum_{t \in R} P(x,t) F(t,0) = F(x,0), \qquad x \in R$$
of P4(b). It follows that
$$\sum_{t \in R} P_2(x,t) F(t,0) = \sum_{t \in R} P(x,t) F(t,0),$$
and by further iterations of the transition operator $P(x,y)$
$$\sum_{t \in R} P_m(x,t) F(t,0) = F(x,0), \qquad m \geq 0, \quad x \in R.$$
Now we take a fixed $x_0 \in R^+$, in order to prove that $F(x_0,0) = 1$. This is done by setting $x = 0$ in the last equation, giving
$$\sum_{t \in R} P_m(0,t) F(t,0) = F(0,0) = F = 1.$$

Since $x_0 \in R^+$, we may select $m_0 \geq 0$ such that $P_{m_0}(0,x_0) > 0$. Then
$$1 = P_{m_0}(0,x_0) F(x_0,0) + \sum_{t \neq x_0} P_{m_0}(0,t) F(t,0)$$
$$\leq P_{m_0}(0,x_0) F(x_0,0) + \sum_{t \neq x_0} P_{m_0}(0,t)$$
$$= 1 + P_{m_0}(0,x_0)[F(x_0,0) - 1],$$
which proves that $F(x_0,0) = 1$.

Furthermore
$$1 = F(x_0,0) = F(0,-x_0),$$
and $F(0,-x_0)$ would be zero if $-x_0$ were not in R^+. Hence $-x_0$ is in R^+ along with x_0, so that $R^+ = \bar{R}$. But then the previous arguments show that $F(0,x_0) = 1$, and it is of course clear that $F(0,x) = 0$ when $x \in R - \bar{R}$, if this set has any elements at all. Thus the proof of P5 is complete.

Before summarizing the results of P2 through P5, it is convenient to simplify matters by focusing attention on the set $\bar{R}$. If $R = \bar{R}$, then the statement in P5 that $R^+ = \bar{R}$ reduces to $R^+ = R$. But even in the transient case that part of R which is not in $\bar{R}$ is quite irrelevant in the following sense. One could always imbed R in a larger group of dimension $d + 1$ without affecting the random walk in the least. So we shall make the definition[3]

D2 *A random walk $P(x,y)$ defined on R is aperiodic if $R = \bar{R}$.*

If a random walk is *periodic* (i.e., not aperiodic), then the problem is *badly posed*. In other words, if $R \neq \bar{R}$, then the random walk is defined on the *wrong group*, or on a coset of the wrong group if it starts at a point x not in $\bar{R}$. It will always be possible to reduce problems where $R \neq \bar{R}$ to the aperiodic case, for the simple reason that in every case $\bar{R}$ happens to be group theoretically isomorphic to R of some dimension $d \geq 0$. (See Chapter II, section 7, for a more detailed discussion of such considerations.)

As a good example of the simplifications due to aperiodicity we have

P6 *For aperiodic recurrent random walk*

$$\lim_{n \to \infty} \frac{G_n(0,x)}{G_n(0,0)} = 1, \qquad x \in R.$$

Proof: By P1.5, the limit is $F(0,x)$. But by P5, $F(0,x) = 1$ for all x in $R^+ = \bar{R}$, and as the random walk is aperiodic this gives $F(0,x) = 1$ in all of R.

Finally we summarize the results of P2 through P5 as

T1 *Let $P(x,y)$ be the transition function of aperiodic random walk. Then there are only two possibilities*
(a) (*Transient case*)

$$G(0,x) < \infty \text{ on } R, \quad F(0,0) < 1, \quad F(0,x) = 0 \text{ on } R - R^+,$$

(b) (*Recurrent case*)

$$G(0,x) = \infty \text{ on } R, \quad F(0,x) = 1 \text{ on } R.$$

There is nothing to prove, all the statements being immediate from D2 applied to P2 and P5. Instead we mention an interesting extension

---

[3] Here again our terminology differs from the conventional one in the theory of Markov chains in [9] and [31]. Our notion of strongly aperiodic random walk, to be introduced in D5.1, is far closer to aperiodicity in Markov chains than the present definition of an aperiodic random walk.

of T1, which, at this point, we do not have the resources to prove. It is proved in P24.9 of Chapter VI, and completes the result of T1 for the transient case. It concerns transient random walk only and asserts that $F(0,x) < 1$ in the transient case for all x, unless the random walk is of a very special type which we shall now define.

D3 *A random walk is called left continuous if it is one dimensional and if*

$$P(0,-1) > 0 \text{ and } P(0,x) = 0 \text{ for } x \leq -2.$$

Similarly it is called right continuous if

$$P(0,1) > 0 \text{ and } P(0,x) = 0 \text{ for } x \geq 2.$$

Note that Bernoulli random walk, with $0 < p = 1 - q < 1$ is both left and right continuous, even when modified so that also $P(0,0) > 0$. The extension of T1, which will be accomplished when we prove P24.9 in Chapter VI, is

Aperiodic transient random walk has $F(0,x) < 1$ for all x in R unless it is
(a) *left continuous, with the additional property that*

$$-\infty < \sum_{x=-\infty}^{\infty} xP(0,x) < 0,$$

in which case $F(0,x) = 1$ for all $x < 0$ and $F(0,x) < 1$ for all $x \geq 0$; or
(b) *right continuous with the additional property that*

$$0 < \sum_{x=-\infty}^{+\infty} xP(0,x) < \infty,$$

in which case $F(0,x) = 1$ for $x > 0$, $F(0,x) < 1$ for $x \leq 0$.

At the present stage of development of the theory it is not even quite obvious that the random walk of type (a) or (b) above is transient. This matter will be settled in T3.1, where some simple sufficient conditions are obtained for a random walk to be transient. Now we turn instead to the task of finding a sufficient condition for recurrence.

Restricting attention to one-dimensional random walk let

D4
$$m = \sum_{x=-\infty}^{\infty} |x|P(0,x) \leq \infty,$$

and, in case $m < \infty$

$$\mu = \sum_{x=-\infty}^{\infty} xP(0,x).$$

22   THE CLASSIFICATION OF RANDOM WALK

We saw in E1.2 that Bernoulli random walk, where μ turns out to have the value $p - q$, is recurrent if and only if $\mu = 0$. We will eventually show that every one-dimensional random walk, with $m < \infty$, has the same property: it is *recurrent if $\mu = 0$ and transient otherwise*. Intuitively the conclusion that random walks with $\mu \neq 0$ are transient seems quite plausible if not obvious. But it turns out that this part of the result is the harder one to prove, and so we shall now consider only the case when $\mu = 0$.

The first step of the proof is nothing but the classical *weak law of large numbers* for sums of identically distributed random variables. Clearly $P(0,x)$ can be taken to be the probability measure of an integer valued random variable. Calling $P(0,x) = p(x)$, one observes that

$$P_2(0,x) = \sum_{y \in R} p(x - y)p(y)$$

is the convolution of two such measures, and similarly $P_n(0,x)$ *is the convolution of the measure $p(x)$ with itself n times*. If $P(0,x)$ has mean μ according to D4 (and $m < \infty$) the usual statement of the weak law becomes

P7 $\lim_{n \to \infty} \sum_{[x] \left| \frac{x}{n} - \mu \right| > \epsilon} P_n(0,x) = 0$ *for every* $\epsilon > 0$.

We shall not give a proof of this remarkable theorem which is due to Khinchin [62], 1929, but only point out that no measure theory is involved as the usual proof[4] proceeds by elementary careful estimation of the values of $P_n(0,x)$ for large n and $|x|$. The first step of the proof of course consists in verifying P7 for the case when $P(0,x)$ satisfies the additional assumption of finite *variance* σ^2, i.e.,

$$\sigma^2 = \sum_{x \in R} |x - \mu|^2 P(0,x) < \infty.$$

In that case

$$\sum_{[x]\left|\frac{x}{n} - \mu\right| > \epsilon} P_n(0,x) \leq \frac{1}{n^2 \epsilon^2} \sum_{[x]\left|\frac{x}{n} - \mu\right| > \epsilon} |x - n\mu|^2 P_n(0,x)$$

$$\leq \frac{1}{n^2 \epsilon^2} \sum_{x \in R} |x - n\mu|^2 P_n(0,x) = \frac{\sigma^2}{n\epsilon^2} \to 0,$$

as $n \to \infty$.

[4] To be found in any good introductory text, such as [31], [34], [84].

Using nothing but P7, Chung and Ornstein [13], 1962, succeeded in showing that every one-dimensional random walk with $m < \infty$ and $\mu = 0$ is recurrent (earlier proofs used the Fourier analytical methods developed in the next chapter). Let us therefore suppose that $P(x,y)$ is a one-dimensional transition function (i.e., R is the set of integers) with $m < \infty$, $\mu = 0$. For every integer N we can assert, using P1.3, that

$$(1) \qquad G_N(0,0) \geq \frac{1}{2M+1} \sum_{|x| \leq M} G_N(0,x)$$

for every positive number M. Furthermore

$$(2) \qquad \sum_{|x| \leq M} G_N(0,x) = \sum_{k=0}^{N} \sum_{|x| \leq M} P_k(0,x) \geq \sum_{k=0}^{N} \sum_{|x| \left|\frac{x}{k}\right| \leq \frac{M}{N}]} P_k(0,x).$$

Choosing $M = aN$, $a > 0$, we may combine (1) and (2) to obtain

$$(3) \qquad G_N(0,0) \geq \frac{1}{2aN+1} \sum_{k=0}^{N} \sum_{|x| \leq ka} P_k(0,x).$$

From P7 we know, since $\mu = 0$ and $\sum_{x \in R} P_k(0,x) = 1$, that

$$\lim_{k \to \infty} \sum_{|x| \leq ka} P_k(0,x) = 1, \qquad \text{when } a > 0,$$

and one may conclude from (3) that

$$(4) \qquad \lim_{\overline{N \to \infty}} G_N(0,0) \geq \frac{1}{2a}.$$

Since a may be taken arbitrarily small, (4) implies

$$G = \lim_{N \to \infty} G_N(0,0) = \sum_{k=0}^{\infty} P_k(0,0) = \infty.$$

Thus we have proved

P8 *If $P(x,y)$ is one dimensional, $m = \sum |x| P(0,x) < \infty$, and $\mu = \sum x P(0,x) = 0$, then the random walk is recurrent.*

Proposition P8 has the partial converse that a random walk with $m < \infty$, $\mu \neq 0$ is transient. As no direct analytic proof seems to be known, we shall use this difficulty as a motivation for developing the correct measure theoretical formulation of the theory of random walk in the next section. One of the most frequently useful measure theoretical results will be the strong law of large numbers (P3.4),

Kolmogorov's [66] (1933) sharpened version of the weak law (P7). In particular, the strong law (P3.4) will make it quite evident that every random walk on the line with $m < \infty$, $\mu \neq 0$ is transient.

3. SOME MEASURE THEORY

First we define what is meant by an arbitrary *probability space*, and then we shall specialize to those probability spaces which are useful in the measure theoretical formulation of random walk problems. A *probability space* is a triple $(\Omega, \mathscr{F}, \mathbf{P})$ where

(a) Ω is an arbitrary space (collection of "points" $\omega \in \Omega$);

(b) $\mathscr{F}$ is a so-called σ-field, i.e., a collection of subsets $A, B, C, \ldots$ of Ω which is closed under the operations of countable unions, intersections, and complementation; in particular the empty set and the whole space Ω are elements of $\mathscr{F}$;

(c) $\mathbf{P}$ is a non-negative set function, defined on $\mathscr{F}$, with the properties that $\mathbf{P}[\Omega] = 1$ and $\sum_{n=1}^{\infty} \mathbf{P}[A_n] = \mathbf{P}[\bigcup_{n=1}^{\infty} A_n]$, when A_n is a sequence of pairwise disjoint sets in $\mathscr{F}$ (i.e., $\mathbf{P}[\cdot]$ is a *countably additive probability measure*). We shall feel free to use (but actually we shall rarely need to do so) a few standard notions and theorems from *measure theory*[5] which apply to an arbitrary probability space:

(1) If A_n is a monotone sequence of sets in $\mathscr{F}$, i.e., if $A_k \supset A_{k+1}$ for every $k \geq 1$, then

$$\lim_{n \to \infty} \mathbf{P}\left[\bigcap_{k=1}^{n} A_k\right] = \mathbf{P}\left[\bigcap_{k=1}^{\infty} A_k\right].$$

(2) A real valued function $f(\omega)$ on Ω (which may also assume the "values" $+\infty$ and $-\infty$) is called *measurable* if the set $[\omega \mid f(\omega) \leq t]$ is in $\mathscr{F}$ for every real number t. A measurable function which assumes only a finite number of distinct values is called *simple*. A sequence of measurable functions $f_n(\omega)$ is said to *converge a.e.* (*almost everywhere*, or *with probability one*) if $[\omega \mid \lim_{n \to \infty} f_n(\omega)$ exists] is a set of measure one. Using this terminology, every measurable function can be represented as the a.e. limit of a sequence of simple functions. And if a sequence of measurable functions converges a.e., then its limit is also measurable.

---

[5] To be found in [37], [73], or [78].

3. SOME MEASURE THEORY

(3) If $\varphi(\omega)$ is a simple function, assuming finite values a_k on the sets $A_k \in \mathscr{F}$, $k = 1, \ldots, n$, its *integral* is defined as

$$\int_\Omega \varphi(\omega) \, d\mathbf{P}(\omega) = \sum_{k=1}^n a_k \mathbf{P}[A_k].$$

An arbitrary measurable function $f(\omega)$ is called *integrable* if

$$\sup_\varphi \int_\Omega |\varphi(\omega)| \, d\mathbf{P}(\omega) < \infty,$$

the supremum being taken over all simple functions such that $|\varphi(\omega)| \leq |f(\omega)|$ almost everywhere. If $f(\omega)$ is integrable, then

$$\int_\Omega f(\omega) \, d\mathbf{P}(\omega) = \lim_{n \to \infty} \int_\Omega \varphi_n(\omega) \, d\mathbf{P}(\omega)$$

exists and has the same value for every sequence $\varphi_n(\omega)$ of simple functions such that $|\varphi_n(\omega)| \leq |f(\omega)|$ and $\varphi_n(\omega)$ converges to $f(\omega)$ almost everywhere.

(4) The usual properties of the Lebesgue integral on a *finite* interval, such as the Lebesgue dominated convergence theorem, apply to the integral defined in (3). Of particular importance is the following special case of Fubini's theorem. Two integrable functions $f(\omega)$ and $g(\omega)$ are said to be *independent* if

$$\mathbf{P}[\omega \mid f(\omega) \leq a, g(\omega) \leq b] = \mathbf{P}[\omega \mid f(\omega) \leq a]\mathbf{P}[\omega \mid g(\omega) \leq b]$$

for all real numbers a and b. If so, then

$$\int_\Omega f(\omega)g(\omega) \, d\mathbf{P}(\omega) = \int_\Omega f(\omega) \, d\mathbf{P}(\omega) \int_\Omega g(\omega) \, d\mathbf{P}(\omega).$$

Now we are ready to specialize to a particular probability space, where Ω will be the set of infinite sequences

$$\omega = (\omega_1, \omega_2, \ldots), \text{ with each } \omega_i \in R.$$

Here R is the state space of a given random walk. The σ-field $\mathscr{F}$ is defined as follows. First we require $\mathscr{F}$ to contain all *cylinder sets*, i.e., sets A_n of the form

$$A_n = [\omega \mid \omega_k = a_k, \quad k = 1, 2, \ldots, n], \qquad n \geq 0.$$

where $a_k \in R$ for each $k = 1, \ldots, n$, and then we define $\mathscr{F}$ to be the *smallest σ-field which contains every cylinder set*. Finally, in order to define the probability measure (in terms of the transition function $P(x,y)$ of a given random walk), we stipulate that

$$\mathbf{P}[A_n] = P(0,a_1)P(0,a_2)\ldots P(0,a_n), \qquad n \geq 0$$

for every cylinder set (the empty product, when $n = 0$, is of course defined to have the value one). It is not difficult to prove the *extension theorem*[6]—that there exists one and only one countably additive probability measure $\mathbf{P}[\cdot]$ defined on the whole of $\mathscr{F}$, which has the desired values on the cylinder sets. That completes the definition of a *probability space* $(\Omega, \mathscr{F}, \mathbf{P})$ corresponding to a given random walk with transition function $P(x,y)$ and state space R. To this definition we now add a list of frequently used probabilistic terminology.

D1 *Measurable functions on the above probability space are called random variables.* In particular we denote the random variables ω_k by

$$\omega_k = X_k(\omega) = \mathbf{X}_k, \qquad k = 1, 2, \ldots,$$

and their sums by

$$\mathbf{S}_0 = 0, \qquad \mathbf{S}_n = \mathbf{X}_1 + \cdots + \mathbf{X}_n, \qquad n \geq 1.$$

If $f(\omega) = \mathbf{f}$ *is a random variable, we write for each* $x \in R$

$$\mathbf{P}[\mathbf{f} = x] = \mathbf{P}[\omega \mid f(\omega) = x]$$

and if $f(\omega) = \mathbf{f}$ *is in addition integrable, then*

$$\int_\Omega f(\omega) \, d\mathbf{P}(\omega) = \mathbf{E}[\mathbf{f}]$$

is called its expectation (expected value). Finally, when $A \in \mathscr{F}$, *the symbol* $\mathbf{E}[\mathbf{f};A]$ *will denote the expectation* $\mathbf{E}[f(\omega)\varphi_A(\omega)]$ *where* $\varphi_A(\omega) = 1$ *for* $\omega \in A$ *and zero otherwise.*

The two most frequently useful corollaries of D1 and of the general theory of measure and integration preceding it are summarized in

P1 (a) *If* $f(\omega) = F(\mathbf{S}_1, \mathbf{S}_2, \ldots, \mathbf{S}_n)$ *is any integrable function of* $\mathbf{S}_1, \ldots, \mathbf{S}_n$, *then its expected value is*

$$\mathbf{E}[\mathbf{f}] = \sum_{x_i \in R, i=1,2,\ldots,n} F(x_1, x_2, \ldots, x_n) \mathbf{P}[\mathbf{S}_1 = x_1, \mathbf{S}_2 = x_2, \ldots, \mathbf{S}_n = x_n]$$

$$= \sum_{x_i \in R, i=1,2,\ldots,n} F(x_1, \ldots, x_n) P(0,x_1) P(x_1,x_2) \ldots P(x_{n-1},x_n).$$

[6] See [37], p. 157.

3. SOME MEASURE THEORY

In particular, let $f(\omega) = 1$ if $\mathbf{S}_n = x \in R$, and 0 otherwise. Then

$$\mathbf{E}[\mathbf{f}] = \mathbf{P}[\mathbf{S}_n = x] = P_n(0,x).$$

(b) *The random variables $\omega_n = \mathbf{X}_n$ are pairwise independent functions. More generally, let I and J be two disjoint subsets of the natural numbers. Let $\mathscr{F}_I$ denote the smallest sub-σ-field which contains all sets of the form $[\omega \mid \omega_k = x]$ for $k \in I$ and $x \in R$, and define $\mathscr{F}_J$ in a similar way. Suppose now that $f(\omega)$ and $g(\omega)$ are integrable on $(\Omega, \mathscr{F}, \mathbf{P})$ and that $f(\omega)$ is measurable with respect to $\mathscr{F}_I$ and $g(\omega)$ with respect to $\mathscr{F}_J$. Then $\mathbf{f}$ and $\mathbf{g}$ are independent functions, and*

$$\mathbf{E}[\mathbf{fg}] = \mathbf{E}[\mathbf{f}]\mathbf{E}[\mathbf{g}].$$

The proof is nothing more or less than an exercise in rewording the content of D1, with two minor exceptions. The last equation in part (a) made use of the definition of $P_n(0,x)$ in D1.1, and the last equation in part (b) is Fubini's theorem as given in equation (4) preceding D1.

To exhibit the advantage of the language of sets (events) and their measures in a familiar setting we shall use P1 to give a proof of P1.2. According to D1.2 we may express $F_n(x,y)$ as

$$F_n(x,y) = \mathbf{P}[x + \mathbf{S}_k \neq y \text{ for } k = 1, 2, \ldots, n-1; x + \mathbf{S}_n = y],$$
$$n \geq 1.$$

If we let $\mathbf{T} = T(\omega)$ denote the first time that $x + \mathbf{S}_n = y$, or more formally

$$\mathbf{T} = \min [k \mid 1 \leq k, x + \mathbf{S}_k = y] \leq \infty$$

then $\mathbf{T} = T(\omega)$ is a random variable (measurable function) on the probability space $(\Omega, \mathscr{F}, \mathbf{P})$. It may or may not assume the value $+\infty$ on a set of positive measure—we do not care. What matters is that we may now write

$$F_n(x,y) = \mathbf{P}[\mathbf{T} = n],$$

so that parts (a) and (b) of P1.2, namely

$$F_n(x,y) = F_n(0, y-x) \quad \text{and} \quad \sum_{k=1}^{n} F_k(x,y) \leq 1$$

are quite obvious. To prove part (c) of P1.2 we decompose the event that $x + \mathbf{S}_n = y$. Using part (a) of P1 and the additivity of $\mathbf{P}[\cdot]$,

$$P_n(x,y) = \mathbf{P}[x + \mathbf{S}_n = y] = \sum_{k=1}^{n} \mathbf{P}[x + \mathbf{S}_n = y; \mathbf{T} = k]^7$$

$$= \sum_{k=1}^{n} \mathbf{P}[\mathbf{T} = k; \mathbf{S}_n - \mathbf{S}_k = 0].$$

(Here we made the observation that $\mathbf{S}_n - \mathbf{S}_k = 0$ when $\mathbf{T} = k$ and $x + \mathbf{S}_n = y$.) Now we may apply part (b) of P1 in the following manner. For each fixed integer $k \geq 1$, let $I = \{1, 2, \ldots, k\}$ and $J = \{k+1, k+2, \ldots, n\}$. Then

$$\mathbf{f}_k = f_k(\omega) = \begin{cases} 1 \text{ if } \mathbf{T} = k \\ 0 \text{ otherwise,} \end{cases}$$

$$\mathbf{g}_k = g_k(\omega) = \begin{cases} 1 \text{ if } \mathbf{S}_n - \mathbf{S}_k = 0 \\ 0 \text{ otherwise,} \end{cases}$$

are a pair of independent random variables, $f_k(\omega)$ being measurable with respect to $\mathscr{F}_I$ and $g_k(\omega)$ with respect to $\mathscr{F}_J$. It follows that $\mathbf{E}[\mathbf{f}_k \mathbf{g}_k] = \mathbf{E}[\mathbf{f}_k]\mathbf{E}[\mathbf{g}_k]$, and resuming our decomposition of $P_n(x,y)$ we obtain

$$P_n(x,y) = \sum_{k=1}^{n} \mathbf{E}[\mathbf{f}_k]\mathbf{E}[\mathbf{g}_k] = \sum_{k=1}^{n} \mathbf{P}[\mathbf{T} = k]\mathbf{P}[\mathbf{S}_n - \mathbf{S}_k = 0]$$

$$= \sum_{k=1}^{n} F_k(x,y) P_{n-k}(0,0),$$

which proves part (c) of P1.2.

As a rule we shall not give such obvious probabilistic arguments in all detail. But occasionally they are sufficiently complicated to warrant careful exposition, and then a more sophisticated notation than that employed in D1 will be helpful. Instead of working with a single probability measure $\mathbf{P}[\cdot]$ and the associated operator $\mathbf{E}[\cdot]$ of integration we introduce the starting point x of the random walk directly into the definition.

D2 *For each $x \in R$, the triple $(\Omega_x, \mathscr{F}_x, \mathbf{P}_x)$ is a probability space defined as follows. The elements ω of Ω_x are infinite sequences*

$$\omega = (\mathbf{x}_0, \mathbf{x}_1, \mathbf{x}_2, \ldots) \text{ with } \mathbf{x}_0 = x.$$

---

[7] $\mathbf{P}[A; B]$ means $\mathbf{P}[A \cap B]$.

If $A_n = [\omega \mid \mathbf{x}_k = a_k, k = 1, \ldots, n]$, $n \geq 0$, $a_k \in R$, then

$$\mathbf{P}_x[A_n] = P(x, a_1)P(a_1, a_2)\ldots P(a_{n-1}, a_n).$$

The σ-field $\mathscr{F}_x$ is the smallest σ-field containing all the cylinder sets A_n, and $\mathbf{P}_x[\cdot]$ is the unique countably additive measure on $\mathscr{F}_x$, having the desired value on cylinder sets. If $f(\omega) = \mathbf{f}$ is integrable on $(\Omega_x, \mathscr{F}_x, \mathbf{P}_x)$, its expectation is defined as

$$\mathbf{E}_x[\mathbf{f}] = \int_{\Omega_x} f(\omega) \, d\mathbf{P}_x(\omega).$$

The triple $(\Omega_x, \mathscr{F}_x, \mathbf{P}_x)$, being completely specified by the transition function $P(x, y)$ and the state space R, will be called "the random walk $\mathbf{x}_n = x_n(\omega)$, starting at the point $\mathbf{x}_0 = x$."

When $x = 0$, it is obvious that the sequence $\mathbf{S}_n$, $n \geq 0$, defined on $(\Omega, \mathscr{F}, \mathbf{P})$ by D1 is the same sequence of random variables as $\mathbf{x}_n$, for $n \geq 0$, defined on $(\Omega_0, \mathscr{F}_0, \mathbf{P}_0)$ by D2. When $x \neq 0$, then $\mathbf{S}_n + x$ on $(\Omega, \mathscr{F}, \mathbf{P})$ corresponds to $\mathbf{x}_n$ on $(\Omega_x, \mathscr{F}_x, \mathbf{P}_x)$, in the sense that

$$\mathbf{P}[x + \mathbf{S}_k = y_k, k = 1, 2, \ldots, n] = \mathbf{P}_x[\mathbf{x}_k = y_k, k = 1, 2, \ldots, n]$$

for every set of points $y_1, y_2, \ldots, y_n$ in R.

The advantage in this slight but important shift in our point of view is particularly important when one works with stopping times. Roughly speaking, a *stopping time* $\mathbf{T}$ is a random variable which depends only on the past of the random walk. Thus the event that $\mathbf{T} = k$ is required to be independent of $\mathbf{x}_{k+1}, \mathbf{x}_{k+2}, \ldots$ In D3 below that will be made precise by saying that the event $\mathbf{T} = k$ lies in the σ-field generated by $\mathbf{x}_1, \mathbf{x}_2, \ldots, \mathbf{x}_k$. Practically all the stopping times we shall ever encounter will be of the same simple type: if A is a subset of R, then $\mathbf{T} = \mathbf{T}_A$, defined as $\mathbf{T} = \min[k \mid k \geq 1, \mathbf{x}_k \in A]$, will be a stopping time; it is called the *hitting time* of the set A. (Our systematic study of hitting times will begin with D10.1 in Chapter III.)

D3 *Let* $\mathbf{T} = T(\omega)$ *be a random variable, i.e., a measurable function on* $(\Omega_x, \mathscr{F}_x, \mathbf{P}_x)$ *for each x in R whose possible values are the non-negative integers and* $+\infty$. *Let* $\mathscr{F}_{k,x}$ *denote the smallest sub-σ-field of* $\mathscr{F}_x$ *containing all the sets* $[\omega \mid \mathbf{x}_n = y]$ *for* $n = 0, 1, 2, \ldots, k$, *and* $y \in R$. *Suppose further that* $[\omega \mid T(\omega) = k] \in \mathscr{F}_{k,x}$ *for all* $k \geq 0$ *and all* $x \in R$. *Then* $\mathbf{T}$ *is called a stopping time.*

P2 *If* **T** *is a stopping time, and* $F(n,x)$ *an arbitrary non-negative function defined for all* $n \geq 0$ *and all* $x \in R$, *and if* $F(\infty, x) = 0$, *then*

$$\mathbf{E}_x[F(\mathbf{T}, \mathbf{x}_{\mathbf{T}+n})] = \mathbf{E}_x\{\mathbf{E}_{\mathbf{x}_\mathbf{T}}[F(\mathbf{T}, \mathbf{x}_n)]\} = \mathbf{E}_x[\sum_{y \in R} P_n(\mathbf{x}_\mathbf{T}, y) F(\mathbf{T}, y)]$$

$$= \sum_{z \in R} \sum_{k=0}^{\infty} \mathbf{P}_x[\mathbf{x}_\mathbf{T} = z; \mathbf{T} = k] \mathbf{E}_z[F(k, \mathbf{x}_n)] \leq \infty.$$

In particular, setting $F(n,x) = 1$ *if* $n = k$ *and* $x = y$, *while* $F(n,x) = 0$ *otherwise,*

$$\mathbf{P}_x[\mathbf{T} = k; \mathbf{x}_{\mathbf{T}+n} = y] = \sum_{z \in R} \mathbf{P}_x[\mathbf{x}_\mathbf{T} = z; \mathbf{T} = k] \mathbf{P}_z[\mathbf{x}_n = y].$$

The proof is omitted—the reduction to a problem involving independent functions is quite natural, and then the result becomes a special case of part (b) of P1. Instead we illustrate the use of P2 by giving a still shorter proof of part (c) of P1.2.

Let $\mathbf{T}_y = \min[k \mid k \geq 1, \mathbf{x}_k = y]$. It is clearly a stopping time in the sense of D3. Now we have

$$P_n(x,y) = \mathbf{P}_x[\mathbf{x}_n = y] = \sum_{k=1}^{n} \mathbf{P}_x[\mathbf{T}_y = k; \mathbf{x}_n = y]$$

$$= \sum_{k=1}^{n} \mathbf{P}_x[\mathbf{T}_y = k; \mathbf{x}_{\mathbf{T}_y + n - k} = y],$$

and since $\mathbf{x}_{\mathbf{T}_y} = y$, the last line of P2 gives

$$P_n(x,y) = \sum_{k=1}^{n} \mathbf{P}_x[\mathbf{T}_y = k] \mathbf{P}_y[\mathbf{x}_{n-k} = y] = \sum_{k=1}^{n} F_k(x,y) P_{n-k}(y,y).$$

As the next illustration of the use of the measure theoretical notation in D1 we shall reformulate the definition of recurrence. Let $A_n = [\omega \mid \mathbf{S}_n = 0]$. Then $\bigcup_{k=1}^{\infty} A_k$ is the event that a return to 0 occurs at some finite time, and

$$\overline{\lim_{n \to \infty}} A_n = \bigcap_{n=1}^{\infty} \bigcup_{k=n}^{\infty} A_k$$

the event that the random walk returns to 0 infinitely often. We shall prove, using the definition of recurrence in D1.5, that

P3 $\mathbf{P}[\overline{\lim}_{n \to \infty} A_n] = 1$ *if the random walk defining the probability measure* $\mathbf{P}[\cdot]$ *on* $(\Omega, \mathscr{F}, \mathbf{P})$ *is recurrent, and* $\mathbf{P}[\overline{\lim}_{n \to \infty} A_n] = 0$ *if it is transient.*

Proof: Since $\mathbf{P}[\cdot]$ is a completely additive measure, the measure of the intersection of a monotone sequence of sets is the limit of their measures (see (1) preceding D1). Since the sets $\bigcup_{k=n}^{\infty} A_k = B_n$ form a monotone sequence, we have

(1) $$\mathbf{P}\left[\varlimsup_{n\to\infty} A_n\right] = \lim_{n\to\infty} \mathbf{P}[B_n].$$

Now consider the sets

$$B_{n,m} = \bigcup_{k=n}^{n+m} A_k$$

which are easily seen to have measure

(2) $$\mathbf{P}[B_{n,m}] = \sum_{t\in R} P_{n-1}(0,t) \sum_{k=1}^{m+1} F_k(t,0), \qquad n \geq 1, \quad m \geq 0.$$

In the *recurrent* case we let $m \to \infty$ in (2), observing that the sets $B_{n,m}$ increase to B_n, so that

(3) $$\mathbf{P}[B_n] = \lim_{m\to\infty} \mathbf{P}[B_{n,m}] = \sum_{t\in R} P_{n-1}(0,t) F(t,0).$$

But we know from P2.5 that $F(t,0) = 1$ for all t such that $P_{n-1}(0,t) > 0$. Therefore $\mathbf{P}[B_n] = 1$ for $n \geq 1$, and equation (1) shows that

$$\mathbf{P}\left[\varlimsup_{n\to\infty} A_n\right] = 1.$$

In the *transient* case one goes back to equation (2), observing that

(4) $$\mathbf{P}[B_{n,m}] \leq \sum_{t\in R} P_{n-1}(0,t) \sum_{k=1}^{m+1} P_k(t,0)$$
$$= \sum_{j=n}^{n+m} P_j(0,0) = G_{n+m}(0,0) - G_{n-1}(0,0),$$
$$n \geq 1.$$

If we let $m \to \infty$, P1.4 tells us that

$$\mathbf{P}[B_n] \leq G(0,0) - G_{n-1}(0,0) < \infty,$$

so that finally

$$\mathbf{P}\left[\varlimsup_{n\to\infty} A_n\right] = \lim_{n\to\infty} \mathbf{P}[B_n] \leq G - G = 0,$$

completing the proof of P3.

32   THE CLASSIFICATION OF RANDOM WALK

The next step brings us to Kolmogorov's strong law of large numbers which we shall formulate, without proof,[8] in the terminology of D1.

P4 *If an arbitrary random walk has*

$$m = \sum_{x \in R} |x| P(0,x) < \infty,$$
$$\mu = \sum_{x \in R} x P(0,x),$$

then the sequence of random variables $\mathbf{S}_n = \mathbf{X}_1 + \cdots + \mathbf{X}_n$ *has the property that*

$$\mathbf{P}\left[\lim_{n \to \infty} \frac{\mathbf{S}_n}{n} = \mu\right] = 1.$$

Remark: (a) The theorem makes perfectly good sense, and is even true, for arbitrary random walk in dimension $d \geq 1$. When $d > 1$ the mean μ is a *vector* and so are the random variables $\mathbf{S}_n$. However, we shall use only P4 when $d = 1$.

(b) It is easy to show that the set

$$\left[\omega \mid \lim_{n \to \infty} \frac{S_n(\omega)}{n} = \mu\right] \in \mathscr{F},$$

by representing it in terms of a countable number of cylinder sets of the form $[\omega \mid |S_n(\omega) - n\mu| > r]$, where r may be any positive rational number.

Our immediate goal is the proof of the sufficient condition for one-dimensional random walk to be transient, which was discussed at the end of section 2, and which served as motivation for introducing measure theory. For this purpose the following seemingly (not actually) weak version of P4 will suffice. We shall assume that $\mu \neq 0$, let

$$C_n = \left[\omega \mid \left|\frac{\mathbf{S}_n}{n} - \mu\right| > \frac{|\mu|}{2}\right],$$

and observe that P4 implies

$$\mathbf{P}\left[\varlimsup_{n \to \infty} C_n\right] = 0.$$

---

[8] The first proof, for the case of Bernoulli random walk, is due to Borel [6], 1909. The general version in P4, announced by Kolmogorov twenty years later, was made possible by his measure theoretic formulation of probability theory [66]. Many modern texts, such as [23] and [73], contain several different proofs (Kolmogorov's as well as proofs based on martingales or ergodic theory).

Now we let $A_n = [\omega \mid S_n = 0]$, and notice that C_n was defined so that $A_n \subset C_n$ for all $n \geq 1$. Consequently

$$\mathbf{P}\left[\varlimsup_{n \to \infty} A_n\right] = 0.$$

Finally, we refer back to P3 where it was shown that the upper limit of the sequence of sets A_n has measure zero if and only if the underlying random walk is transient.

Hence every random walk with $m < \infty$ and $\mu \neq 0$ is transient! (Our proof is perfectly valid in dimension $d > 1$.) Combining this fact, when $d = 1$, with P2.8, we can assert that

T1 *One-dimensional random walk with finite absolute mean m is recurrent if and only if the mean μ is zero.*

The ideas of this section are now applied to obtain an amusing generalization of certain properties of Bernoulli random walk.

E1 It was shown in E1.2 that *Bernoulli random walk* with $p > q$ has the property that

(1) $$G(0,x) = \frac{1}{p-q} \text{ for } x \geq 0.$$

This result can be extended to arbitrary *right-continuous random walk with positive mean* in a surprisingly interesting fashion; we shall assume that

$$P(0,x) = 0 \text{ for } x \geq 2, \quad 0 < \mu = \sum_{x=-\infty}^{1} xP(0,x)$$

and prove that *for such a random walk*

(2) $$G(0,0) = G(0,1) = G(0,2) = \cdots = \frac{1}{\mu}.$$

This is a clear-cut generalization of (1) where μ happens to have the value $p - q$.

First we observe that this random walk is transient according to T1. Also $S_n \to +\infty$ (or $\mathbf{x}_n \to +\infty$) with probability one, according to the strong law of large numbers (P4). Consequently, if

$$\mathbf{T}_y = \min [k \mid 1 \leq k \leq \infty, \mathbf{x}_k = y],$$

we may conclude that

$$\mathbf{P}_0[\mathbf{T}_y < \infty] = 1 \text{ for } y \geq 1,$$

as right-continuous random walk cannot "skip" a point in going to the right. Thus

(3) $$F(0,y) = \sum_{k=1}^{\infty} F_k(0,y) = 1, \quad y \geq 1.$$

Our next observation is that, when $x \neq y$,

(4) $$G(x,y) = F(x,y)G(y,y) = F(x,y)G(0,0)$$

which was contained in P1.5. Calling $G(0,0) = G$ and setting $x = 0$ in (4), we get from (3)

(5) $$G(0,y) = G \quad \text{for } y \geq 1.$$

Thus it remains to show only that the constant G has the value μ^{-1}. For this purpose we use the identity

(6) $$G(0,x) = \sum_{y \in R} G(0,y)P(y,x) + \delta(0,x), \quad x \in R.$$

When $x \geq 1$, this identity reduces to the trivial result $G = G$ so we must consider also negative values of x, although $G(0,x)$ is then unknown. One can write

$$\mu = P(0,1) - \sum_{y=1}^{\infty} \sum_{x=-\infty}^{0} P(y,x)$$

which suggests summing x in (6) from $-\infty$ to 0. For $n \geq 1$

$$\sum_{x=-n}^{0} G(0,x) - 1 = \sum_{y \in R} G(0,y) \sum_{x=-n}^{0} P(y,x)$$

$$= \sum_{y=1}^{\infty} G(0,y) \sum_{x=-n}^{0} P(y,x) + \sum_{y=-\infty}^{0} G(0,y) \sum_{x=-n}^{0} P(y,x)$$

Hence, letting $n \to +\infty$,

$$G\mu = 1 + GP(0,1) - \sum_{y=-\infty}^{0} G(0,y)\left[1 - \sum_{x=-\infty}^{0} P(y,x)\right]$$

$$= 1 - \sum_{y=-\infty}^{-1} G(0,y)\left[1 - \sum_{x=-\infty}^{0} P(y,x)\right] = 1,$$

since

$$\sum_{x=-\infty}^{0} P(y,x) = 1 \quad \text{when } y \leq -1.$$

That completes the proof of (2). In Chapter VI (P24.6) we shall prove a further generalization of (2). Adhering to the assumption that $m < \infty$, $\mu > 0$, while discarding the hypothesis of right continuity, one can no longer assert (2). But in P24.6 equation (2) will be shown to remain true asymptotically, in the sense that

$$\lim_{x \to +\infty} G(0,x) = \frac{1}{\mu}.$$

4. THE RANGE OF A RANDOM WALK[9]

To illustrate the phenomenon of recurrence from a novel point of view we shall study a random variable associated in a natural way with every random walk. The transition function $P(x,y)$ and the dimension of R will be completely arbitrary. The most natural way to proceed is in terms of the measure space of D3.1. We had $\mathbf{S}_0 = 0$, $\mathbf{S}_n = \mathbf{X}_1 + \cdots + \mathbf{X}_n$, where the $\mathbf{X}_i$ are identically distributed independent random variables with distribution $\mathbf{P}[\mathbf{X}_i = x] = P(0,x)$. Thus $\mathbf{S}_n$ represents the position of the random walk at time n.

D1 *For $n \geq 0$, $\mathbf{R}_n$ is the cardinality of the (random) subset of R which consists of the points $0 = \mathbf{S}_0, \mathbf{S}_1, \ldots, \mathbf{S}_n$.*

Thus $\mathbf{R}_n$ (called the *range* of the random walk in time n) is clearly a measurable function on $(\Omega, \mathscr{F}, \mathbf{P})$ being simply the number of distinct elements in the sequence $\mathbf{S}_0, \mathbf{S}_1, \ldots, \mathbf{S}_n$, or *the number of distinct points visited by the random walk in time n.*

Note that it would be perfectly possible to define the probability law of $\mathbf{R}_n$ (and that is all we shall be concerned with) in terms of the simple analytic setup of section 1. For example,

$$\mathbf{P}[\mathbf{R}_2 = 1] = [P(0,0)]^2$$
$$\mathbf{P}[\mathbf{R}_2 = 2] = \sum_{x \neq 0} P(0,x)[P(x,x) + P(x,0)] + P(0,0) \sum_{x \neq 0} P(0,x),$$

but evidently this is not a very convenient representation.

We shall be concerned with a weak law of large numbers for the sequence $\mathbf{R}_n$ and we shall say that $\mathbf{R}_n/n$ converges *in measure* to the constant c if

$$\lim_{n \to \infty} \mathbf{P}\left[\left|\frac{\mathbf{R}_n}{n} - c\right| > \epsilon\right] = 0$$

for every $\epsilon > 0$. We shall prove the theorem

T1 *If a random walk is transient, then $\mathbf{R}_n/n$ converges in measure to $G^{-1} = 1 - F > 0$ and, if it is recurrent, $\mathbf{R}_n/n$ converges in measure to 0 (which equals $1 - F$ in this case too!).*

Proof: First we calculate $\mathbf{E}[\mathbf{R}_n]$. The formula

$$\mathbf{E}[\mathbf{R}_n] = \sum_{k=1}^{n+1} k\mathbf{P}[\mathbf{R}_n = k]$$

---

[9] The results of this section have recently been extended in several directions. Central limit and iterated logarithm theorems have been proved for $\mathbf{R}_n$ in the transient case, ([S11], [S12], [S14], [S15]) and a strong law of large numbers for $\mathbf{R}_n/\mathbf{E}(\mathbf{R}_n)$ in the recurrent case [S13].

is not useful, so that it is better to go back to D1, in order to represent $\mathbf{R}_n$ as a sum of random variables, each of which has value 0 or 1 according as $\mathbf{S}_n$ equals one of the preceding partial sums or not. We can write

$$\mathbf{R}_n = \sum_{k=0}^{n} \boldsymbol{\varphi}_k,$$

where $\boldsymbol{\varphi}_0 = 1$, $\boldsymbol{\varphi}_k(\mathbf{S}_1, \ldots, \mathbf{S}_k) = 1$ if $\mathbf{S}_k \neq \mathbf{S}_\nu$ for all $\nu = 0, 1, \ldots, k-1$, and $\boldsymbol{\varphi}_k = 0$ otherwise. Then, even though the random variables $\boldsymbol{\varphi}_k$ are not independent,

$$\mathbf{E}[\mathbf{R}_n] = \sum_{k=0}^{n} \mathbf{E}[\boldsymbol{\varphi}_k].$$

We have, for $k \geq 1$,

$$\begin{aligned}
\mathbf{E}[\boldsymbol{\varphi}_k] &= \mathbf{P}[\mathbf{S}_k - \mathbf{S}_{k-1} \neq 0, \mathbf{S}_k - \mathbf{S}_{k-2} \neq 0, \ldots, \mathbf{S}_k \neq 0] \\
&= \mathbf{P}[\mathbf{X}_k \neq 0, \mathbf{X}_k + \mathbf{X}_{k-1} \neq 0, \ldots, \mathbf{X}_k + \cdots + \mathbf{X}_1 \neq 0] \\
&= \mathbf{P}[\mathbf{X}_1 \neq 0, \mathbf{X}_1 + \mathbf{X}_2 \neq 0, \ldots, \mathbf{X}_1 + \cdots + \mathbf{X}_k \neq 0] \\
&= \mathbf{P}[\mathbf{S}_j \neq 0 \text{ for } j = 1, 2, \ldots, k] \\
&= 1 - \sum_{j=1}^{k} F_j(0,0),
\end{aligned}$$

in the notation of D1.2. By D1.5

$$\lim_{n \to \infty} \mathbf{E}[\boldsymbol{\varphi}_n] = 1 - F \geq 0,$$

where $1 - F = 0$ if and only if the random walk is recurrent. It follows that the averages of $\mathbf{E}[\boldsymbol{\varphi}_n]$ also converge to $1 - F$, or

$$\lim_{n \to \infty} \frac{1}{n} \mathbf{E}[\mathbf{R}_n] = \lim_{n \to \infty} \frac{1}{n} \sum_{k=0}^{n} \mathbf{E}[\boldsymbol{\varphi}_k] = 1 - F.$$

Finally, to show that $\mathbf{R}_n/n$ converges in measure to $1 - F$, we have to distinguish between the cases of recurrent and transient random walk. The easy case is the *recurrent* one. There we have for arbitrary $\epsilon > 0$

$$\begin{aligned}
\mathbf{P}\left[\frac{\mathbf{R}_n}{n} > \epsilon\right] &= \sum_{[k \mid k > n\epsilon]} \mathbf{P}[\mathbf{R}_n = k] \leq \frac{1}{n\epsilon} \sum_{k=0}^{\infty} k \mathbf{P}[\mathbf{R}_n = k] \\
&= \frac{1}{n\epsilon} \mathbf{E}[\mathbf{R}_n] \to \frac{1 - F}{\epsilon} = 0,
\end{aligned}$$

as $n \to \infty$, so that $\mathbf{R}_n/n \to 0$ in measure.

In the *transient* case, a similar estimate (by the method of Chebychev's inequality) gives

$$\mathbf{P}\left[\left|\frac{\mathbf{R}_n}{n} - (1-F)\right| > \epsilon\right] \le \frac{1}{n^2\epsilon^2} \mathbf{E}[|\mathbf{R}_n - n(1-F)|^2]$$

$$= \frac{1}{n^2\epsilon^2} \sigma^2(\mathbf{R}_n) + \frac{1}{\epsilon^2}\left\{1 - F - \mathbf{E}\left[\frac{\mathbf{R}_n}{n}\right]\right\}^2.$$

Here $\sigma^2(\mathbf{R}_n) = \mathbf{E}[\{\mathbf{R}_n - \mathbf{E}[\mathbf{R}_n]\}^2]$ is the variance of $\mathbf{R}_n$, and since we have shown that $\mathbf{E}[\mathbf{R}_n/n] \to 1 - F$, the proof will be complete if we can show that

$$\lim_{n\to\infty} \frac{1}{n^2} \sigma^2(\mathbf{R}_n) = 0.$$

One calculates as follows:

$$\sigma^2(\mathbf{R}_n) = \mathbf{E}\left[\sum_{j=0}^{n}\sum_{k=0}^{n} \boldsymbol{\varphi}_j\boldsymbol{\varphi}_k\right] - \left[\sum_{j=0}^{n} \mathbf{E}[\boldsymbol{\varphi}_j]\right]^2$$

$$= \sum_{j=0}^{n}\sum_{k=0}^{n} \{\mathbf{E}[\boldsymbol{\varphi}_j\boldsymbol{\varphi}_k] - \mathbf{E}[\boldsymbol{\varphi}_j]\mathbf{E}[\boldsymbol{\varphi}_k]\}$$

$$\le 2\sum_{0\le j<k\le n} \{\mathbf{E}[\boldsymbol{\varphi}_j\boldsymbol{\varphi}_k] - \mathbf{E}[\boldsymbol{\varphi}_j]\mathbf{E}[\boldsymbol{\varphi}_k]\} + \sum_{j=0}^{n} \mathbf{E}[\boldsymbol{\varphi}_j].$$

Now we observe that when $j < k$

$$\begin{aligned}\mathbf{E}[\boldsymbol{\varphi}_j\boldsymbol{\varphi}_k] &= \mathbf{P}[\mathbf{S}_j \ne \mathbf{S}_\alpha \text{ for } \alpha < j \text{ and } \mathbf{S}_k \ne \mathbf{S}_\beta \text{ for } \beta < k]\\ &\le \mathbf{P}[\mathbf{S}_j \ne \mathbf{S}_\alpha \text{ for } \alpha < j \text{ and } \mathbf{S}_k \ne \mathbf{S}_\beta \text{ for } j \le \beta < k]\\ &= \mathbf{P}[\mathbf{X}_j \ne 0, \mathbf{X}_j + \mathbf{X}_{j-1} \ne 0, \ldots, \mathbf{X}_j + \cdots + \mathbf{X}_1 \ne 0;\\ &\qquad \mathbf{X}_k \ne 0, \mathbf{X}_k + \mathbf{X}_{k-1} \ne 0, \ldots, \mathbf{X}_k + \cdots \mathbf{X}_{j+1} \ne 0]\\ &= \mathbf{P}[\mathbf{X}_1 \ne 0, \ldots, \mathbf{X}_1 + \cdots + \mathbf{X}_j \ne 0]\mathbf{P}[\mathbf{X}_1 \ne 0, \ldots,\\ &\qquad \mathbf{X}_1 + \cdots + \mathbf{X}_{k-j} \ne 0],\end{aligned}$$

so that for $j < k$

$$\mathbf{E}[\boldsymbol{\varphi}_j\boldsymbol{\varphi}_k] \le \mathbf{E}[\boldsymbol{\varphi}_j]\mathbf{E}[\boldsymbol{\varphi}_{k-j}].$$

Going back to the estimation of $\sigma^2(\mathbf{R}_n)$,

$$\sigma^2(\mathbf{R}_n) \le 2\sum_{j=0}^{n} \mathbf{E}[\boldsymbol{\varphi}_j] \sum_{k=j+1}^{n} [\mathbf{E}\boldsymbol{\varphi}_{k-j} - \mathbf{E}\boldsymbol{\varphi}_k] + \mathbf{E}[\mathbf{R}_n].$$

Given any monotone nonincreasing sequence $a_1 \ge a_2 \ge a_3 \ldots$, it is easy to see that

$$\sum_{k=j+1}^{n} [a_{k-j} - a_k] = (a_1 + a_2 + \cdots + a_{n-j}) - (a_{j+1} + \cdots + a_n)$$

assumes its maximum when $j = [n/2]$, the greatest integer in $n/2$. Setting $a_k = \mathbf{E}\boldsymbol{\varphi}_k$ and continuing the estimation, one therefore obtains

$$\sigma^2(\mathbf{R}_n) \leq 2 \sum_{j=0}^{n} \mathbf{E}[\boldsymbol{\varphi}_j]\mathbf{E}[\mathbf{R}_{n-[n/2]} + \mathbf{R}_{[n/2]} - \mathbf{R}_n] + \mathbf{E}[\mathbf{R}_n].$$

Since we know that $\mathbf{E}[\mathbf{R}_n/n]$ tends to $1 - F$ as $n \to \infty$, it follows that

$$\lim_{n \to \infty} \frac{\sigma^2(\mathbf{R}_n)}{n^2} \leq 2(1 - F) \lim_{n \to \infty} \mathbf{E}\left[\frac{\mathbf{R}_{n-[n/2]} + \mathbf{R}_{[n/2]} - \mathbf{R}_n}{n}\right]$$

$$= 2(1 - F)\left[\frac{1-F}{2} + \frac{1-F}{2} - (1-F)\right] = 0.$$

That completes the proof of T1.

The method by which we proved T1 was devised by Dvoretzky and Erdös ([27], 1951) in the course of a detailed study of the range $\mathbf{R}_n$ for simple random walk. One of their results was the strong law

(1) $$\mathbf{P}\left[\lim_{n \to \infty} \frac{\mathbf{R}_n}{n} = 1 - F\right] = 1$$

for simple random walk, which of course implies T1. Actually, this is *a general theorem, valid for arbitrary random walk*. We shall sketch a simple proof of (1) which is beyond the scope of this chapter as it depends on Birkhoff's *ergodic theorem*.

E1 In order to establish (1) above, for arbitrary random walk, we shall introduce the upper and lower bounds

(2) $$\mathbf{D}_n \leq \mathbf{R}_n \leq \mathbf{R}_{n,M}, \qquad n \geq 0,$$

defined as follows.

We choose a (large) positive integer M and define $\mathbf{R}_{n,M}$ as the number of distinct points counted by "wiping the slate clean" at times M, $2M$, $3M$, ... To formalize this idea, let $\mathbf{Z}_k(M)$ be the independent random variables defined by: $\mathbf{Z}_k(M) =$ the number of distinct partial sums among $\mathbf{S}_{kM}, \mathbf{S}_{kM+1}, \ldots, \mathbf{S}_{(k+1)M-1}$, for $k = 0, 1, 2, \ldots$. Then let

$$\mathbf{R}_{n,M} = \sum_{k=0}^{[n/M]+1} \mathbf{Z}_k(M).$$

Clearly this makes $\mathbf{R}_n \leq \mathbf{R}_{n,M}$. We may apply the strong law of large numbers, P3.4, to the averages of the sequence $\mathbf{Z}_k(M)$ obtaining

$$\varlimsup_{n \to \infty} \frac{\mathbf{R}_n}{n} \leq \varlimsup_{n \to \infty} \frac{1}{n} \sum_{k=0}^{[n/M]+1} \mathbf{Z}_k(M) = \frac{1}{M} \mathbf{E}[\mathbf{Z}_0(M)] = \frac{1}{M} \mathbf{E}[\mathbf{R}_M].$$

Now we let $M \to \infty$, and as it was shown in the proof of T1 that
$$M^{-1}\mathbf{E}[\mathbf{R}_M] \to 1 - F,$$
we have
(3)
$$\varlimsup_{n \to \infty} \frac{\mathbf{R}_n}{n} \leq 1 - F,$$
with probability one.

Keeping in mind the definition of $\mathbf{R}_n = \mathbf{R}_n(\mathbf{X}_1, \mathbf{X}_2, \ldots, \mathbf{X}_n)$ as a random variable on the measure space $(\Omega, \mathscr{F}, \mathbf{P})$ where $\omega_k = X_k(\omega)$, and the $\mathbf{X}_k$ are independent with distribution $P(0,x) = \mathbf{P}[\mathbf{X}_k = x]$, we define $\mathbf{D}_n$ as the *number of points visited in time n which are never revisited at any later time*. Thus $\mathbf{D}_n$ will be a function on the same measure space $(\Omega, \mathscr{F}, \mathbf{P})$, and its precise definition is

$$\mathbf{D}_n = \sum_{k=0}^n \psi_k,$$

where
$$\psi_k = 1 \text{ if } \mathbf{X}_{k+1} + \cdots + \mathbf{X}_{k+\nu} \neq 0 \text{ for } \nu = 1, 2, \ldots$$
$$= 0 \text{ otherwise.}$$

It should be obvious that $\mathbf{D}_n$ satisfies (2); every point which is counted in determining $\mathbf{D}_n$ is certainly visited in time n.

We are now ready to apply the individual ergodic theorem (see [23], p. 464 or [73], p. 421). Let T be the shift operator on Ω defined by $(T\omega)_k = \omega_{k+1} = \mathbf{X}_{k+1}$. It is measure preserving, i.e.,
$$\mathbf{P}[T^{-1}A] = \mathbf{P}[A], \qquad A \in \mathscr{F},$$
so that the limit
$$\lim_{n \to \infty} \frac{1}{n} \sum_{k=0}^n f(T^k\omega)$$
exists with probability one (it may be a random variable) for every integrable function $\mathbf{f}$ on Ω. Of course our choice for $\mathbf{f}$ is $f(\omega) = \psi_0(\omega)$, so that $f(T^k\omega) = \psi_k(\omega)$ and the limit
$$\lim_{n \to \infty} \frac{1}{n} \sum_{k=0}^n \psi_0(T^k\omega) = \lim_{n \to \infty} \frac{\mathbf{D}_n}{n}$$
exists with probability one. Since $(\Omega, \mathscr{F}, \mathbf{P})$ is a product measure space, much more can be said: the above limit must in fact be a constant (this is the so-called zero-one law of Kolmogorov, [23], p. 102), and this constant is
$$\mathbf{E}[f(\omega)] = \mathbf{E}[\psi_0(\omega)] = \mathbf{P}[\mathbf{X}_1 + \cdots + \mathbf{X}_\nu \neq 0 \text{ for all } \nu \geq 1] = 1 - F.$$

Therefore we now have, with probability one,
(4)
$$\varliminf_{n \to \infty} \frac{\mathbf{R}_n}{n} \geq \lim_{n \to \infty} \frac{\mathbf{D}_n}{n} = 1 - F,$$
and equations (3) and (4) together imply (1).

Remark: The above proof was discovered jointly with Kesten and Whitman (unpublished). Not only is it simpler than the proof of T1 but it also yields interesting generalizations without any further work. In problem 14 of Chapter VI it is shown how $\mathbf{R}_n$, the number of points swept out in time n, may be replaced by the *number of points swept out by a finite set*. The limit in (1) is then the *capacity* of the set in question. Thus the study of $\mathbf{R}_n$ takes us into the realm of *potential theory*, which will be discussed briefly in sections 13 (Chapter III) and 25 (Chapter VI). To anticipate a little, let us state the analogue of T1 and E1 for a *three-dimensional Brownian motion process* $\mathbf{x}(t)$, the continuous time analogue of simple random walk in three space (see Feller [31], Vol. I, Ch. XIV). Suppose that S is a compact subset of three-space. Let $\mathbf{x}(\tau) + S$ denote the translate of S by the random process $\mathbf{x}(\tau)$. The set swept out by S in time t is the union of the sets $\mathbf{x}(\tau) + S$ over $0 \le \tau \le t$, and we denote the Lebesgue measure (volume) of the set thus swept out as

$$\mathbf{R}_t(S) = \left| \bigcup_{0 \le \tau \le t} \{\mathbf{x}(\tau) + S\} \right|.$$

This definition makes sense since Brownian motion, when properly defined, is continuous almost everywhere on its probability space. On this space one can then prove, following the method of E1, that

$$\lim_{t \to \infty} \frac{\mathbf{R}_t(S)}{t} = C(S)$$

exists almost everywhere. The limit $C(S)$ is nothing but the ordinary *electrostatic capacity of the set S*.

We shall find it convenient, in section 25, to associate a notion of capacity with each random walk, and it will be seen (see E25.1) that the limit $1 - F$ in equation (1) is again simply the *capacity of a set consisting of a single point*.

5. THE STRONG RATIO THEOREM

Here we are concerned with ratios of transition probabilities of the form

$$\frac{P_n(0,x)}{P_n(0,0)} \quad \text{and} \quad \frac{P_{n+1}(0,0)}{P_n(0,0)},$$

which will be shown to converge (to the limit one) as $n \to \infty$, under suitable assumptions about the transition function $P(x,y)$. This result, in T1 below, evidently constitutes a significant improvement over the weak ratio theorem P1.5. It was first proved by Chung and Erdös [11], 1951.

This rather deep theorem will in fact "lie fallow" until we reach the last section of Chapter VII. There we shall be concerned with ratios of the form

$$J_n(x) = \frac{\sum_{k=n}^{\infty} F_k(x,0)}{\sum_{k=n}^{\infty} F_k(0,0)},$$

for recurrent random walk, and the proof of the convergence of $J_n(x)$ as $n \to \infty$ will make use of T1 which concerns the far simpler ratio $P_{n+1}(0,0)/P_n(0,0)$. While $P_{n+1}(0,0)/P_n(0,0)$ and even $P_n(0,x)/P_n(0,0)$ tend to the "uninteresting" limit one, this is not true of

(1) $$\lim_{n \to \infty} J_n(x) = J(x).$$

Since this fact lies at the heart of the theory of recurrent random walk, a few remarks, even out of their proper context are called for as motivation for later work. Long before we are able to prove (in T32.1) that the ratios $J_n(x)$ converge, we shall prove (in Chapters III and VII) that the series

(2) $$\sum_{n=0}^{\infty} [P_n(0,0) - P_n(x,0)] = a(x)$$

converges for arbitrary aperiodic random walk. This result is of course related to the problem of this section as it is much stronger than P1.5 while unfortunately being too weak to imply that $P_n(0,x)/P_n(0,0)$ converges. Thus our plan is the following. In this section we prove, by a rather "brute force" attack, that $P_n(0,x)/P_n(0,0) \to 1$, under suitable conditions (T1). Later (in Chapter III for dimension 2 and in Chapter VII for dimension 1) we establish (2), and finally, in section 32 (Chapter VII), we use (2) and T1 to show that (1) holds, *with the same limit function as in equation (2). Thus, we shall find that $a(x) = J(x)$ for $x \neq 0$,* or

(3) $$\delta(x,0) + \sum_{n=1}^{\infty} [P_n(0,0) - P_n(x,0)] = \lim_{n \to \infty} \frac{\sum_{k=n}^{\infty} F_k(x,0)}{\sum_{k=n}^{\infty} F_k(0,0)}, \quad x \in R$$

for arbitrary aperiodic random walk, recurrent or transient, regardless of dimension.

It is clear that any set of conditions for the convergence of ratios like $P_n(0,x)/P_n(0,0)$ must ensure the positivity of $P_n(0,0)$, at least for all sufficiently large integers n. Observe that a random walk may be aperiodic, and even recurrent, and still fail to satisfy this condition. (For simple random walk in one dimension, $P_n(0,0) = 0$ for all odd n.) Thus we have to strengthen the condition of aperiodicity and we adopt a definition which may at first sight be complicated but which will be seen to accomplish what we want. At the same time it will be weak enough that it will impose no essential restriction. P1 will show that at least every recurrent random walk may be modified (by an appropriate change in its space and time scale) so as to be *strongly aperiodic*.

D1 *A random walk with transition function $P(x,y)$ on R is called strongly aperiodic if it has the property that for each x in R, the smallest subgroup of R which contains the set*

$$x + \Sigma = [y \mid y = x + z, \text{ where } P(0,z) > 0]$$

is R itself.

Since we shall be interested primarily in recurrent random walk in this section, it will suffice to study the implications of D1 under this restriction. (In Chapter II (P7.8) we shall derive a Fourier analytic characterization of arbitrary strongly aperiodic random walk, which is used in P7.9 to obtain information concerning the asymptotic behavior of $P_n(0,x)$ as $n \to \infty$. With this exception strong aperiodicity will not be of any real interest to us, until we get to the aforementioned ratio theorem in Chapter VII.)

P1 *For aperiodic recurrent random walk there are two possibilities.*

(a) *The random walk is strongly aperiodic. In this case, given any x in R, there is some integer $N = N(x)$ such that $P_n(0,x) > 0$ for all $n \geq N$.*

(b) *The random walk is not strongly aperiodic. Then there is an integer $s \geq 2$ (the period) such that $P_{ns}(0,0) > 0$ for all sufficiently large n, while $P_k(0,0) = 0$ when k is not a multiple of s.*

Remark: The proof will show that in case (b), when $x \neq 0$, $P_n(0,x) = 0$ unless n belongs to an arithmetic progression of the form $n = ks + r$, where r depends on x.

Proof: We begin by observing that the set of positive integers $\mathcal{N} = [n \mid P_n(0,0) > 0]$ is a semigroup; for if $P_n(0,0) > 0$, $P_m(0,0) > 0$, then $P_{n+m}(0,0) \geq P_n(0,0)P_m(0,0) > 0$. Since the random walk is recurrent (this assumption is much stronger than necessary), $\mathcal{N}$ is nonempty. Hence there are only two possibilities:
 (i) the greatest common divisor of the elements of $\mathcal{N}$ is 1,
 (ii) the greatest common divisor is $s > 1$.
The proof of P1 will consist in showing that case (i) corresponds to case (a) in P1 and case (ii) to case (b).

In case (i) $P_n(0,0)$ is obviously positive for sufficiently large n. Similarly, given any $x \neq 0$, there exists an integer m such that $P_m(0,x) > 0$. Therefore

$$P_{n+m}(0,x) \geq P_n(0,0)P_m(0,x) > 0$$

for all sufficiently large n. In case (i) it therefore remains only to show that the random walk is strongly aperiodic. Thus, choosing two arbitrary points x and y in R, we must show that y is in the group generated by $x + \Sigma$. Since $P_n(0,y)$ as well as $P_n(0,0)$ are positive for all sufficiently large n, it follows that one can find some $n > 0$ such that

$$y = \sigma_1 + \sigma_2 + \cdots + \sigma_n \quad \text{and} \quad 0 = -\sigma_{n+1} - \sigma_{n+2} - \cdots - \sigma_{2n}$$

where $\sigma_1, \ldots, \sigma_{2n}$ are elements (not necessarily distinct) of Σ. This enables us to represent y as

$$y = \sum_{k=1}^{n} (x + \sigma_k) - \sum_{k=1}^{n} (x + \sigma_{n+k})$$

which is the desired representation of y as an element of the group generated by $x + \Sigma$. Therefore the random walk is strongly aperiodic.

Finally, suppose we are in case (ii). Let us call $-\Sigma_k$ the set of points of the form $-(\sigma_1 + \cdots + \sigma_k)$, $\sigma_i \in \Sigma$. Observe that if $x \in -\Sigma_1$, then $P_n(0,x) = 0$ unless n is of the form $n = ks - 1$. This is quite clear as the two conditions $x \in -\Sigma_1$, and $P_n(0,x) > 0$ imply that $P_{n+1}(0,0) \geq P_n(0,x)P(x,0) > 0$. Thus we have shown that *points in $-\Sigma_1$ can be visited only at times in the progression $ks - 1$*. Similarly points in $-\Sigma_2$ can be visited only at times in the progression $ks - 2$, and so forth, until we come to the set $-\Sigma_s$ which can be visited only at times ks. Furthermore, each set $-\Sigma_{js+r}$, $j \geq 0$, $1 \leq r \leq s$, clearly shares with $-\Sigma_r$ the property of being visited only at times which are in the progression $ks - r$. Since the random walk with transition function $P(x,y)$ is recurrent and aperiodic, we know from T2.1 that $\bar{R} = R = R^+$, which is the semigroup generated by $-\Sigma$. Hence every point of R is in some $-\Sigma_{js+r}$. Thus R is decomposed into s equivalence classes of points, two points x and y being equivalent if the arithmetic progression of integers n on which $P_n(0,x) > 0$ for large enough n is the same as that for y. It follows from $P_{m+n}(0,0) \geq P_m(0,x)P_n(x,0)$ that the points of $-\Sigma_{js+r}$ are in the same equivalence class as those of $\Sigma_{js+(s-r)}$. Therefore the equivalence class containing the origin is a group H. (The other equivalence classes are its cosets!) To complete the proof of P1 we simply select x from Σ_{s-1}. Then $x + \Sigma \subset \Sigma_s$ and since Σ_s generates the proper subgroup H of R, we have proved that a random walk with property (ii) cannot be strongly aperiodic.

It will not suffice to know that the n-step transition function of strongly aperiodic recurrent random walk is positive for sufficiently large n. We shall require an explicit lower bound which is given by

P2 *Let $P(x,y)$ be the transition function of a recurrent strongly aperiodic random walk. Given any point x in R and any ϵ, $0 < \epsilon < 1$, there is some $N = N(x,\epsilon)$ such that*

$$P_n(0,x) \geq (1 - \epsilon)^n \text{ for } n \geq N.$$

Proof: The random walk is recurrent, so that $\sum P_n(0,0) = \infty$, which implies that the power series $\sum z^n P_n(0,0)$ has radius of convergence one. Hence

$$\varlimsup_{n \to \infty} [P_n(0,0)]^{1/n} = 1.$$

It follows that, given any ϵ with $0 < \epsilon < 1$,

(1) $$P_n(0,0) \geq (1 - \epsilon)^n$$

for infinitely many positive integers n. We also observe that if (1) holds for $n = m$, then

(2) $$P_{km}(0,0) \geq [P_m(0,0)]^k \geq (1 - \epsilon)^{mk}$$

for every integer $k \geq 1$. Now we pick an arbitrary point x in R, and from the infinite sequence of integers m for which (2) holds we select m, as we may by use of part (a) of P1, such that

(3) $$\min_{m < j \leq 2m} P_j(0,x) = A > 0.$$

5. THE STRONG RATIO THEOREM

When $n > 2m$, we write $n = (k+1)m + r$, $0 < r \leq m$, so that

$$P_n(0,x) \geq P_{m+r}(0,x) P_{km}(0,0),$$

and using (2) and (3),

$$P_n(0,x) \geq A(1-\epsilon)^{mk} > A(1-\epsilon)^n.$$

Even though $A < 1$, we have shown that

$$\lim_{n \to \infty} P_n(0,x)(1-\epsilon)^{-n} > 0, \qquad x \in R,$$

for every ϵ between 0 and 1 which implies that P2 is true.

The next lemma concerns a well-known[10] exponential lower bound for the Bernoulli distribution. In terms of Bernoulli random walk it becomes

P3 *Let* $\mathbf{S}_n = \mathbf{X}_1 + \cdots + \mathbf{X}_n$, *in the notation of* D3.1, *denote Bernoulli random walk with*

$$0 < p = P(0,1) = \mathbf{P}[\mathbf{X}_k = 1] = 1 - P(0,-1) < 1.$$

There exists a constant $a > 0$, depending on p, but independent of n, such that

$$\mathbf{P}\left[\left|\frac{\mathbf{S}_n}{n} - (2p-1)\right| \geq \epsilon\right] \leq 2e^{-a\epsilon^2 n}$$

for all $n \geq 1$ and all $\epsilon > 0$.

Proof: For convenience we call $\mathbf{T}_n = n^{-1}[\mathbf{S}_n - n(2p-1)]$. Since $|\mathbf{T}_n| \leq 2$, P3 is automatically true when $\epsilon > 2$ so that we may assume that $0 < \epsilon \leq 2$. We may also simplify the problem by proving only that

(1) $\qquad \mathbf{P}[\mathbf{T}_n \geq \epsilon] \leq e^{-a\epsilon^2 n}, \qquad n \geq 1, \ 0 < \epsilon \leq 2.$

for some $a > 0$. The proof of (1) with $\mathbf{T}_n$ replaced by $-\mathbf{T}_n$ will be the same, since it concerns the Bernoulli random walk $-\mathbf{S}_n$. In other words, if we prove (1) with $a = a(p)$, for arbitrary p, $0 < p < 1$, then (1) will hold for $-\mathbf{T}_n$ with $a = a(1-p)$ and P3 will hold with $a = \min[a(p), a(1-p)]$.

Using the notation of D3.1, we obtain, for every $t > 0$ and $\epsilon \geq 0$,

(2) $\qquad \mathbf{P}[\mathbf{T}_n \geq \epsilon] \leq e^{-t\epsilon} \mathbf{E}[e^{t\mathbf{T}_n}; \mathbf{T}_n \geq \epsilon] \leq e^{-t\epsilon} \mathbf{E}[e^{t\mathbf{T}_n}].$

---

[10] This is S. Bernstein's estimate, [84], pp. 322–326.

Using the independence of the random variables $\mathbf{X}_k$ (P3.1)

(3) $$\mathbf{E}[e^{t\mathbf{T}_n}] = \mathbf{E}\left[e^{\frac{t}{n}\sum_{k=1}^{n}(\mathbf{X}_k - 2p+1)}\right] = \prod_{k=1}^{n}\mathbf{E}\left[e^{\frac{t}{n}(\mathbf{X}_k - 2p+1)}\right]$$
$$= \left\{\mathbf{E}\left[e^{\frac{t}{n}(\mathbf{X}_1 - 2p+1)}\right]\right\}^n = \left[pe^{\frac{2t}{n}(1-p)} + (1-p)e^{\frac{-2t}{n}p}\right]^n.$$

Expanding the function
$$f(x) = pe^{x(1-p)} + (1-p)e^{-xp}$$
in a Taylor series about the origin, one readily verifies that $f(0) = 1$, $f'(0) = 0$, $f''(0) > 0$. Hence there exist two constants $k_1 > 0$, $k_2 > 0$, depending on p such that
$$f(x) \leq 1 + k_1 x^2 \leq e^{k_1 x^2}$$
whenever $|x| \leq k_2$. Therefore it follows from (2) and (3) that
$$\mathbf{P}[\mathbf{T}_n \geq \epsilon] \leq e^{-t\epsilon}\left[f\left(\frac{2t}{n}\right)\right]^n \leq e^{\{-t\epsilon + (4k_1 t^2)/n\}}$$
when $|2t| \leq nk_2$. Now we let $t = cn\epsilon$ where $c > 0$ is still to be determined. Then
$$\mathbf{P}[\mathbf{T}_n \geq \epsilon] \leq e^{-n\epsilon^2(c - 4k_1 c^2)}$$
when $|2c\epsilon| \leq k_2$. Since $\epsilon \leq 2$ we may indeed choose a small enough positive value for c so that $2c\epsilon \leq k_2$ and simultaneously $c - 4k_1 c^2 > 0$. If we then set $a = a(p) = c - 4k_1 c^2$, it is clear that we have proved (1) and hence P3.

We are now in a position to prove the strong ratio theorem (T1). The proof of Orey [64] uses the elegant device of first imposing the condition that $P(0,0) > 0$. In P4 we therefore consider only random walk with this property, and the subsequent proof of T1 will show how the assumption that $P(0,0) > 0$ may be removed.

P4 *For every aperiodic recurrent random walk with $0 < P(0,0) = 1 - \alpha < 1$*

(a) $$\lim_{n \to \infty} \frac{P_n(0,0)}{P_{n+1}(0,0)} = 1,$$

(b) $$\lim_{n \to \infty} \frac{P_n(0,x)}{P_n(0,0)} = 1 \text{ for all } x \in R.$$

Proof: The assumption that $P(0,0) > 0$ implies that $P_n(0,0) > 0$ for all $n \geq 0$. Therefore we know from P1 that the random walk in

5. THE STRONG RATIO THEOREM

P4 is strongly aperiodic. That is crucial because it will enable us to apply P3 in the estimation procedure which follows.

We may decompose

(1) $$P_n(0,0) = \mathbf{P}_0[\mathbf{x}_n = 0]$$
$$= \sum_{k=0}^{n} \mathbf{P}_0\left[\mathbf{x}_n = 0; \; n - \sum_{j=1}^{n} \delta(\mathbf{x}_{j-1}, \mathbf{x}_j) = k\right]$$

according to the number of *jumps* of the random walk in time n, if we say that a jump occurs at time $j-1$ if $\mathbf{x}_{j-1} \neq \mathbf{x}_j$. Since $P(0,0) = 1 - \alpha$, the probability of k jumps in time n is given by the binomial distribution

(2) $$b(n,k,\alpha) = \binom{n}{k}\alpha^k(1-\alpha)^{n-k}.$$

But if a jump occurs at time j and if $\mathbf{x}_j = x$, then the probability that $\mathbf{x}_{j+1} = y$ is

(3) $$\begin{aligned} Q(x,y) &= \alpha^{-1}P(x,y) \quad \text{for } x \neq y \\ &= 0 \quad \text{for } x = y. \end{aligned}$$

Equation (3) defines a perfectly legitimate transition function $Q(x,y)$. We may call its iterates $Q_n(x,y)$ and conclude from the identity

$$P(x,y) = (1-\alpha)\delta(x,y) + \alpha Q(x,y), \qquad x,y \in R$$

that

(4) $$P_n(0,0) = \sum_{k=0}^{n} Q_k(0,0)b(n,k,\alpha), \qquad n \geq 0.$$

Now we shall decompose the sum in equation (4) into two parts ($\sum = \sum_{n,\epsilon}' + \sum_{n,\epsilon}''$), $\sum_{n,\epsilon}'$ being over the set of integers k such that $0 \leq k \leq n$ and in addition $|k - n\alpha| < \epsilon n$. Here ϵ is an arbitrary positive number which will tend to zero later, and $\sum_{n,\epsilon}''$ is simply the sum over the remaining values of k between 0 and n. We shall use the exponential estimate in P3 in the evidently correct form

(5) $$\sum_{n,\epsilon}'' b(n,k,\alpha) \leq 2e^{-A\epsilon^2 n}, \qquad n \geq 1$$

for some positive constant A which may be chosen independent of ϵ.

Finally we use (4) to decompose the ratios

(6) $$\frac{P_n(0,0)}{P_{n+1}(0,0)} = \frac{\sum_{n,\epsilon}' Q_k(0,0)b(n,k,\alpha)}{P_{n+1}(0,0)} + \frac{\sum_{n,\epsilon}'' Q_k(0,0)b(n,k,\alpha)}{P_{n+1}(0,0)}.$$

In view of (5) and P2, the last term in (6) is bounded above by

$$\frac{2e^{-A\epsilon^2 n}}{P_{n+1}(0,0)} \leq \frac{2e^{-A\epsilon^2 n}}{(1-\delta)^{n+1}} \text{ for all } n \geq N(\delta)$$

whenever $0 < \delta < 1$, so that this term tends to zero as $n \to \infty$. It follows that

$$(7) \quad \varlimsup_{n\to\infty} \frac{P_n(0,0)}{P_{n+1}(0,0)} \leq \varlimsup_{n\to\infty} \frac{\sum_{n,\epsilon}' Q_k(0,0)b(n,k,\alpha)}{P_{n+1}(0,0)}$$

$$\leq \varlimsup_{n\to\infty} \left[\frac{\sum_{n+1,\epsilon}' Q_k(0,0)b(n,k,\alpha)}{P_{n+1}(0,0)} + R_{n,\epsilon} \right].$$

Here $R_{n,\epsilon}$ is an error term which tends to zero as $n \to \infty$. (It is of the form $[P_{n+1}(0,0)]^{-1} \sum''' Q_k(0,0)b(n,k,\alpha)$ where $\sum'''$ extends over all values of k which are included in the summation $\sum_{n,\epsilon}'$ but not in $\sum_{n+1,\epsilon}'$. Hence

$$|R_{n,\epsilon}| \leq \frac{\sum_{n,\epsilon/2}'' b(n,k,\alpha)}{P_{n+1}(0,0)}$$

for all sufficiently large n, and tends to zero by previous arguments using P2 and P3.) Resuming the estimation from (7),

$$(8) \quad \varlimsup_{n\to\infty} \frac{P_n(0,0)}{P_{n+1}(0,0)} \leq \varlimsup_{n\to\infty} \frac{\sum_{n+1,\epsilon}' Q_k(0,0)b(n,k,\alpha)}{\sum_{n+1,\epsilon}' Q_k(0,0)b(n+1,k,\alpha)}$$

$$\leq \varlimsup_{n\to\infty} \max_{[k\,|\,|k-(n+1)\alpha|\leq\epsilon(n+1)]} \frac{b(n,k,\alpha)}{b(n+1,k,\alpha)}$$

$$= \varlimsup_{n\to\infty} \max_{[k\,|\,|k-(n+1)\alpha|\leq\epsilon(n+1)]} \frac{n+1-k}{(n+1)(1-\alpha)}$$

$$\leq \lim_{n\to\infty} \frac{(n+1)(1-\alpha+\epsilon)}{(n+1)(1-\alpha)} = 1 + \frac{\epsilon}{1-\alpha}.$$

Since ϵ can be taken arbitrarily small, we have shown that

$$\varlimsup_{n\to\infty} \frac{P_n(0,0)}{P_{n+1}(0,0)} \leq 1.$$

Exactly the same method of proof may be applied to the reciprocal ratio $P_{n+1}(0,0)/P_n(0,0)$ which of course also has upper limit one, so that part (a) of P4 is proved.

To prove part (b) of P4, use P1.2 to write

$$P_n(0,x) = \sum_{k=0}^{n} F_k(0,x) P_{n-k}(0,0), \quad n \geq 1.$$

For every positive integer m

$$\lim_{n \to \infty} \frac{P_n(0,x)}{P_n(0,0)} \geq \lim_{n \to \infty} \sum_{k=0}^{m} F_k(0,x) \frac{P_{n-k}(0,0)}{P_n(0,0)} = \sum_{k=0}^{m} F_k(0,x),$$

where we applied part (a) of P4 to evaluate the lower limit. Since the random walk is recurrent, the last term tends to one as $m \to \infty$, so that

(9) $$\lim_{n \to \infty} \frac{P_n(0,x)}{P_n(0,0)} \geq 1.$$

To complete the proof of P4, suppose that (b) is false. Then there is some $x_0 \in R$ where

$$\overline{\lim_{n \to \infty}} \frac{P_n(0,x_0)}{P_n(0,0)} = 1 + \delta > 1.$$

We may choose an integer k such that $P_k(x_0, 0) > 0$. Then

(10) $$\frac{P_{n+k}(0,0)}{P_n(0,0)} = \sum_{y \neq x_0} \frac{P_n(0,y)}{P_n(0,0)} P_k(y,0) + \frac{P_n(0,x_0)}{P_n(0,0)} P_k(x_0,0).$$

The left-hand side has limit one as $n \to \infty$, so that

$$1 \geq \sum_{y \neq x_0} P_k(y,0) \lim_{n \to \infty} \frac{P_n(0,y)}{P_n(0,0)} + (1 + \delta) P_k(x_0,0)$$
$$\geq 1 + \delta P_k(x_0,0) > 1,$$

providing the contradiction that completes the proof of P4.

Finally we shall drop the hypothesis that $P(0,0) > 0$, to obtain

T1 *For every strongly aperiodic recurrent random walk*

(a) $$\lim_{n \to \infty} \frac{P_n(0,0)}{P_{n+1}(0,0)} = 1,$$

(b) $$\lim_{n \to \infty} \frac{P_n(0,x)}{P_n(0,0)} = 1 \text{ for all } x \in R.$$

Proof: First observe that it will suffice to prove part (a) only. For once this is done, the proof of part (b) in P4 may be used to obtain (b) here too—it did not depend on the positivity of $P(0,0)$ at all. Now let s denote the smallest positive integer such that $P_s(0,0) > 0$. Actually any positive integer with this property would do, and we may of course assume that $s > 1$, as the case of $s = 1$ was settled in P4. It is clear from P1 that such a value of s exists, and also that the transition function

$$Q(x,y) = P_s(x,y)$$

is again the transition function of an aperiodic recurrent random walk. It satisfies the hypotheses of P4, so that

$$(1) \qquad \lim_{n \to \infty} \frac{Q_n(0,0)}{Q_{n+1}(0,0)} = \lim_{n \to \infty} \frac{P_{ns}(0,0)}{P_{(n+1)s}(0,0)} = 1.$$

Therefore, as is easy to see, we shall have proved part (a) of T1 if we show that

$$(2) \qquad \lim_{n \to \infty} \frac{P_{ns+r}(0,0)}{P_{ns}(0,0)} = 1 \text{ for } 1 \leq r < s.$$

One obtains a lower estimate for the ratio in (2) from

$$(3) \qquad P_{ns+r}(0,0) = \sum_{x \in R} P_r(0,x) P_{ns}(x,0).$$

If we apply P4(b) to the ratios

$$\frac{P_{ns}(x,0)}{P_{ns}(0,0)} = \frac{Q_n(x,0)}{Q_n(0,0)},$$

then (3) leads to

$$\lim_{n \to \infty} \frac{P_{ns+r}(0,0)}{P_{ns}(0,0)} \geq \sum_{|x| \leq M} P_r(0,x) \lim_{n \to \infty} \frac{Q_n(x,0)}{Q_n(0,0)}$$

$$= \sum_{|x| \leq M} P_r(0,x)$$

for every $M > 0$, so that

$$(4) \qquad \lim_{n \to \infty} \frac{P_{ns+r}(0,0)}{P_{ns}(0,0)} \geq 1.$$

The upper bound will come from

$$P_n(0,0) = \sum_{k=0}^{n} F_k(0,0) P_{n-k}(0,0), \qquad n \geq 1,$$

with $F_k(0,0)$ as defined in D1.2. It is first necessary to verify that $F_k(0,0) > 0$ for all sufficiently large values of k, say for all $k \geq N$

(that follows quite easily from the recurrence and strong aperiodicity of the random walk). Then we choose a positive integer p such that $p(s-1) \geq N$, and let $n \to \infty$ in

$$\text{(5)} \qquad \frac{P_{(n+p)s}(0,0)}{P_{ns}(0,0)} = \sum_{k=0}^{(n+p)s} F_k(0,0) \frac{P_{(n+p)s-k}(0,0)}{P_{ns}(0,0)}.$$

If T1 were false, then (2) would have to fail, and in view of (4) there would be some value of r, $0 \leq r < s$ such that the ratio $P_{ns+r}(0,0)/P_{ns}(0,0)$ has upper limit $1 + \delta > 1$. Let us suppose this to be so. The coefficient of this ratio on the right-hand side in (5) is $F_{ps-r}(0,0)$ which is positive by our choice of p. As the left side in (5) tends to one, we obtain

$$\text{(6)} \qquad 1 \geq (1 + \delta) F_{ps-r}(0,0)$$
$$+ \sum_{[k \mid k \leq M;\ k \neq ps-r]} F_k(0,0) \lim_{n \to \infty} \frac{P_{(n+p)s-k}(0,0)}{P_{ns}(0,0)}$$

for every $M > 0$. Using (4)

$$1 \geq (1 + \delta) F_{ps-r}(0,0) + \sum_{k \neq ps-r} F_k(0,0) = 1 + \delta F_{ps-r}(0,0) > 1$$

which is the desired contradiction. Hence $P_{ns+r}(0,0)/P_{ns}(0,0)$ has upper limit at most one for every r, equation (2) holds, and T1 is proved.

Remark: Although the dimension of the random walk did in no way *seem* to affect the proof of T1, we shall see in the next chapter (T8.1) that T1 has only been proved when $d \leq 2$. The reason is that *no aperiodic recurrent random walk exists when $d \geq 3$*. Of course there are random walks of dimension $d \geq 3$ where all or part of the conclusion of T1 are valid (for instance simple random walk in any dimension satisfies (a) of T1). On the other hand, one can easily exhibit a transient strongly aperiodic random walk that violates T1. Any one-dimensional random walk with positive mean will serve as an example, if $P(0,x) > 0$ only at a finite number of points. It can be shown that $P_n(0,0)$ then decreases geometrically, so that not only T1 is false, but also the crucial estimate in P2 is violated.

Problems

1. For an arbitrary random walk, let

$$E_{m,n} = G_{n+m}(0,0) - G_m(0,0)$$

denote the expected number of visits to the origin in the time interval $[m, m+n]$. Prove that

$$E_{0,n} \geq E_{m,n} \text{ for all } m \geq 0,\ n \geq 0.$$

2. Again for arbitrary random walk prove that

$$\sum_{x \in R} F_1(0,x) = 1, \text{ while } \sum_{x \in R} F_n(0,x) = 1 - \sum_{k=1}^{n-1} F_k(0,0)$$

when $n \geq 2$.

3. A two-dimensional random walk is defined (in complex notation) by

$$0 < P(0,1) = P(0,-1) = p < 1,\ P(0,i) = 1 - 2p,$$

so that $P(0,z) = 0$ for all other $z = m + ni$. Show that this random walk is transient and calculate F and G.

4. Find the simplest possible expression, in terms of binomial coefficients, for $P_n(0,0)$ for simple random walk in the plane. Apply Stirling's formula to show that this random walk is recurrent. This fact—together with the result of the next problem, constitutes a famous theorem of Polya [83], 1921.

5. Use Stirling's formula to show that simple random walk is transient in dimension $d \geq 3$.

6. For an arbitrary aperiodic recurrent random walk $\mathbf{x}_n$ starting at $\mathbf{x}_0 = 0$, calculate the expected number of visits to the point $x \neq 0$, before the first return to 0. Show that the answer is *independent of x* (and of the random walk in question).

7. Repeat the calculation in problem 6 for transient random walk.

8. Explain why (6) and (7) were easy, and why it is harder to calculate
  (a) The expected number of returns to 0 before the first visit to x;
  (b) The probability of a visit to x before the first return to 0;
  (c) The probability of return to 0 before the first visit to x.[11]

9. Express the probability $\mathbf{P}[\mathbf{R}_n > \mathbf{R}_{n-1}]$ that a *new* point is visited at time n in terms of the sequence $F_k(0,0)$. Use your result to prove the statement of problem 2.

10. The state space R of a random walk is subjected to the following indignity: the points of R are *painted red* with probability α and *green* with probability $1 - \alpha$. The colors of distinct points are independent. Now we wish to define $\mathbf{T}_\alpha$ as the first time that the random walk lands on a red point. Define an appropriate probability space $(\Omega, \mathscr{F}, \mathbf{P})$ on which $\mathbf{T}_\alpha = \min [n \mid n \geq 0;\ \mathbf{x}_n \in \text{set of red points}]$ is a bona fide random variable, and on which also $\mathbf{x}_\alpha = \mathbf{x}_{\mathbf{T}_\alpha}$ is a random variable.

---

[11] This problem is intended to give a glimpse of the difficulties in store for us. Large parts of Chapters III and VII are indeed devoted to the three questions (a), (b), and (c).

11. Continuing (10), prove that

$$P[T_\alpha > n] = E[(1 - \alpha)^{R_n}], \qquad n \geq 0,$$

$$P[\mathbf{x}_\alpha = x] = \frac{\alpha}{1 - \alpha} P[T < T_\alpha], \qquad x \in R.$$

Here $T = \min[n \mid n \geq 0, \mathbf{x}_n = x]$, and $\mathbf{x}_0 = 0$, while $\mathbf{R}_n$ is the *range* of T4.1.

12. Continuing (11), prove that for transient random walk

$$\lim_{\alpha \to 0} P[\alpha T_\alpha > t] = e^{-(1-F)t}, \qquad t \geq 0.$$

13. Continuing (11), prove that for *simple random walk* in one dimension

$$\lim_{\alpha \to 0} P[\alpha \mathbf{x}_\alpha \leq y] = F(y)$$

for all real y, where $F(y)$ is the distribution function with density

$$f(y) = F'(y) = |y| \int_{|y|}^\infty \frac{e^{-s} \, ds}{s^2}, \qquad y \neq 0.$$

14. Let A denote the *random subset* of R constructed, as in problem 10, by selecting each point of R with probability α. If φ_A is the characteristic function of the set A then

$$\mathbf{N}_n(\alpha) = \sum_{k=0}^n \varphi_A(\mathbf{x}_k)$$

may be interpreted as the *occupation time* of the random set A up to time n. Show that $E[\mathbf{N}_n(\alpha)] = (n+1)\alpha$, and that $\sigma^2[\mathbf{N}_n(\alpha)]/n^2 \to 0$, so that $\mathbf{N}_n(\alpha)/n \to \alpha$ in measure. (A proof of convergence with probability one may be based on Birkhoff's ergodic theorem since the sequence $\varphi_A(\mathbf{x}_k)$, $k = 0, 1, 2, \ldots$ forms a so-called strictly stationary process.) Although all this goes through for arbitrary random walk, prove that the limit of $\sigma^2[\mathbf{N}_n(\alpha)]/n$ is finite or infinite exactly according as the random walk is transient or recurrent. It has the value

$$\alpha(1 - \alpha)(1 + F)/(1 - F).$$

15. Continuation. For arbitrary transient random walk derive a central limit theorem for the occupation time $\mathbf{N}_n(\alpha)$ of a random subset of the state space with density α. As shown by Whitman,

$$\lim_{n \to \infty} P[\mathbf{N}_n(\alpha) - n\alpha \leq x\{n\alpha(1-\alpha)(1+F)/(1-F)\}^{1/2}]$$

$$= \frac{1}{\sqrt{2\pi}} \int_{-\infty}^x e^{-t^2/2} \, dt, \qquad -\infty < x < \infty.$$

Hint: Verify that the limiting moments are those of the normal distribution. This will suffice according to P23.3 in Chapter V.

Chapter II

HARMONIC ANALYSIS

6. CHARACTERISTIC FUNCTIONS AND MOMENTS[1]

As in the last chapter R will denote the d-dimensional group of lattice points $x = (x^1, x^2, \ldots, x^d)$ where the x^i are integers. To develop the usual notions of Fourier analysis we must consider still another copy of Euclidean space, which we shall call E. It will be the whole of Euclidean space (not just the integers) and of the same dimension d as R. For convenience we shall use Greek letters to denote elements of E, and so, if R is α-dimensional, the elements of E will be $\theta = (\theta_1, \theta_2, \ldots, \theta_d)$, where each θ_i is a real number for $i = 1, 2, \ldots, d$. The following notation will be convenient.

D1 *For $x \in R$, $\theta \in E$,*

$$|x|^2 = \sum_1^d (x^i)^2, \qquad |\theta|^2 = \sum_1^d (\theta_i)^2, \qquad x \cdot \theta = \sum_1^d x^i \theta_i.$$

Given, as in Chapter I, a random walk which is completely specified by its transition function $P(x,y)$ on a d-dimensional state space R, we make the definition

D2 *The characteristic function of the random walk is*

$$\phi(\theta) = \sum_{x \in R} P(0,x) e^{ix \cdot \theta}, \qquad \theta \in E.$$

Thus $\phi(\theta)$ is nothing but a special kind of Fourier series, special in that its Fourier coefficients are non-negative and their sum over R is

---

[1] The material of this section is well known. More detailed expositions are in [18], [23], [35], [73], and [84], for the one-dimensional case, and the extension to dimension $d \geq 2$ is nowhere difficult. For a deeper study of the theory of Fourier series, Zygmund's treatise [106] is recommended.

one. The term "characteristic function" is customarily used in probability theory for Fourier series, or for Fourier integrals of probability measures, which are of this special type.

Many important properties of characteristic functions are in fact general properties of Fourier series. After establishing the orthogonality of the exponential functions in P1, we shall record, in P2, a very weak form of Parseval's theorem, which in its usual strong form amounts to the assertion that the exponential functions form a complete orthonormal set. Then in P3 we shall derive the convolution and inversion theorems for Fourier series which are elementary as well as general. From that point on, most of this chapter will deal with special properties of characteristic functions (properties which do depend on the non-negativity of the coefficients). Only in section 9, P9.1, will we return to a general principle, the *Riemann Lebesgue Lemma*, which will find important applications in several later chapters.

Our first task is to set up convenient notation for integration on E, and in particular on the cube C in E with center at the origin and sides of length 2π.

D3 $\qquad C = [\theta \mid \theta \in E, |\theta_i| \leq \pi \text{ for } i = 1, 2, \ldots, d]$

and for complex valued functions $f(\theta)$ which are Lebesgue integrable on C, the integral over C is denoted by

$$\int f \, d\theta = \int_C f \, d\theta = \int_{-\pi}^{\pi} \cdots \int_{-\pi}^{\pi} f(\theta) \, d\theta_1 \ldots d\theta_d.$$

Thus $d\theta$ will always denote the volume element (Lebesgue measure in E). To achieve corresponding economy of notation for summation over R, we shall write

$$\sum g(x) = \sum_{x \in R} g(x) \qquad \text{when } \sum_{x \in R} |g(x)| < \infty.$$

In most of our work $f(\theta)$ will in fact be a continuous function on C so that the integral $\int f \, d\theta$ is then just the ordinary d-tuple Riemann integral. The basic result underlying our Fourier analysis will of course be the orthogonality of the exponential functions

P1 $\quad (2\pi)^{-d} \int e^{i\theta \cdot (x-y)} \, d\theta = \delta(x,y),$ *for every pair x, y in R.*

The obvious proof proceeds by reduction to the one-dimensional case, since the integral in P1 may be written as the product of d

one-dimensional integrals over the real interval from $-\pi$ to π. Thus the integral is

$$\prod_{k=1}^{d} \frac{1}{2\pi} \int_{-\pi}^{\pi} e^{i\theta_k(x^k-y^k)} d\theta_k = \prod_{k=1}^{d} \delta(x^k, y^k) = \delta(x,y),$$

which depends only on the fact that for every integer n

$$\frac{1}{2\pi} \int_{-\pi}^{\pi} e^{in\theta} d\theta = \delta(n,0).$$

Next we give a very weak form of the Parseval Theorem. Let $a_1(x)$ and $a_2(x)$ be two summable complex valued functions on R, i.e., we suppose that

$$\sum |a_k(x)| < \infty \text{ for } k = 1, 2.$$

Then it makes sense to define their Fourier series, $f_1(\theta)$ and $f_2(\theta)$, for $\theta \in E$, by

$$f_k(\theta) = \sum a_k(x) e^{ix\cdot\theta}, \qquad k = 1,2.$$

Since the $a_k(x)$ are summable, the series defining $f_k(\theta)$ converge absolutely, so that in fact each $f_k(\theta)$ is a continuous function on E. Then Parseval's Theorem is (in the notation of D3)

P2 $$(2\pi)^{-d} \int f_1(\theta)\overline{f_2(\theta)}\, d\theta = \sum a_1(x)\overline{a_2(x)}.$$

Proof: Using the summability condition one can interchange summation and integration on the left so that

$$(2\pi)^{-d} \int f_1 \bar{f_2}\, d\theta = \sum_{x \in R} \sum_{y \in R} a_1(x)\overline{a_2(y)} (2\pi)^{-d} \int e^{i(x-y)\cdot\theta}\, d\theta,$$

and using the orthogonality relation P1 we immediately get P2.

Remark: Actually P2 is valid also under the weaker condition that

$$\sum |a_k(x)|^2 < \infty \text{ for } k = 1, 2.$$

Under this condition the functions $|f_1|^2$, $|f_2|^2$ and $f_1\bar{f_2}$ are all Lebesgue integrable on C, so that P2 makes perfect sense. In particular, let

$$f_1(\theta) = f_2(\theta) = f(\theta),$$
$$a_1(x) = a_2(x) = a(x) = (2\pi)^{-d} \int e^{-ix\cdot\theta} f(\theta)\, d\theta.$$

Then the two conditions

$$\sum |a(x)|^2 < \infty \quad \text{and} \quad \int |f(\theta)|^2 \, d\theta < \infty$$

are equivalent, and if either of them holds, then

$$(2\pi)^{-d} \int |f(\theta)|^2 \, d\theta = \sum |a(x)|^2.$$

This is *Parseval's identity* which we shall never have occasion to use. All we shall ever need in our work is *Bessel's inequality*, to the effect that

$$(2\pi)^{-d} \int |f(\theta)|^2 \, d\theta \geq \sum |a(x)|^2,$$

which will be derived in the course of the proof of P9.1.

Returning now to characteristic functions of a random walk, we prove

P3 *If $\phi(\theta)$ is the characteristic function of the transition function $P(x,y)$, then for every integer $n \geq 0$*

(a) $\quad\quad\quad \phi^n(\theta) = \sum P_n(0,x) e^{ix\cdot\theta} \quad \text{for all } \theta \text{ in } E,$

(b) $\quad\quad\quad P_n(0,y) = (2\pi)^{-d} \int e^{-iy\cdot\theta} \phi^n(\theta) \, d\theta \quad \text{for all } y \text{ in } R.$

Proof: It is instructive to prove the first part using the probability interpretation of the problem. In the notation of D3.1

$$P_n(0,x) = \mathbf{P}[\mathbf{S}_n = x]$$

where $\mathbf{S}_n = \mathbf{X}_1 + \cdots + \mathbf{X}_n$ is the sum of independent d-dimensional random variables which describes the position of the random walk at time n. Using P3.1 we obtain

$$\sum P_n(0,x) e^{i\theta\cdot x} = \mathbf{E}[e^{i\theta\cdot\mathbf{S}_n}] = \mathbf{E}[e^{i\theta\cdot(\mathbf{X}_1 + \cdots + \mathbf{X}_n)}]$$
$$= \prod_{k=1}^n \mathbf{E}[e^{i\theta\cdot\mathbf{X}_k}] = \phi^n(\theta).$$

Here we used the independence of the $\mathbf{X}_k$, and the fact that they each have the characteristic function $\phi(\theta)$. This proves part (a), and part (b) follows by multiplying each side in (a) by $e^{-iy\cdot\theta}$ and by integrating over C.

To illustrate the use of characteristic functions we now turn to a somewhat sketchy discussion of the behavior of characteristic functions near $\theta = 0$, emphasizing those results which will be most immediately useful in deciding whether a random walk is recurrent or not. Some related results, such as P5, are included because they are indispensable for work in later chapters. We begin with one-dimensional random walk, where we pay particular attention to the relations between the moments of a transition function and the derivatives of its characteristic function at the origin. It is an old and familiar fact in Fourier analysis that the behavior of a sequence (or function), far out, is reflected in the behavior of its Fourier series (or transform) near the origin. In probability theory these connections become particularly sharp and interesting, because we are dealing with a very restricted class of Fourier series.

D4 *For one-dimensional random walk,*

$$m_k = \sum |x|^k P(0,x), \qquad \mu_k = \sum x^k P(0,x), \qquad k \geq 1.$$
$$\mu = \mu_1, \qquad m = m_1, \qquad \sigma^2 = \mu_2 - \mu^2.$$

We only define μ_k when $m_k < \infty$.

P4 *In one dimension, when $m_k < \infty$, the kth derivative $\phi^{(k)}(\theta)$ is a continuous function, and $\phi^{(k)}(0) = (i)^k \mu_k$. The converse is false, i.e. $\phi^{(1)}(0)$ may exist without m_1 being finite, but if $\phi^{(2)}(0)$ exists, then $m_2 = -\phi^{(2)}(0) < \infty$.*

Proof: If $m_1 < \infty$, then

$$\frac{1}{h}[\phi(\theta + h) - \phi(\theta)] = \frac{1}{h}\sum (e^{ihx} - 1)e^{i\theta x}P(0,x).$$

Note that, according to D3, the sum extends over all of R. Now

$$|e^{ihx} - 1| = \left|\int_0^{hx} e^{it}\, dt\right| \leq \int_0^{|hx|} dt = |hx|,$$

so that

$$\frac{1}{h}\sum |e^{ihx} - 1| \cdot |e^{i\theta x}| P(0,x) \leq m_1.$$

Therefore one can interchange this summation over R with the limiting operation $h \to 0$. This is simply the discrete analogue for series of the *dominated convergence theorem* in the theory of Lebesgue integration. By dominated convergence then,

$$\phi^{(1)}(\theta) = \sum \frac{d}{dh}(e^{ihx})e^{i\theta x}P(0,x) = i\sum x e^{i\theta x}P(0,x)$$

and $\phi^{(1)}(0) = i\mu_1$. Clearly $\phi^{(1)}(\theta)$ is continuous, the defining series being uniformly convergent. Exactly the same argument applies to give the continuity of $\phi^{(k)}(\theta)$ when $m_k < \infty$ so that the relation $\phi^{(k)}(0) = (i)^k \mu_k$ holds.

Where the converse is concerned we offer the following well known example, leaving the verification to the reader. Let $P(0,x) = c|x|^{-2}(\ln |x|)^{-1}$ when $|x| \geq 2$, $P(0,x) = 0$ when $|x| \leq 1$, where the constant c is chosen so that $\sum P(0,x) = 1$. Then evidently $m_1 = \infty$, while a somewhat lengthy calculation will show that $\phi^{(1)}(0)$ exists.

Finally, suppose that

$$\phi^{(2)}(0) = \lim_{h \to 0} \frac{\phi^{(1)}(h) - \phi^{(1)}(0)}{h}$$

exists. Then it turns out that

$$\phi^{(2)}(0) = \lim_{\theta \to 0} \frac{1}{\theta^2} [\phi(\theta) + \phi(-\theta) - 2],$$

which is a more convenient representation for the second derivative at the origin. It implies

$$-\phi^{(2)}(0) = \lim_{\theta \to 0} \sum \left(\frac{2}{\theta} \sin \frac{x\theta}{2}\right)^2 P(0,x) = \lim_{\theta \to 0} \sum \left(\frac{\sin \theta x}{\theta x}\right)^2 x^2 P(0,x).$$

Consequently, for each $n \geq 0$,

$$\sum_{x=-n}^{n} x^2 P(0,x) = \lim_{\theta \to 0} \sum_{x=-n}^{n} \left(\frac{\sin \theta x}{\theta x}\right)^2 x^2 P(0,x) \leq -\phi^{(2)}(0).$$

This shows that $m_2 < \infty$ and the proof of P4 is completed by observing that the first part of P4 gives $m_2 = -\phi^{(2)}(0)$.

The extension of the converse part of P4 to arbitrary even moments will never be used and is relegated to problem 1. It is interesting, and useful, however, that assertions stronger than those in P4 can be made under further restrictive assumptions on $P(0,x)$. We shall now consider the case when $P(0,x) = 0$ for $x < 0$ and show that one can then learn something about m_1 from the first derivative of the characteristic function.

P5 *If $P(0,x) = 0$ for $x < 0$ and if*

$$0 \leq -i\phi^{(1)}(0) = \lim_{\theta \to 0} i \frac{1 - \phi(\theta)}{\theta} = \alpha < \infty,$$

then $\sum_{x=0}^{\infty} xP(0,x) = \alpha.$

Proof: To simplify the notation let $p_n = P(0,x)$ for $x = n \geq 0$. Then

(1) $$\sum_0^\infty p_k \frac{1 - \cos k\theta}{\theta} \to 0, \qquad \sum_0^\infty p_k \frac{\sin k\theta}{\theta} \to \alpha$$

as $\theta \to 0$. Now the plan of the proof is to use this information to show that $m = \sum_0^\infty k p_k < \infty$.

The rest of the result will then be an automatic consequence of P4. We start by breaking up the second sum in (1) into two pieces. The choice of pieces is dictated by a convenient estimate of $\sin x$ to the effect that

$$\frac{\sin x}{x} \geq \frac{2}{\pi} \quad \text{for } |x| \leq \frac{\pi}{2}.$$

Let us use the symbol $\hat{\theta}$ to denote $[\pi/2\theta]$, the greatest integer in $\pi/2\theta$. When $0 \leq k \leq \hat{\theta}$ the above estimate shows that $\sin k\theta \geq 2\theta k/\pi$. Therefore

(2) $$\sum_0^\infty p_k \frac{\sin k\theta}{\theta} \geq \frac{2}{\pi} \sum_{k=0}^{\hat{\theta}} k p_k + \sum_{k=\hat{\theta}+1}^\infty p_k \frac{\sin k\theta}{\theta}.$$

As $\theta \to 0$, $\hat{\theta} \to \infty$, the left-hand side in the above inequality tends to α by (1) and so we can conclude that $\sum_0^\infty k p_k < \infty$, provided we show that the last sum in (2) stays bounded as $\theta \to 0$. This in turn will be true if we can show that

$$A(\theta) = \frac{1}{\theta} \sum_{k=\hat{\theta}+1}^\infty p_k$$

stays bounded as $\theta \to 0$, and we shall in fact show more, namely that $A(\theta) \to 0$ as $\theta \to 0$.

When $k \geq \hat{\theta} + 1$, $k\theta \geq \pi/2 > 1$. Therefore there is a positive number b such that

$$1 - \frac{\sin k\theta}{k\theta} \geq b > 0 \qquad \text{when } k \geq \hat{\theta} + 1.$$

This gives

$$|A(\theta)| \leq \frac{1}{b|\theta|} \sum_{k=0}^\infty \left[1 - \frac{\sin k\theta}{k\theta}\right] p_k = \frac{1}{b|\theta|} \sum_{k=0}^\infty p_k \frac{1}{\theta} \int_0^\theta (1 - \cos kt) \, dt$$

$$\leq \frac{1}{b\theta} \int_0^\theta \left[\sum_{k=0}^\infty p_k \frac{1 - \cos kt}{|t|}\right] dt.$$

The last term tends to zero in view of the first part of (1), and this shows that $A(\theta) \to 0$ and completes the proof.

The next result concerns a necessary and sufficient condition for the finiteness of the first moment m which is quite different from the condition in P4. One can extend it to moments of arbitrary odd order but this matter is again relegated to the problem section (problem 2).

P6 $$m_1 = \frac{1}{2\pi} \int_{-\pi}^{\pi} \frac{\text{Re } [1 - \phi(\theta)]}{1 - \cos \theta} d\theta \leq \infty,$$

and $m_1 < \infty$ *if and only if the function* $\theta^{-2} \text{ Re } [1 - \phi(\theta)]$ *is (Lebesgue) integrable on the interval* $[-\pi,\pi]$.

Proof: The real part of the characteristic function is

$$\text{Re } \phi(\theta) = \sum P(0,x) \cos x\theta,$$

and therefore $\text{Re } [1 - \phi(\theta)] \geq 0$, and the integral in P6, whether finite or not, equals

$$\sum P(0,x) \frac{1}{2\pi} \int_{-\pi}^{\pi} \frac{1 - \cos x\theta}{1 - \cos \theta} d\theta$$

in the sense that this series diverges if and only if the integral in P6 is infinite. To complete the proof it suffices to verify that

$$\frac{1}{2\pi} \int_{-\pi}^{\pi} \frac{1 - \cos x\theta}{1 - \cos \theta} d\theta = |x|, \quad x \in R,$$

This is easily done by using the trigonometric identity

$$\frac{1 - \cos x\theta}{1 - \cos \theta} = \left(\frac{\sin \frac{x\theta}{2}}{\sin \frac{\theta}{2}} \right)^2 = \left| \sum_{k=1}^{|x|} e^{ik\theta} \right|^2, \quad x \in R.$$

For higher dimensional random walk ($d \geq 2$) we shall be content to give the analogue of P4 for the first and second moments only, but it will first be convenient to adopt a sensible notation, which agrees with that of D4.

D5 $m_1 = \sum |x| P(0,x), \qquad \mu = \sum x P(0,x)$ if $m_1 < \infty$,
$m_2 = \sum |x|^2 P(0,x), \qquad Q(\theta) = \sum (x \cdot \theta)^2 P(0,x)$ if $m_2 < \infty$.

Note that while m_1 and m_2 are scalars, μ *is of course a d-dimensional vector*. $Q(\theta)$ is called the *moment quadratic form* of the random walk, for reasons which will promptly be discernible.

If $\mathbf{X} = (\mathbf{X}_1, \ldots, \mathbf{X}_d)$ is a vector random variable such that $\mathbf{X} = x$ with probability $P(0,x)$, then evidently

$$Q(\theta) = \sum_{i=1}^{d} \sum_{j=1}^{d} E[\mathbf{X}_i \mathbf{X}_j] \theta_i \theta_j$$

when the expected values $E|\mathbf{X}_i\mathbf{X}_j|$ all exist. But their existence is assured by the assumption that $m_2 = E|\mathbf{X}|^2 < \infty$, as one can see from the Schwarz inequality

$$(\theta \cdot \mathbf{X})^2 \leq |\theta|^2 |\mathbf{X}|^2.$$

This inequality implies that $Q(\theta) < \infty$ for all θ, and the fact that

$$E|\mathbf{X}_i\mathbf{X}_j| \leq \{E|\mathbf{X}_i|^2 E|\mathbf{X}_j|^2\}^{1/2} \leq E|\mathbf{X}|^2$$

gives another way of seeing that the coefficients in the quadratic form $Q(\theta)$ are well defined.

To illustrate what kind of results can be obtained, we prove

P7 *For d-dimensional random walk, $d \geq 1$, suppose that $m_1 < \infty$. If α is a vector in E, then*

$$\lim_{h \to 0} \frac{\phi(h\alpha) - 1}{h} = i\alpha \cdot \mu.$$

If, in addition, $\mu = 0$ and $m_2 < \infty$, then

$$\lim_{h \to 0} \frac{1 - \phi(h\alpha)}{h^2} = \tfrac{1}{2} Q(\alpha).$$

Proof: Imitating the proof of P6 up to a point,

$$|\phi(h\alpha) - 1| = \left|\sum (e^{ih\alpha \cdot x} - 1) P(0,x)\right|$$
$$\leq \sum |e^{ih\alpha \cdot x} - 1| P(0,x) \leq h \sum |\alpha \cdot x| P(0,x) \leq h m_1 |\alpha|.$$

Therefore the dominated convergence theorem gives

$$\lim_{h \to 0} \frac{1}{h}[\phi(h\alpha) - 1] = \sum \lim_{h \to 0} \left(\frac{e^{ih\alpha \cdot x} - 1}{h}\right) P(0,x) = i\alpha \cdot \mu.$$

In the case when $\mu = 0$ and $m_2 < \infty$ we can write

$$\phi(h\alpha) - 1 = \sum (e^{ih\alpha \cdot x} - 1 - ih\alpha \cdot x) P(0,x).$$

But for complex z with real part zero there is some $c > 0$ so that

$$|e^z - 1 - z| \le c|z|^2, \qquad \text{Re}(z) = 0.$$

Hence

$$\sum |e^{i h \alpha \cdot x} - 1 - i h \alpha \cdot x| P(0,x) \le ch^2 \sum (\alpha \cdot x)^2 P(0,x),$$

and by the Schwarz inequality

$$\sum (\alpha \cdot x)^2 P(0,x) \le |\alpha|^2 m_2 < \infty.$$

Again therefore, one can interchange limits to obtain the desired result.

To conclude this section it is only proper to point out that the *Central Limit Theorem* certainly belongs to the domain of ideas discussed here. Although it will play a decidedly minor role in the theory of random walk we shall sketch the essential ingredients of the traditional Fourier analytical proof of the Central Limit Theorem for identically distributed independent random variables.

A monotone nondecreasing function $F(t)$, $-\infty < t < \infty$ is a *distribution function* if $F(-\infty) = 0$, $F(+\infty) = 1$. A sequence of distribution functions $F_n(t)$ is said to *converge weakly* to the distribution function $F(t)$ if

$$\lim_{n \to \infty} \int_{-\infty}^{\infty} g(t) \, dF_n(t) = \int_{-\infty}^{\infty} g(t) \, dF(t)$$

for every bounded continuous function $g(t)$, $-\infty < t < \infty$. An equivalent definition of weak convergence is the statement that $F_n(t)$ has the limit $F(t)$ for every real t which is a point of continuity of $F(t)$.

The Fourier transform of a distribution function $F(t)$,

$$\phi(\lambda) = \int_{-\infty}^{\infty} e^{i\lambda t} \, dF(t), \qquad -\infty < \lambda < \infty,$$

is called the characteristic function of F. If ϕ_n are the characteristic functions of a sequence of distributions F_n, and if the sequence F_n converges weakly to the distribution F, then ϕ_n converges to the characteristic function ϕ of F. Conversely (this is the important *continuity theorem* of Lévy, 1925) if the sequence ϕ_n of characteristic functions converges pointwise to a function ϕ which is continuous at $\lambda = 0$, then F_n converges weakly to the (unique) distribution function F which has ϕ as its characteristic function.

We shall illustrate these facts by proving

P8 *If $P(x,y)$ is the transition function of one-dimensional random walk with mean $\mu = 0$, and variance $\sigma^2 = m_2 < \infty$, then*

$$\lim_{n \to \infty} \sum_{x < \sqrt{n}\sigma t} P_n(0,x) = F(t),$$

where

$$F(t) = \frac{1}{\sqrt{2\pi}} \int_{-\infty}^{t} e^{-x^2/2}\, dx, \qquad -\infty < t < \infty.$$

Proof: If

$$F_n(t) = \sum_{x < \sqrt{n}\sigma t} P_n(0,x),$$

which clearly is a sequence of distribution functions, the proof of P8 will follow from the Lévy continuity theorem, provided we can show that

$$\lim_{n \to \infty} \int_{-\infty}^{\infty} e^{i\lambda t}\, dF_n(t) = \frac{1}{\sqrt{2\pi}} \int_{-\infty}^{\infty} e^{i\lambda t} e^{-t^2/2} dt = e^{-\lambda^2/2},$$

for all real λ. To this end let

$$\phi(\theta) = \sum P(0,x) e^{ix\theta}, \qquad -\infty < \theta < \infty,$$

be the characteristic function of the random walk, defined in D2. Then a simple calculation yields

$$\phi_n(\lambda) = \int_{-\infty}^{\infty} e^{i\lambda t}\, dF_n(t) = \phi^n\left(\frac{\lambda}{\sigma\sqrt{n}}\right).$$

In view of P4,

$$\phi\left(\frac{\lambda}{\sigma\sqrt{n}}\right) = 1 - \frac{\lambda^2}{2n} + \frac{\epsilon(\lambda,n)}{n},$$

where, for every fixed λ, the error term $\epsilon(\lambda,n)$ tends to zero as n tends to infinity. Consequently

$$\lim_{n \to \infty} \int_{-\infty}^{\infty} e^{i\lambda t}\, dF_n(t) = \lim_{n \to \infty} \left[1 - \frac{\lambda^2}{2n} + \frac{\epsilon(\lambda,n)}{n}\right]^n = e^{-\lambda^2/2},$$

which proves P8.

7. PERIODICITY

In the terminology of Chapter I, a random walk with d-dimensional state space R is aperiodic if the group $\bar{R}$ and the group R are the same group. Of course, given any transition function we could artificially increase the dimension of R (by imbedding R in a space of higher dimension) and then extend $P(0,x)$ by defining it to be zero, where it was not previously defined. Therefore it is important to make our terminology absolutely unambiguous. We shall speak of a d-dimensional random walk only when R has dimension d, and when $P(0,x)$ is defined for all x in R. This random walk is then said to be aperiodic if $\bar{R} = R$.

Having pointed out how one can trivially make an aperiodic random walk periodic, by artificially enlarging its state space, we shall now show that it is almost as easy to replace a periodic random walk by an aperiodic one, which retains all properties which could possibly be of any interest. This possibility depends on a simple lemma from linear algebra.

P1 *If R_d is the group of d-dimensional integers, and if $\bar{R}$ is a proper subgroup of R_d containing more elements than just the origin, then there exists an integer k, $1 \leq k \leq d$, and k linearly independent points in R_d, namely $x_1, x_2, \ldots, x_k$, such that $\bar{R}$ is the additive group generated by $x_1, x_2, \ldots, x_k$. The integer k is uniquely determined by $\bar{R}$, and $\bar{R}$ is isomorphic to the group R_k of k-dimensional integers.*

Proof: A subgroup of R_d is a so-called vector-module over the ring of integers (not a vector space, since the integers do not form a field). Hence it should be no surprise that the usual proof for vector spaces ([38], p. 18) breaks down. Indeed the theorem is false for modules over arbitrary rings, but true for modules over principal ideal rings—rings which like the integers have the property that every ideal is generated by a single integer. In this case the usual proof ([99], p. 149) proceeds by induction. Suppose that P1 holds for $d \leq n - 1$ (it obviously does when $d = 1$), and let X be a submodule (subgroup) of R_n. If the last coordinate of every element of X is 0, then we are obviously finished, in view of the induction hypothesis. If not, let L denote the set of all integers which occur as last coordinates of elements of X. Clearly L forms an ideal, and therefore it consists of all multiples of a positive integer p.

Now we may choose an element $x \in X$ whose last coordinate $x^n = p$. For every $y \in X$ there is some integer r (depending on y, of course) such that the last coordinate of $y - rx$ is zero. The set Y of all points $y - rx$ obtained by varying y over X forms a submodule of R_{n-1}, and by the induction hypothesis, one can find $k - 1 \leq n - 1$ linearly independent points $x_1, x_2, \ldots, x_{k-1}$ which generate Y. But then the set $\{x_1, x_2, \ldots, x_{k-1}, x\}$ will clearly generate X. It is a linearly independent set since the last coordinate of x is not zero, so it forms a basis for X of dimension k. This implies that X is isomorphic to R_k with $k \leq n$; just map x_1 on the first unit vector of R_k, x_2 on the second, ..., x_{k-1} on the $(k-1)$ st, and x on the last unit vector. The induction is therefore complete.

Lemma P1 may be used in the following way. Suppose a random walk $P(x,y)$ on the state space $R = R_d$ is given and happens to be periodic. If $\bar{R} = \{0\}$, then $P(x,y) = \delta(x,y)$ and the random walk is of no interest. Otherwise $\bar{R}$ is isomorphic to R_k, with basis $x_1, x_2, \ldots, x_k$ for some $k \leq d$. Then we can define an operator T which maps $\bar{R}$ onto R_k linearly and one-to-one so that $Tx_i = \xi_i$ for $i = 1, 2, \ldots, k$, where the ξ_i are the unit vectors in R_k. On R_k we now define the function $Q(x,y)$ so that for x,y in R_k

$$Q(x,y) = P(T^{-1}x, T^{-1}y).$$

It should be clear that $Q(x,y)$ is a transition function. Since $P \geq 0$, we have $Q \geq 0$. Since T is linear, $Q(x,y) = Q(0, y - x)$. Finally, then,

$$\sum_{x \in R_k} Q(0,x) = \sum_{x \in R_k} P(0, T^{-1}x) = \sum_{y \in R} P(0,y) = 1,$$

as $P(0,y) = 0$ when y is not in $\bar{R}$.

By our construction the random walk determined by Q on R_k is aperiodic, and it has all the essential properties of the periodic random walk defined by P on R. For example,

$$\sum_{n=0}^{\infty} Q_n(0,0) = \sum_{n=0}^{\infty} P_n(0,0) \leq \infty$$

so that *if one of the random walks is recurrent, then so is the other.*

E1 One can of course apply similar ideas to stochastic processes which are almost, but not quite, random walks by our definition. Consider, for example, the *triangular lattice* in the plane which consists of all points (complex numbers) z of the form

$$z = m + ne^{\pi i/3},$$

where m and n are arbitrary positive or negative integers or zero. This lattice, it is easily seen, forms an additive group G, but G is not a subset of the plane of lattice points $R = R_2$. Nevertheless we shall try to define, with G as state space, a stochastic process which, in a reasonable sense, has transition probabilities of *one third from each point z to the neighbors*

$$z + 1, \quad z + e^{2\pi i/3}, \quad z + e^{-2\pi i/3}.$$

Observing that each point z in G has a unique representation as

$$z = m + ne^{\pi i/3}, \quad m, n = 0, \pm 1, \pm 2, \ldots,$$

we may define a linear operator T mapping G on R_2 in a one-to-one fashion by

$$T(z) = T(m + ne^{\pi i/3}) = m + ni.$$

This is clearly a group isomorphism of G onto R_2. In particular

$$T(1) = 1, \quad T(e^{2\pi i/3}) = -1 + i, \quad T(e^{-2\pi i/3}) = -i.$$

This isomorphism therefore transforms the given process into the bona fide random walk with transition probabilities

$$P(0,1) = P(0,-1+i) = P(0,-i) = 1/3,$$

and $P(0,x) = 0$ for all other x in $R = R_2$. This is a random walk in the plane with mean vector $\mu = 0$ and finite second moment m_2. As we shall see in section 8, T8.1, such a random walk is recurrent. The sense in which this makes the given process on G recurrent is obviously the following. Define $Q(0,z) = 1/3$ for $z = 1$, $z = e^{2\pi i/3}$, $z = e^{-2\pi i/3}$, and zero elsewhere on G. Then define $Q(z_1, z_2) = Q(0, z_2 - z_1)$ for all pairs z_1, z_2 in G. Finally, imitating P1.1, let

$$Q_0(z_1, z_2) = \delta(z_1, z_2), \quad Q_1(z_1, z_2) = Q(z_1, z_2),$$
$$Q_{n+1}(z_1, z_2) = \sum_{z \in G} Q(z_1, z) Q_n(z, z_2), \quad n \geq 0.$$

Then the recurrence of the P-random walk on R_2 implies that

$$\sum_{n=0}^{\infty} Q_n(0,0) = \infty.$$

It is obvious that one should let this statement serve as the definition of recurrence for the process with transition function Q.

Now we return to Fourier analysis in order to establish a simple criterion for aperiodicity, in terms of the characteristic function of a random walk.

T1 *A random walk on R (of dimension d), is aperiodic if and only if its characteristic function $\phi(\theta)$, defined for θ in E (of dimension d) has the following property: $\phi(\theta) = 1$ if and only if all the coordinates of θ are integer multiples of 2π.*

Proof:[2] To fix the notation, let

$$\bar{E} = [\theta \mid \theta \in E; \phi(\theta) = 1]$$

and let

$$E_0 = [\theta \mid \theta \in E; (2\pi)^{-1}\theta_k = \text{integer for } k = 1, 2, \ldots, d].$$

Then the theorem we have to prove takes on the simple form

(1) $\qquad \bar{E} = E_0$ if and only if $\bar{R} = R$.

The easy part of (1) is the implication $\bar{R} = R \Rightarrow \bar{E} = E_0$. It is obvious from the definition of $\phi(\theta)$ that $E_0 \subset \bar{E}$, and therefore it suffices to prove that $\bar{E} \subset E_0$. Suppose that θ is in $\bar{E}$. This means that $(2\pi)^{-1}\theta \cdot x$ is an integer for all x such that $P(0,x) > 0$, or for all x in Σ. (See D2.1 for the definition of Σ, R^+, and $\bar{R}$.) By the definition of R^+, $(2\pi)^{-1}\theta \cdot x$ is then an integer for x in R^+, and by the definition of $\bar{R}$, $(2\pi)^{-1}\theta \cdot x$ is an integer for all x in $\bar{R}$. But we have assumed that $\bar{R} = R$, and therefore $(2\pi)^{-1}\theta \cdot x$ is an integer for each unit vector in R. By letting x run through the d unit vectors we see that each component of θ is a multiple of 2π. Thus θ is in E_0, and since θ was an arbitrary point in $\bar{E}$ we have shown that $R = \bar{R} \Rightarrow \bar{E} \subset E_0$.

The other implication, $\bar{E} = E_0 \Rightarrow \bar{R} = R$, will lead us to a famous problem in number theory. When the common dimension d of E_0 and R is zero, there is nothing to prove. So we shall now assume that $\bar{R}$ is a proper subgroup of R and that $d = \dim(E_0) = \dim(R) \geq 1$, and attempt to construct a point θ_0 in $\bar{E}$ which is not in E_0. Such a point θ_0 has to have the property that $(2\pi)^{-1}\theta_0 \cdot x$ is an integer for all x in $\bar{R}$, and that $(2\pi)^{-1}\theta_0$ has some coordinate which is not an integer. Using P1 we construct a basis $a_1, a_2, \ldots, a_k$ of $\bar{R}$, with $k \leq d$. If it turns out that $k = \dim(\bar{R}) < d$, we are in the easiest case; for then we let β be any vector in the d-dimensional space E which is perpendicular to each basis vector, and such that $(2\pi)^{-1}\beta$ has some non-integer coordinate. Now we take $\theta_0 = \beta$ which gives $(2\pi)^{-1}\theta_0 \cdot x = 0$ when $x = a_1, a_2, \ldots, a_k$, and hence when x is an arbitrary point in $\bar{R}$.

Finally suppose that $k = \dim(\bar{R}) = d$. Let

$$a_i = (a_{i1}, a_{i2}, \ldots, a_{id})$$

be the basis vectors of $\bar{R}$ in the cartesian coordinates of R, and let

$$\mathscr{A} = [x \mid x = \sum_{i=1}^{d} \xi_i a_i, 0 \leq \xi_i \leq 1].$$

Thus $\mathscr{A}$ is a subset of Euclidean space which we shall call the

---

[2] A brief abstract proof can be based on the Pontryagin duality theorem for locally compact Abelian groups [74].

fundamental parallelogram of $\bar{R}$. Its *vertices* (those points such that each $\xi_i = 0$ or 1 for $i = 1, \ldots, d$) belong to $\bar{R}$, in fact $\bar{R}$ is the group generated by these vertices. Let us call a point of $\mathscr{A}$ which is not a vertex an *interior point*. Since, by hypothesis $\bar{R} \neq R$, it is clear that some interior point of $\mathscr{A}$ must belong to R. This implies (as will be shown in P2 below, and this is the crux of the proof) that the *volume V* of $\mathscr{A}$ is greater than one.

We shall proceed on the assumption that P2 holds, and begin by observing that

$$1 < V = |\det A|$$

where $A = (a_{ij})$, $i,j = 1, 2, \ldots, d$, is the matrix whose entries are the components of the basis vectors a_i. Since these basis vectors are linearly independent, the matrix A must be nonsingular. Hence the determinant of A^{-1} is $(\det A)^{-1}$ which is not zero and which is *less than one in magnitude*. This fact leads to the conclusion that *not every element of A^{-1} can be an integer*. Suppose therefore that a noninteger element has been located in the pth column of A^{-1}. In this case we let θ_0 be this column multiplied by 2π, i.e.,

$$\theta_0 = 2\pi(A^{-1}{}_{1p}, A^{-1}{}_{2p}, \ldots, A^{-1}{}_{dp}).$$

Since $A \cdot A^{-1} = I$, we have $(2\pi)^{-1} a_k \cdot \theta_0 = \delta(k,p)$, so that $(2\pi)^{-1} x \cdot \theta_0$ is an integer for $x = a_1, a_2, \ldots, a_d$, and by linearity, for all x in $\bar{R}$. Thus θ_0 satisfies the requirements; it is a point in $\bar{E}$ which is not in E_0, and the proof of T1 is complete if we show that

P2 *Every fundamental parallelogram which contains an interior lattice point must have volume greater than one.*

Proof: Let us take a large cube, in d-dimensional space, with edge of length M. The fundamental parallelogram $\mathscr{A}$ fits into this cube approximately M^d/V times, if V is the volume of $\mathscr{A}$. Since each congruent (translated) version of $\mathscr{A}$ accounts for at least two lattice points (one boundary point and at least one interior one), it follows that the number of lattice points in the large cube is *at least* of the order of $2 M^d V^{-1}$. But the volume M^d of the large cube is also approximately the number of lattice points in the large cube. Hence one has the "approximate inequality" $2M^d V^{-1} \leq M^d$, and it is quite trivial to make the argument sufficiently precise, letting $M \to \infty$, to conclude that $2V^{-1} \leq 1$. Hence $V > 1$.

That completes the proof of P2 and hence of T1, but we yield to the temptation of pointing out that the converse of P2 is also true.

P3 *If a fundamental parallelogram has volume $V > 1$, then it must contain an interior lattice point.*

This is the lemma of Minkowski (1896). As it is valuable for its applications to number theory rather than probability, the proof is left to the reader.

It is very convenient to use T1 to obtain further estimates concerning characteristic functions and transition functions. We shall find it expedient to use some basic facts about quadratic forms. For θ in E

$$\theta \cdot A\theta = \sum_{i=1}^{d} \sum_{j=1}^{d} a_{ij}\theta_i\theta_j$$

is called a *quadratic form* if $a_{ij} = a_{ij}$, i.e., if the matrix $A = (a_{ij})$, $i,j = 1, 2, \ldots, d$ is symmetric. Such a form is called *positive definite* if

$$\theta \cdot A\theta \geq 0 \text{ for all } \theta \text{ in } E,$$

and $\theta \cdot A\theta = 0$ only when $\theta = 0$. A positive definite quadratic form has positive real *eigenvalues*. These are simply the eigenvalues of the matrix A and we shall often use the well known estimate (see [38], Ch. III)

P4 $\qquad \lambda_1|\theta|^2 \leq \theta \cdot A\theta \leq \lambda_d|\theta|^2, \qquad \theta \in E,$

if A is positive definite and has eigenvalues $0 < \lambda_1 \leq \lambda_2 \leq \cdots \leq \lambda_d$.

With the aid of P4 one obtains the following estimate for the real part of the characteristic function of a random walk. (Remember that by D6.3, C denotes the cube with side 2π about the origin in E.)

P5 *For d-dimensional aperiodic random walk, $d \geq 1$, with characteristic function $\phi(\theta)$ there exists a constant $\lambda > 0$ such that*

$$1 - \text{Re } \phi(\theta) \geq \lambda|\theta|^2$$

for all θ in C.

Proof: Since $R = \bar{R}$, of dimension $d \geq 1$, one can find d linearly independent vectors $a_1, a_2, \ldots, a_d$ in the set $\Sigma = [x \mid P(0,x) > 0]$. Suppose that the longest of these vectors has length $L = \max |a_k|$. Then we can conclude that the quadratic form

$$Q_L(\theta) = \sum_{|x| \leq L} (x \cdot \theta)^2 P(0,x)$$

is positive definite. This is so because

$$Q_L(\theta) \geq \sum_{k=1}^{d} (a_k \cdot \theta)^2 P(0,a_k).$$

Since $a_k \in \Sigma$, $k = 1, \ldots, d$, the right-hand side could vanish only if θ were perpendicular to $a_1, a_2, \ldots, a_d$. This is impossible as they are linearly independent, so $Q_L(\theta)$ is positive definite.

We have

$$\text{Re}\,[1 - \phi(\theta)] = \sum_{x \in R} [1 - \cos x \cdot \theta] P(0,x)$$

$$= 2 \sum_{x \in R} \sin^2\left(\frac{x \cdot \theta}{2}\right) P(0,x) \geq 2 \sum_{|x| \leq L} \sin^2\left(\frac{x \cdot \theta}{2}\right) P(0,x)$$

Since $\left|\sin \dfrac{x \cdot \theta}{2}\right| \geq \pi^{-1}|x \cdot \theta|$ when $|x \cdot \theta| \leq \pi$,

$$\text{Re}\,[1 - \phi(\theta)] \geq \frac{2}{\pi^2} \sum_{[x||x| \leq L;\,|x \cdot \theta| \leq \pi]} (x \cdot \theta)^2 P(0,x).$$

But under the restriction on x that $|x| \leq L$ we know that $|x \cdot \theta| \leq \pi$ whenever $|\theta| \leq \pi L^{-1}$. Therefore

$$\text{Re}\,[1 - \phi(\theta)] \geq \frac{2}{\pi^2} \sum_{|x| \leq L} (x \cdot \theta)^2 P(0,x) = \frac{2}{\pi^2} Q_L(\theta)$$

for all θ such that $|\theta| \leq \pi L^{-1}$. If λ_1 is the smallest eigenvalue of $Q_L(\theta)$, then P4 yields

(1) $\qquad 1 - \text{Re}\,\phi(\theta) \geq \dfrac{2}{\pi^2} \lambda_1 |\theta|^2$ for $|\theta| \leq \pi L^{-1}$.

The sphere defined by $|\theta| \leq \pi L^{-1}$ is a subset of the cube C, since $L^{-1} \leq 1$. But now we use T1 which tells us that $1 - \text{Re}\,\phi(\theta) > 0$ when $\theta \in C$ and $|\theta| \geq \pi L^{-1}$. As $1 - \text{Re}\,\phi(\theta)$ is a continuous function on C we have

$$m = \min_{[\theta|\theta \in C;\,|\theta| \geq \pi L^{-1}]} [1 - \text{Re}\,\phi(\theta)] > 0,$$

which implies

(2) $\qquad 1 - \text{Re}\,\phi(\theta) \geq \dfrac{mL^2}{\pi^2} |\theta|^2$ for $\theta \in C$, $|\theta| \geq \pi L^{-1}$.

Combining the two inequalities (1) and (2) we find that P5 has been proved with $\lambda = \min\,[2\pi^{-2}\lambda_1,\,m\pi^{-2}L^2] > 0$.

The estimate provided by P5 will turn out to be very useful. With rather little effort it can be converted into statements concerning the asymptotic behavior of the transition function $P_n(0,x)$ as $n \to \infty$. The simplest statements of this type would be

(1) $$\lim_{n \to \infty} P_n(0,0) = 0, \text{ when } P(0,0) \neq 1,$$

and a slightly more sophisticated version, namely

(2) $$\lim_{n \to \infty} \sup_{x \in R} P_n(0,x) = 0,$$

the latter being correct in all cases except when $P(0,x) = 1$ for some $x \in R$. Note that even (1), although obvious for transient random walk, is not too easy to prove in the recurrent case, say by the methods of Chapter I. But, using Fourier analytical devices, we shall be able to obtain even sharper results than (2), in the form of an upper bound on the supremum in (2) which will depend on the dimension d of the random walk. Since the dimension is the crucial thing, but not aperiodicity, it will be useful to state

D1 *a random walk is genuinely d-dimensional if the group $\bar{R}$ associated with the random walk is d-dimensional.*

Thus genuinely d-dimensional random walk need not be aperiodic, but $\bar{R}$ must have a basis, according to P1, of d linearly independent vectors $a_1, \ldots, a_d$ in R. We shall prove the following result

P6 *If $P(x,y)$ is the transition function of a genuinely d-dimensional random walk $(d \geq 1)$ with the property that*[3]

(3) $$Q(x,y) = \sum_{t \in R} P(x,t) P(y,t)$$

is also the transition function of a genuinely d-dimensional random walk, then there exists a constant $A > 0$, such that

(4) $$P_n(0,x) \leq A n^{-d/2}, \quad x \in R, \quad n \geq 1.$$

Furthermore, every genuinely d-dimensional random walk $(d \geq 1)$, with the single exception of those where $P(0,x) = 1$ for some $x \in R$, satisfies

(5) $$P_n(0,x) \leq A n^{-1/2}, \quad x \in R, \quad n \geq 1,$$

for some $A > 0$.

Proof: It seems best to begin by explaining the usefulness of the transition function $Q(x,y)$ in (3). It clearly is a transition function, and if

---

[3] This class of random walks has the following equivalent algebraic characterization: the support Σ of $P(0,x)$ is not contained in any coset of any k-dimensional subgroup of R, for any $k < d$.

7. PERIODICITY

$$\phi(\theta) = \sum_{x \in R} P(0,x) e^{ix \cdot \theta}$$

is the characteristic function of the random walk, then $Q(x,y)$ has the characteristic function

$$\psi(\theta) = \sum_{x \in R} Q(0,x) e^{ix \cdot \theta} = |\phi(\theta)|^2.$$

Suppose now that we wish to get an upper bound for $P_n(0,x)$. Clearly

$$(2\pi)^d \sup_{x \in R} P_{2n}(0,x) = \sup_{x \in R} \int e^{-ix \cdot \theta} \phi^{2n}(\theta) \, d\theta \le \int |\phi(\theta)|^{2n} \, d\theta = \int \psi^n(\theta) \, d\theta,$$

and the same upper bound is obtained if $P_{2n}(0,x)$ is replaced by $P_{2n+1}(x)$. Therefore (4) will be true if we can exhibit a constant $B > 0$ such that

(6) $$(2\pi)^d Q_n(0,0) = \int_C \psi^n(\theta) \, d\theta \le B n^{-d/2}, \qquad n \ge 1.$$

The stage is now almost set for the application of P5. We must observe only that we may assume $Q(x,y)$ to be aperiodic. If it were not, then, as remarked after D1, one might change coordinates to replace $Q(x,y)$ by a symmetric, d-dimensional, aperiodic transition function $Q'(x,y)$ such that $Q'_n(0,0) = Q_n(0,0)$. (The characteristic function of Q' will be real but it might no longer be non-negative, and we shall want to use the non-negativity of ψ; but that is a decidedly minor point, as it suffices to prove (6) for even values of n.)

As ψ is real, P5 gives

$$1 - \psi(\theta) \ge \lambda |\theta|^2, \qquad \theta \in C,$$

for some $\lambda > 0$. Thus

$$0 \le \psi(\theta) \le 1 - \lambda|\theta|^2 \le e^{-\lambda|\theta|^2}, \qquad \theta \in C,$$

$$\int_C \psi^n(\theta) \, d\theta \le \int_C e^{-\lambda|\theta|^2 n} \, d\theta \le \int_E e^{-\lambda|\theta|^2 n} \, d\theta$$

$$= n^{-d/2} \int_E e^{-\lambda|\alpha|^2} \, d\alpha = B n^{-d/2}, \qquad n \ge 1.$$

The constant B is finite, being the integral of the Gaussian function $\exp(-\lambda|\alpha|^2)$ over all of Euclidean d-dimensional space. That completes the proof of (4).

As an illustration of the necessity of a condition of the type we imposed, consider the *two-dimensional* aperiodic random walk defined, in complex notation, by

$$P(0,1) = P(0,i) = 1/2.$$

It is easy to verify that $P_{2n}(0, n - in)$ is of the order of $n^{-1/2}$ rather than n^{-1}, the reason being that $Q(x,y)$ is *one-dimensional* in this case.

The proof of (5) is now immediate. As we saw, the true dimension of the state space of $Q(x,y)$ influences the asymptotic behavior of $P_n(0,x)$. Thus (5) will hold provided $Q(x,y)$ is at least genuinely one dimensional, i.e., if $Q(0,x) > 0$ for at least one $x \neq 0$. But an examination of (3) shows this to be always the case, except when $P(0,x) = 1$ for some $x \in R$, in which case Q is zero dimensional ($Q(0,0) = 1$).

This section is concluded by showing that P6 is the best possible result of its type, because there are random walks for which the upper bound is actually attained, in the sense that

$$\lim_{n \to \infty} n^{d/2} P_n(0,x)$$

exists and is positive for every x in R of dimension d. P7 and P8 will serve as preliminaries to the explicit calculation of such limits in P9 and P10.

P7 *For aperiodic random walk of arbitrary dimension $d \geq 1$, with mean vector $\mu = 0$, second absolute moment $m_2 < \infty$, $Q(\theta) = \Sigma_{x \in R} (x \cdot \theta)^2 P(0,x)$ is positive definite and*

$$\lim_{\theta \to 0} \frac{1 - \phi(\theta)}{Q(\theta)} = \frac{1}{2}.$$

Proof: Since $\mu = 0$ we may write

$$1 - \phi(\theta) = \tfrac{1}{2}Q(\theta) + \sum_{x \in R} [1 + i\theta \cdot x - \tfrac{1}{2}(\theta \cdot x)^2 - e^{i\theta \cdot x}] P(0,x).$$

Observe now that $Q(\theta)$ is positive definite (as shown in the proof of P5 for $Q_L(\theta) \leq Q(\theta)$) and that by P4 $Q(\theta) \geq \lambda |\theta|^2$ for some $\lambda > 0$. Therefore it will suffice to prove that

$$\lim_{\theta \to 0} |\theta|^{-2} \sum_{x \in R} [1 + i\theta \cdot x - \tfrac{1}{2}(\theta \cdot x)^2 - e^{i\theta \cdot x}] P(0,x) = 0.$$

Clearly each term of the sum over R, divided by $|\theta|^2$ tends to zero as $\theta \to 0$. But to interchange the limits we need to establish dominated convergence. There is some $A > 0$ such that

$$|1 + it - \tfrac{1}{2}t^2 - e^{it}| \leq At^2$$

for all real t. Therefore

$$|\theta|^{-2} \left| 1 + i\theta \cdot x - \frac{1}{2}(\theta \cdot x)^2 - e^{i\theta \cdot x} \right| \leq \frac{A(\theta \cdot x)^2}{|\theta|^2} \leq A|x|^2$$

which yields dominated convergence and completes the proof. (Observe that P7 represents a considerable refinement of P6.7.)

The next lemma is the analogue of the criterion in T1 for *strongly aperiodic* instead of *aperiodic* random walk.

P8 *Strongly aperiodic random walk of arbitrary dimension $d \geq 1$ has the property that $|\phi(\theta)| = 1$ only when each coordinate of θ is a multiple of 2π. Conversely, every random walk with this property is strongly aperiodic.*

Proof: For the direct part suppose that $|\phi(\theta_0)| = 1$, so that $\phi(\theta_0) = e^{it}$ for some real t. It follows that $x \cdot \theta_0 = t + 2n\pi$ for all x such that $P(0,x) > 0$, where of course the integer n may depend on x. Continuing to work with this fixed value of θ_0, we select a point z_0 in R such that $z_0 \cdot \theta_0 = t + 2m\pi$, where m is an integer. Now consider the transition function defined by $Q(0,x) = P(0, x + z_0)$, where $P(0,x)$ is the transition function of the given, strongly aperiodic random walk. We know from D5.1 that $Q(0,x)$ is the transition function of an aperiodic random walk. Its characteristic function is

$$\psi(\theta) = \sum_{x \in R} Q(0,x) e^{ix \cdot \theta} = \sum_{x \in R} P(0, x + z_0) e^{ix \cdot \theta} = e^{-iz_0 \cdot \theta} \phi(\theta).$$

Hence

$$\psi(\theta_0) = e^{-iz_0 \cdot \theta_0} \phi(\theta_0) = e^{-iz_0 \cdot \theta_0} e^{it} = 1,$$

but since $\psi(\theta)$ is the characteristic function of an aperiodic random walk we conclude from T1 that θ_0 is a point all of whose coordinates are integer multiples of 2π.

To prove the converse, suppose that a random walk fails to be strongly aperiodic. Then we have to exhibit some θ_0 in E such that $|\phi(\theta_0)| = 1$, and such that not all of its coordinates are multiples of 2π. By D5.1 we can find a point z_0 in R such that $P(0, x + z_0)$, as a function of x, fails to be aperiodic. This random walk has the characteristic function $\exp(-iz_0 \cdot \theta)\phi(\theta)$. By T1 this function equals one at some point θ_0 in E, not all of whose coordinates are multiples of 2π. Hence $|\phi(\theta_0)| = 1$, i.e., θ_0 satisfies our requirements, and the proof of P8 is complete.

P9 *For strongly aperiodic random walk of dimension $d \geq 1$ with mean $\mu = 0$ and finite second moments,*

$$\lim_{n \to \infty} (2\pi n)^{d/2} P_n(0,x) = |Q|^{-1/2}, \quad x \in R,$$

where $|Q|$ is the determinant of the quadratic form

$$Q(\theta) = \sum_{x \in R} (x \cdot \theta)^2 P(0,x).$$

Proof:[4] According to P6.3,

$$(2\pi n)^{d/2} P_n(0,x) = n^{d/2}(2\pi)^{-d/2} \int_C \phi^n(\theta) e^{-ix\cdot\theta}\, d\theta.$$

We perform the change of variables $\theta\sqrt{n} = \alpha$, so that

$$(2\pi n)^{\frac{d}{2}} P_n(0,x) = (2\pi)^{-\frac{d}{2}} \int_{\alpha\in\sqrt{n}C} \phi^n\left(\frac{\alpha}{\sqrt{n}}\right) e^{-\frac{ix\cdot\alpha}{\sqrt{n}}}\, d\alpha.$$

With $A > 0$, and $0 < r < \pi$, one decomposes this integral to obtain

(1) $\quad (2\pi n)^{\frac{d}{2}} P_n(0,x) = (2\pi)^{-\frac{d}{2}} \int_E e^{-\frac{1}{2}Q(\alpha)} e^{-\frac{ix\cdot\alpha}{\sqrt{n}}}\, d\alpha$
$$+ I_1(n,A) + I_2(n,A) + I_3(n,A,r) + I_4(n,r).$$

The last four integrals, which will play the role of error terms, are

$$I_1(n,A) = (2\pi)^{-\frac{d}{2}} \int_{|\alpha|\leq A} \left[\phi^n\left(\frac{\alpha}{\sqrt{n}}\right) - e^{-\frac{1}{2}Q(\alpha)}\right] e^{-\frac{ix\cdot\alpha}{\sqrt{n}}}\, d\alpha,$$

$$I_2(n,A) = -(2\pi)^{-\frac{d}{2}} \int_{|\alpha|>A} e^{-\frac{1}{2}Q(\alpha)} e^{-\frac{ix\cdot\alpha}{\sqrt{n}}}\, d\alpha,$$

$$I_3(n,A,r) = (2\pi)^{-\frac{d}{2}} \int_{A<|\alpha|\leq r\sqrt{n}} \phi^n\left(\frac{\alpha}{\sqrt{n}}\right) e^{-\frac{ix\cdot\alpha}{\sqrt{n}}}\, d\alpha,$$

$$I_4(n,r) = (2\pi)^{-\frac{d}{2}} \int_{|\alpha|>r\sqrt{n};\, \alpha\in\sqrt{n}C} \phi^n\left(\frac{\alpha}{\sqrt{n}}\right) e^{-\frac{ix\cdot\alpha}{\sqrt{n}}}\, d\alpha.$$

Our first task is to show that the principal term gives the desired limit. If $0 < \lambda_1 \leq \lambda_2 \leq \cdots \leq \lambda_d$ are the eigenvalues of the positive definite quadratic form $Q(\theta)$, then a rotation of the coordinate system gives

$$I_0 = (2\pi)^{-\frac{d}{2}} \int_E e^{-\frac{1}{2}Q(\alpha)}\, d\alpha$$
$$= (2\pi)^{-\frac{d}{2}} \int_E e^{-\frac{1}{2}\sum_{k=1}^d \lambda_k \alpha_k^2}\, d\alpha$$
$$= (2\pi)^{-\frac{d}{2}} \prod_{k=1}^d \int_{-\infty}^{\infty} e^{-\frac{1}{2}\lambda_k \alpha_k^2}\, d\alpha_k = \left[\prod_{k=1}^d \lambda_k\right]^{-\frac{1}{2}}.$$

---

[4] We reproduce the proof of Gnedenko [35], pp. 232–235, with the obvious modifications when $d \geq 2$.

But the product of the eigenvalues of Q is the determinant $|Q|$ so that $I_0 = |Q|^{-1/2}$. Since this integral is finite, it is clear that

$$\lim_{n\to\infty} (2\pi)^{-\frac{d}{2}} \int_E e^{-\frac{1}{2}Q(\alpha)} e^{-\frac{ix\cdot\alpha}{\sqrt{n}}} \, d\alpha = |Q|^{-\frac{1}{2}},$$

and therefore the proof of P9 will be complete as soon as we show that the sum of the four error terms tends to zero as $n \to \infty$.

To estimate $I_1(n,A)$ we use P7 which implies that

$$\lim_{n\to\infty} \phi^n\left(\frac{\alpha}{\sqrt{n}}\right) = e^{-\frac{1}{2}Q(\alpha)}$$

for each α in E. Thus $I_1(n,A)$ tends to zero as $n \to \infty$, for every $A > 0$. Next we estimate $I_4(n,r)$ by remarking that

$$|I_4(n,r)| \le n^{d/2}(2\pi)^{-d/2} \int_{[\theta|\theta\in C; |\theta|>r]} |\phi(\theta)|^n \, d\theta.$$

We know from P8 that $|\phi(\theta)| < 1 - \delta$ for some $\delta = \delta(r)$ on the set of integration. Therefore $I_4(n,r)$ tends to zero as $n \to \infty$, whenever $0 < r < \pi$. Now we have to worry only about I_2 and I_3. Using P7 again we can choose r small enough so that

$$\left|\phi^n\left(\frac{\alpha}{\sqrt{n}}\right)\right| \le e^{-\frac{1}{4}Q(\alpha)}$$

when $|\alpha| \le r\sqrt{n}$. Then

$$|I_3(n,A,r)| \le (2\pi)^{-d/2} \int_{|\alpha|>A} e^{-\frac{1}{4}Q(\alpha)} \, d\alpha$$

for all n. Similarly

$$|I_2(n,A)| \le (2\pi)^{-d/2} \int_{|\alpha|>A} e^{-\frac{1}{2}Q(\alpha)} \, d\alpha$$

is independent of n. Therefore the sum of I_2 and I_3 can be made arbitrarily small by choosing r small enough and A large enough, and that completes the proof of P9.

Remark: The following, slightly stronger form of P9 is sometimes useful:

Under the conditions in P9

(2) $$\lim_{n\to\infty} \left[(2\pi n)^{\frac{d}{2}} P_n(0,x) - |Q|^{-\frac{1}{2}} e^{-\frac{1}{2n}(x\cdot Q^{-1}x)}\right] = 0,$$

uniformly for all $x \in R$.

This is really an immediate corollary of the method of proof of P9. It follows from the way the four error terms I_1, I_2, I_3, and I_4 were estimated in the proof of P9 that their sum tends to zero uniformly in x. Comparing equations (1) and (2), it therefore suffices to show that

$$(3) \qquad (2\pi)^{-\frac{d}{2}} \int_E e^{-\frac{1}{2}Q(\alpha)} e^{-\frac{ix\cdot\alpha}{\sqrt{n}}} \, d\alpha = |Q|^{-\frac{1}{2}} e^{-\frac{1}{2n}(x\cdot Q^{-1}x)}.$$

In (3) as well as (2), $x \cdot Q^{-1} x$ denotes the inverse quadratic form of $Q(\alpha) = \alpha \cdot Q\alpha$. It is easy to verify (3) by making an orthogonal transformation from the α-coordinate system to one where the quadratic form $Q(\alpha)$ is of the diagonal type. Due to the presence of the exponential involving x on the left in (3), this calculation is slightly more complicated than the evaluation of I_0 in the proof of P9, but no new ideas are involved.

E2 Simple random walk of dimension $d \geq 1$ is not strongly aperiodic. Nevertheless it is quite easy to modify the proof of P9 appropriately. The characteristic function, for dimension d, is

$$\phi(\theta) = \sum P(0,x) e^{ix\cdot\theta} = \frac{1}{d}[\cos\theta_1 + \cdots + \cos\theta_d].$$

Although $P_n(0,0) = 0$ when n is odd, we get

$$(1) \qquad P_{2n}(0,0) = (2\pi)^{-d} \int_C \left[\frac{1}{d}\sum_{k=1}^d \cos\theta_k\right]^{2n} d\theta.$$

Since the integrand in (1) is periodic in each coordinate θ_k with period 2π, we may translate the cube C by the vector $v = (\pi/2)(1, 1, \ldots, 1)$. Calling $C + v = C'$, we see that $P_{2n}(0,0)$ is still given by (1), if C is replaced by C'.

The point of this translation from C to C' was that the integrand in (1) assumes its maximum (the value one) at two *interior* points of C', namely at the origin and at the point $\pi(1, 1, \ldots, 1)$. The contributions from these two points to the asymptotic behavior of $P_{2n}(0,0)$ are the same, since $|\phi(\theta)| = |\phi(\theta + w)|$ when $w = \pi(1, 1, \ldots, 1)$. Also the proof of P9 has shown that the asymptotic behavior of the integral in (1) is unchanged if we integrate only over arbitrarily small spheres about the points where $|\phi(\theta)| = 1$. Letting S denote such a sphere, of radius $r < \pi/2$, one has

$$(2) \qquad P_{2n}(0,0) \sim 2(2\pi)^{-d} \int_S \left[\frac{1}{d}\sum_{k=1}^d \cos\theta_k\right]^{2n} d\theta.$$

For small values of $|\theta|$

$$(3) \qquad \frac{1}{d}\sum_{k=1}^d \cos\theta_k = \frac{1}{d}\sum_{k=1}^d \left[1 - 2\sin^2\left(\frac{\theta_k}{2}\right)\right]$$

$$\sim 1 - \frac{|\theta|^2}{2d} \sim e^{-\frac{|\theta|^2}{2d}}, \qquad |\theta| \to 0.$$

Applying the estimate (3) to (2), one obtains for large n

$$P_{2n}(0,0) \sim 2(2\pi)^{-d} \int_S e^{-\frac{|\theta|^2}{d}n} d\theta$$

$$\sim 2(2\pi)^{-d} \int_E e^{-\frac{|\theta|^2}{d}n} d\theta = 2\left[\frac{1}{\sqrt{2\pi}} \int_{-\infty}^{\infty} e^{-\frac{x^2}{2}} dx\right]^d \left(\frac{d}{4n\pi}\right)^{\frac{d}{2}}.$$

The integral of the Gaussian density is one, so that we have shown, for d-dimensional simple random walk, that

$$P_{2n}(0,0) \sim 2\left(\frac{d}{4n\pi}\right)^{d/2}, \qquad \text{as } n \to \infty.$$

The strong form of P9 given in the remark following its proof is known as the *Local Central Limit Theorem*. Although it is an extremely useful tool, in many applications of probability theory, there are occasions when one needs even better error estimates (P9 is really quite poor for large values of $|x|$, say when $|x|$ is of the order of $\sqrt{n}$ or larger). There is a variant of P9 (due to Smith [89], 1953, in the one-dimensional case) which provides sharper estimates for large $|x|$. For us it will be indispensable in section 26, where it will be combined with P9 to shed light on the asymptotic behavior of the Green function $G(x,y)$ for three-dimensional random walk.

P10   *For strongly aperiodic random walk of dimension $d \geq 1$, with mean $\mu = 0$ and finite second moments*

$$\lim_{n \to \infty} \frac{|x|^2}{n} \left[(2\pi n)^{\frac{d}{2}} P_n(0,x) - |Q|^{-\frac{1}{2}} e^{-\frac{1}{2n}(x \cdot Q^{-1}x)}\right] = 0,$$

uniformly for all $x \in R$.

Proof: Just as in the proof of P9 we write

$$\frac{|x|^2}{n}(2\pi n)^{d/2} P_n(0,x) = \left(\frac{n}{2\pi}\right)^{d/2} \frac{|x|^2}{n} \int_C \phi^n(\theta) e^{-ix\cdot\theta} d\theta.$$

It is now possible to bring the term $|x|^2$ in under the integral sign on the right. If $f(\theta)$ is a function with continuous second derivatives, we have by Green's theorem

$$|x|^2 \int_C f(\theta) e^{-ix\cdot\theta} d\theta = -\int_C f(\theta) \Delta_\theta [e^{-ix\cdot\theta}] d\theta$$

$$= -\int_C e^{-ix\cdot\theta} \Delta_\theta f(\theta) d\theta + \text{boundary terms.}$$

Here

$$\Delta_\theta = \sum_{i=1}^{d} \frac{\partial^2}{\partial \theta_i^2}$$

is the Laplace operator. We may apply this idea to $\phi^n(\theta)$ whose second derivatives are continuous by the same proof as in P6.4. Furthermore the boundary terms vanish since $\phi(\theta)$ is periodic. In this way

(1) $\quad \dfrac{|x|^2}{n}(2\pi n)^{\frac{d}{2}}P_n(0,x) = -\left(\dfrac{n}{2\pi}\right)^{\frac{d}{2}}\dfrac{1}{n}\int_C e^{-ix\cdot\theta}\,\underset{\theta}{\Delta}[\phi^n(\theta)]\,d\theta$

$\qquad = -(2\pi)^{-\frac{d}{2}}\displaystyle\int_{\sqrt{n}\,C} e^{-\frac{ix\cdot\alpha}{\sqrt{n}}}\underset{\alpha}{\Delta}\left[\phi^n\left(\dfrac{\alpha}{\sqrt{n}}\right)\right]d\alpha,$

after a dilation of coordinates to $\alpha = \sqrt{n}\,\theta$.

From now on the proof proceeds much as did the one of P9, the principal difference being that we *do not* rely on

(2) $\quad \displaystyle\lim_{n\to\infty}\phi^n\left(\dfrac{\alpha}{\sqrt{n}}\right) = e^{-\frac{1}{2}Q(\alpha)}, \qquad \alpha \in E,$

but *use instead*

(3) $\quad \displaystyle\lim_{n\to\infty}\underset{\alpha}{\Delta}\left[\phi^n\left(\dfrac{\alpha}{\sqrt{n}}\right)\right] = \underset{\alpha}{\Delta}e^{-\frac{1}{2}Q(\alpha)}, \qquad \alpha \in E.$

The proof of (3) depends on

(4) $\quad \underset{\alpha}{\Delta}\left[\phi^n\left(\dfrac{\alpha}{\sqrt{n}}\right)\right]$

$\qquad = n\phi^{n-1}\left(\dfrac{\alpha}{\sqrt{n}}\right)\underset{\alpha}{\Delta}\left[\phi\left(\dfrac{\alpha}{\sqrt{n}}\right)\right] + n(n-1)\phi^{n-2}\left(\dfrac{\alpha}{\sqrt{n}}\right)\left|\underset{\alpha}{\mathrm{grad}}\,\phi\left(\dfrac{\alpha}{\sqrt{n}}\right)\right|^2.$

A calculation based on P6.4 and P6.7 shows that

(5) $\quad \displaystyle\lim_{n\to\infty} n\,\underset{\alpha}{\Delta}\phi\left(\dfrac{\alpha}{\sqrt{n}}\right) = -m_2 = -\sum_{x\in R}|x|^2 P(0,x),$

$\qquad \displaystyle\lim_{n\to\infty} n^2\left|\underset{\alpha}{\mathrm{grad}}\,\phi\left(\dfrac{\alpha}{\sqrt{n}}\right)\right|^2 = (Q\alpha\cdot Q\alpha) = |Q\alpha|^2,$

where Q is the covariance matrix. Applying (2) and (5) to (4) one obtains (3).

Equation (3) suggests the decomposition

(6) $\quad \dfrac{|x|^2}{n}(2\pi n)^{\frac{d}{2}}P_n(0,x)$

$\qquad = -(2\pi)^{-\frac{d}{2}}\displaystyle\int_E e^{-\frac{ix\cdot\alpha}{\sqrt{n}}}\underset{\alpha}{\Delta}e^{-\frac{1}{2}Q(\alpha)}\,d\alpha + I_1 + I_2 + I_3 + I_4,$

where the I_k are exactly the same type of error terms as in the proof of P9, but with the Laplacian in the integrand. Thus

$$-I_1 = (2\pi)^{-\frac{d}{2}} \int_{|\alpha| \leq A} \Delta_\alpha \left[\phi^n\left(\frac{\alpha}{\sqrt{n}}\right) - e^{-\frac{1}{2}Q(\alpha)} \right] e^{-\frac{ix\cdot\alpha}{\sqrt{n}}} d\alpha,$$

$$I_2 = (2\pi)^{-\frac{d}{2}} \int_{|\alpha| > A} \Delta_\alpha \left[e^{-\frac{1}{2}Q(\alpha)} \right] e^{-\frac{ix\cdot\alpha}{\sqrt{n}}} d\alpha,$$

$$-I_3 = (2\pi)^{-\frac{d}{2}} \int_{A < |\alpha| \leq r\sqrt{n}} \Delta_\alpha \left[\phi^n\left(\frac{\alpha}{\sqrt{n}}\right) \right] e^{-\frac{ix\cdot\alpha}{\sqrt{n}}} d\alpha,$$

$$-I_4 = (2\pi)^{-\frac{d}{2}} \int_{|\alpha| > r\sqrt{n};\ \alpha \in \sqrt{n}C} \Delta_\alpha \left[\phi^n\left(\frac{\alpha}{\sqrt{n}}\right) \right] e^{-\frac{ix\cdot\alpha}{\sqrt{n}}} d\alpha.$$

The principal term on the right-hand side in (6) is easily seen to have the right value, i.e.,

$$\frac{|x|^2}{n} |Q|^{-\frac{1}{2}} e^{-\frac{1}{2n}(x\cdot Q^{-1} x)}$$

To show that the sum of the four error terms tends to zero uniformly in x one replaces each integral I_k by the integral of the absolute value of the integrand. That eliminates the dependence on x. The order in which one eliminates the error terms is the same as in P9. An examination of (3), (4), and (5) shows the integrand in I_1 can be majorized by a bounded function on every finite interval. I_1 therefore tends to zero for every $A > 0$, uniformly in x. To deal with I_4 one obtains from (4) and (5) that

$$\left| \Delta_\alpha \phi^n\left(\frac{\alpha}{\sqrt{n}}\right) \right| \leq k_1(1 - \delta)^{n-1} + k_2(1 - \delta)^{n-2},$$

where $1 - \delta$ is the upper bound of $\phi(\theta)$ when $\theta \in C$ and $|\theta| \geq r$, and where k_1 and k_2 are positive constants depending on r, but independent of n. Thus $I_4 \to 0$ for each r, $0 < r < \pi$, uniformly in x. Next one disposes of I_3 by choosing r small enough so that

$$\left| \Delta_\alpha \phi^n\left(\frac{\alpha}{\sqrt{n}}\right) \right| \leq M e^{-\frac{1}{4}Q(\alpha)}$$

for some $M > 0$, when $|\alpha| \leq r\sqrt{n}$. This shows that I_3 can be made arbitrarily small, uniformly in n and x, by taking A large enough. Finally, one shows just as in P9 that I_2 tends to zero as $A \to \infty$, uniformly in n and x. That completes the proof.

8. RECURRENCE CRITERIA AND EXAMPLES

The harmonic analysis of transition functions provides elegant methods for deciding whether a random walk is recurrent or transient. We begin with a general criterion, proved by Chung and Fuchs [12], 1951, in the more general and more difficult context of arbitrary (noninteger valued) random walk in Euclidean space.

P1 *If $\phi(\theta)$ is the characteristic function of a random walk, then*

$$\lim_{t \nearrow 1} \int_C \mathrm{Re}\left[\frac{1}{1 - t\phi(\theta)}\right] d\theta < \infty$$

if and only if the random walk is transient.

Proof: The random walk is transient, according to P1.4 if and only if

$$G = \sum_{n=0}^{\infty} P_n(0,0) < \infty.$$

But

$$G = \lim_{t \nearrow 1} \sum_{n=0}^{\infty} t^n P_n(0,0),$$

whether this limit is finite or not. Using P6.3 we have

$$(2\pi)^d G = \lim_{t \nearrow 1} \sum_{n=0}^{\infty} t^n \int \phi^n(\theta) \, d\theta = \lim_{t \nearrow 1} \int \frac{d\theta}{1 - t\phi(\theta)}.$$

Observing that, for $0 \leq t < 1$,

$$\int \frac{d\theta}{1 - t\phi(\theta)} = \mathrm{Re} \int \frac{d\theta}{1 - t\phi(\theta)} = \int \mathrm{Re}\left[\frac{1}{1 - t\phi(\theta)}\right] d\theta,$$

we find

$$(2\pi)^d G = \lim_{t \nearrow 1} \int \mathrm{Re}\left[\frac{1}{1 - t\phi(\theta)}\right] d\theta,$$

which proves P1.

This criterion may be used to obtain simple sufficient conditions for recurrence or for transience. By analogy with T3.1, which states that one-dimensional random walk is recurrent if $m < \infty$ and $\mu = 0$, one may ask if two-dimensional random walk is recurrent under suitable assumptions about its moments.

T1 *Genuinely d-dimensional random walk is recurrent if*

(a) $\qquad d = 1 \quad \text{and} \quad m_1 < \infty, \qquad \mu = 0,$

or if

(b) $\qquad d = 2 \quad \text{and} \quad m_2 < \infty, \qquad \mu = 0.$

(c) *It is always transient when $d \geq 3$.*

Proof: Part (a) was already proved in Chapter I, but in E1 below we shall also sketch the proof of Chung and Fuchs [12], which uses P1. Concerning part (b) we know from P7.9 that

$$P_n(0,0) \sim Cn^{-1}, \qquad \text{as } n \to \infty,$$

for some positive constant C, provided that the random walk is strongly aperiodic. Since the harmonic series diverges, every strongly aperiodic random walk in the plane, with mean vector $\mu = 0$, and finite second moments, must be recurrent. But the condition of strong aperiodicity can be eliminated quite easily: we must construct a new random walk, with transition function $Q(x,y)$ which is strongly aperiodic, satisfies the hypotheses in part (b), and for which

$$\sum_{n=0}^{\infty} Q_n(0,0) \leq \sum_{n=0}^{\infty} P_n(0,0).$$

This construction is based on the observation that *a random walk with $P(0,0) > 0$ is aperiodic if and only if it is* strongly aperiodic. (If $P(0,0) > 0$, then $|\phi(\theta)| = 1$ if and only if $\phi(\theta) = 1$. Thus the conclusion follows from T7.1 and P7.8.) If $P(x,y)$ satisfies the hypotheses in (b) we can certainly find an integer $t \geq 1$ such that $P_t(0,0) > 0$. If the random walk with transition function $P_t(x,y)$ is aperiodic, then it is strongly aperiodic. If not, then we use P7.1 to construct an isomorphism T of the d-dimensional group generated by the points x in R where $P_t(0,x) > 0$ onto R, and define

$$Q(x,y) = P_t(Tx,Ty).$$

Now $Q(x,y)$ is strongly aperiodic,

$$Q_n(0,0) = P_{nt}(0,0), \qquad n \geq 0,$$

so that

$$\sum_{n=0}^{\infty} Q_n(0,0) \leq \sum_{n=0}^{\infty} P_n(0,0).$$

By P7.9 the series on the left diverges, and that proves part (b) of T1.

To prove part (c) we may, if necessary by relabeling the state space, assume that the random walk is aperiodic as well as genuinely d-dimensional. Now we observe that

$$\operatorname{Re} \frac{1}{1 - t\phi(\theta)} \leq \frac{t^{-1}}{\operatorname{Re}[1 - \phi(\theta)]}, \quad \theta \in E,$$

and conclude from P7.5 that

$$(2\pi)^d G = \lim_{t \nearrow 1} \int_C \operatorname{Re} \frac{1}{1 - t\phi(\theta)} \, d\theta$$

$$\leq \int_C \frac{1}{\operatorname{Re}[1 - \phi(\theta)]} \, d\theta \leq \lambda^{-1} \int_C \frac{d\theta}{|\theta|^2}$$

where λ is the positive constant in P7.5. Since the integral of $|\theta|^{-2}$ over C is finite when $d \geq 3$ (introduce polar coordinates!) the proof of T1 is complete.

Remark: If it were possible to justify the interchange of limits in P1, then one would have a slightly more elegant criterion:

"*d-dimensional random walk is transient or recurrent according as the real part of* $[1 - \phi(\theta)]^{-1}$ *is Lebesgue integrable on the d-dimensional cube C or not.*"

This statement is in fact correct. By Fatou's lemma[5]

$$\int_C \operatorname{Re} \frac{1}{1 - \phi(\theta)} \, d\theta = \int_C \lim_{t \nearrow 1} \operatorname{Re} \frac{1}{1 - t\phi(\theta)} \, d\theta$$

$$\leq \lim_{t \nearrow 1} \int_C \operatorname{Re} \frac{1}{1 - t\phi(\theta)} \, d\theta = (2\pi)^d G \leq \infty.$$

Therefore $\operatorname{Re}[1 - \phi(\theta)]^{-1}$ is integrable on C when $G < \infty$, i.e., when the random walk is transient. But unfortunately no direct (Fourier analytic) proof is known for the converse. The only proof known at this time (1963) depends on a number of facts from the theory of recurrent random walk—to be developed in Chapters III and VII. Nevertheless we shall present this proof here—it is quite brief and will serve as a preview of methods to be developed in later chapters. We shall freely use the necessary facts from these later chapters, without fear of dishonesty, as the resulting recurrence criterion (T2 below) will never again be used in the sequel. To avoid trivial but tedious reformulations we shall work only with aperiodic random walk and prove

[5] Cf. Halmos [37], p. 113.

T2 *Aperiodic d-dimensional random walk is transient if and only if* $\operatorname{Re}[1 - \phi(\theta)]^{-1}$ *is integrable on the d-dimensional cube C.* (*In view of T1 this is always the case when* $d \geq 3$.)

Proof: As shown by the argument involving Fatou's lemma we need only prove that $G < \infty$ when $\operatorname{Re}[1 - \phi(\theta)]^{-1}$ is integrable. Let us therefore assume that $G = \infty$, and work toward a contradiction. The random walk is then recurrent and aperiodic, and in this case it is known that the series

$$(1) \qquad a(x) = \sum_{n=0}^{\infty} [P_n(0,0) - P_n(x,0)]$$

converges for all $x \in R$. (That is shown in T28.1 of Chapter VII.) It is further known that

$$(2) \quad a(x) + a(-x) = 2(2\pi)^{-d} \int_C \frac{1 - \cos x \cdot \theta}{1 - \phi(\theta)} d\theta$$

$$= 2(2\pi)^{-d} \int_C [1 - \cos x \cdot \theta] \operatorname{Re} \frac{1}{1 - \phi(\theta)} d\theta, \quad x \in R.$$

(That follows from P28.4 when $d = 1$; when $d = 2$ it is proved earlier, in P12.1.) Finally it will be shown, in P29.4, that the function $a(x) + a(-x)$ has an interesting probability interpretation: it represents the expected number of visits of the random walk $\mathbf{x}_n$, $n \geq 0$, starting at $\mathbf{x}_0 = x$, to the point x, before the first visit to the origin. Thus, in the notation of P29.4,

$$(3) \qquad a(x) + a(-x) = g_{\{0\}}(x,x) = \sum_{k=0}^{\infty} \mathbf{P}_x[\mathbf{x}_k = x; \mathbf{T} > k],$$

where $\mathbf{T} = \min[k \mid k \geq 0; \mathbf{x}_k = 0]$.

The proof of T2 will now be terminated by studying the asymptotic behavior of $a(x) + a(-x)$ for large values of x. The Riemann Lebesgue lemma (which will be proved in P9.1) may be applied to equation (2) to yield the conclusion that

$$(4) \qquad \lim_{|x| \to \infty} [a(x) + a(-x)] = 2(2\pi)^{-d} \int_C \operatorname{Re} \frac{1}{1 - \phi(\theta)} d\theta < \infty.$$

On the other hand, we may use equation (3) to obtain the desired contradiction, in the form

$$(5) \qquad \lim_{|x| \to \infty} [a(x) + a(-x)] = +\infty.$$

To achieve this select a positive integer N, and observe that $\mathbf{T} < \infty$ with probability one for recurrent random walk. It follows that

$$a(x) + a(-x) \geq \sum_{k=0}^{N} \mathbf{P}_x[\mathbf{x}_k = x; \mathbf{T} > k] = \sum_{k=0}^{N} \mathbf{P}_x[\mathbf{x}_k = x]$$

$$- \sum_{k=0}^{N} \mathbf{P}_x[\mathbf{x}_k = x; \mathbf{T} \leq k] = \sum_{k=0}^{N} P_k(0,0) - \sum_{k=0}^{N} \mathbf{P}_x[\mathbf{x}_k = x; \mathbf{T} \leq k].$$

For each fixed integer k

$$\mathbf{P}_x[\mathbf{x}_k = x; \mathbf{T} \leq k] \leq \mathbf{P}_x[\mathbf{T} \leq k]$$

$$= \sum_{j=0}^{k} \mathbf{P}_x[\mathbf{T} = j] \leq \sum_{j=0}^{k} \mathbf{P}_x[\mathbf{x}_j = 0] = \sum_{j=0}^{k} P_j(x,0),$$

which tends to zero as $|x| \to \infty$. Therefore

$$\lim_{|x| \to \infty} [a(x) + a(-x)] \geq \sum_{k=0}^{N} P_k(0,0)$$

for arbitrary N. As the random walk was assumed recurrent it follows that (5) holds. The contradiction between (4) and (5) shows that aperiodic random walk cannot be recurrent when $\text{Re } [1 - \phi(\theta)]^{-1}$ is integrable, and hence the proof of T2 is complete.

E1 *To prove part* (a) *of* T1 *by use of* P1, *observe that*

(1) $$\int_{-\pi}^{\pi} \text{Re} \left[\frac{1}{1 - t\phi(\theta)} \right] d\theta \geq \int_{-\alpha}^{\alpha} \text{Re} \left[\frac{1}{1 - t\phi(\theta)} \right] d\theta$$

when $0 \leq t < 1$ and $0 < \alpha < \pi$, since the real part of $[1 - t\phi(\theta)]^{-1}$ is non-negative. Now

(2) $$\text{Re} \left[\frac{1}{1 - t\phi(\theta)} \right] \geq \frac{1 - t}{[\text{Re }(1 - t\phi)]^2 + t^2[\text{Im }\phi]^2},$$

and, given any $\epsilon > 0$, we may use P6.4 to choose α small enough so that

(3) $$|\text{Im } \phi(\theta)| \leq \epsilon |\theta|,$$

(4) $$[\text{Re }(1 - t\phi(\theta))]^2 \leq 2(1 - t)^2 + 2t^2 [\text{Re }(1 - \phi(\theta))]^2$$
$$\leq 2(1 - t)^2 + 2t^2 \epsilon^2 \theta^2$$

when $|\theta| \leq \alpha$. Consequently, combining (2), (3), and (4),

$$\int_{-\alpha}^{\alpha} \text{Re} \left[\frac{1}{1 - t\phi(\theta)} \right] d\theta \geq (1 - t) \int_{-\alpha}^{\alpha} \frac{d\theta}{2(1 - t)^2 + 3t^2 \epsilon^2 \theta^2}$$

$$\geq \frac{1}{3} \int_{-\frac{\alpha}{1-t}}^{\frac{\alpha}{1-t}} \frac{dx}{1 + \epsilon^2 x^2}.$$

It follows from (1) that

(5) $$\lim_{t \nearrow 1} \int_{-\pi}^{\pi} \mathrm{Re}\left[\frac{1}{1 - t\phi(\theta)}\right] d\theta \geq \frac{1}{3} \int_{-\infty}^{\infty} \frac{dx}{1 + \epsilon^2 x^2} = \frac{\pi}{3\epsilon}.$$

By letting ϵ tend to zero we see that the limit in (5) is infinite, and by P1 the random walk in question is recurrent.

The Fourier analytical proof of part (b) of T1 is so similar to that of part (a) that we go on to a different problem. We shall prove (without using the strong law of large numbers that was used in T3.1) that *one-dimensional random walk is transient if $m < \infty$, $\mu \neq 0$.* Curiously it seems to be impossible to give a proof using only P1 and P6.4. But if in addition we use P6.6, to the effect that the function $\theta^{-2} \mathrm{Re}\,[1 - \phi(\theta)]$ is Lebesgue integrable on the interval $[-\pi,\pi]$, then the proof is easy. As we may as well assume the random walk to be aperiodic, we have to show that

(6) $$\lim_{t \nearrow 1} \int_{-\alpha}^{\alpha} \mathrm{Re}\left[\frac{1}{1 - t\phi(\theta)}\right] d\theta < \infty$$

for some $\alpha > 0$. We use the decomposition

(7) $$\mathrm{Re}\left[\frac{1}{1 - t\phi}\right] = \frac{\mathrm{Re}\,(1 - \phi) + (1 - t)\,\mathrm{Re}\,\phi}{[\mathrm{Re}\,(1 - t\phi)]^2 + t^2\,(\mathrm{Im}\,\phi)^2}$$

$$\leq \frac{\mathrm{Re}\,(1 - \phi)}{t^2(\mathrm{Im}\,\phi)^2} + \frac{1 - t}{|1 - t\phi|^2}.$$

Choosing α sufficiently small so that

$$|\mathrm{Im}\,\phi(\theta)| \geq \left|\frac{\mu\theta}{2}\right|$$

when $|\theta| < \alpha$ (this can be done by P6.4), we see that

$$\lim_{t \nearrow 1} \int_{-\alpha}^{\alpha} \frac{\mathrm{Re}\,[1 - \phi(\theta)]}{t^2(\mathrm{Im}\,\phi)^2} d\theta \leq c \int_{-\pi}^{\pi} \frac{\mathrm{Re}\,[1 - \phi(\theta)]}{|\theta|^2} d\theta < \infty$$

for some $c > 0$. Now it remains only to estimate the integral of the last term in (7). This integral is decomposed into the sum

$$\int_{|\theta| \leq 1-t} \frac{1 - t}{|1 - t\phi|^2} d\theta + \int_{1-t < |\theta| \leq \alpha} \frac{1 - t}{|1 - t\phi|^2} d\theta$$

$$\leq (1 - t)^{-1} \int_{|\theta| \leq 1-t} d\theta + \frac{1 - t}{t^2} \cdot \frac{4}{\mu^2} \int_{1-t < |\theta| \leq \alpha} \frac{d\theta}{\theta^2},$$

which remains bounded as $t \nearrow 1$, so that (6) is proved.

E2 Consider *one-dimensional symmetric random walk* with the property that for some $\alpha > 0$

$$0 < \lim_{|x| \to \infty} |x|^{1+\alpha} P(0,x) = c_1 < \infty.$$

It is known from T1 that such a random walk is recurrent when $\alpha > 1$ (for then the first moment m will be finite). We shall now prove that the random walk in question is in fact *recurrent when $\alpha \geq 1$ and transient when $\alpha < 1$*. This will be accomplished by investigating the asymptotic behavior of $1 - \phi(\theta)$ as $\theta \to 0$. When $\alpha > 2$, the second moment exists and we know from P6.4 that $1 - \phi(\theta) \sim c_2 \theta^2$ for some positive c_2 as $\theta \to 0$. Thus we may confine the discussion to the case when $0 < \alpha \leq 2$. We shall prove that

$$\text{(1)} \qquad \lim_{\theta \to 0} \frac{1 - \phi(\theta)}{|\theta|^\alpha} = c_1 \int_{-\infty}^{\infty} \frac{1 - \cos x}{|x|^{\alpha+1}} \, dx < \infty$$

when $0 < \alpha < 2$, and once this is done it is trivial to conclude from P1 that the random walk is transient if and only if $0 < \alpha < 1$. To prove (1) we write

$$1 - \phi(\theta) = \sum_{n=-\infty}^{\infty} [1 - \cos n\theta] P(0,n),$$

$$\text{(2)} \qquad \frac{1 - \phi(\theta)}{|\theta|^\alpha} = \sum_{n=-\infty}^{\infty} |n|^{1+\alpha} P(0,n) \left|\frac{1}{n\theta}\right|^{\alpha+1} |\theta| [1 - \cos n\theta].$$

Letting

$$f(x) = |x|^{-(1+\alpha)}(1 - \cos x), \qquad -\infty < x < \infty,$$

observe that (2) becomes

$$\text{(3)} \qquad \frac{1 - \phi(\theta)}{|\theta|^\alpha} = c_1 \sum_{n=-\infty}^{\infty} |\theta| f(n\theta) + \sum_{n=-\infty}^{\infty} f(n\theta) |\theta| \epsilon_n$$

where $\epsilon_n = |n|^{1+\alpha} P(0,n) - c_1 \to 0$ as $n \to \infty$. Since

$$\int_{-\infty}^{\infty} f(x) \, dx = \int_{-\infty}^{\infty} \frac{1 - \cos x}{|x|^{\alpha+1}} \, dx,$$

which exists as an improper Riemann integral for every positive $\alpha < 2$, we may conclude from (3) that (1) holds if

$$\lim_{|\theta| \to 0} \sum_{n=-\infty}^{\infty} |\theta| f(n\theta) = \int_{-\infty}^{\infty} f(x) \, dx.$$

But that is true because the sequence of sums above are just the ordinary Riemann approximating sums to the integral of $f(x)$.

Observe that when $\alpha = 2$ the above argument permits us only to assert (and even then one has to be careful) that (1) holds in the sense that the integral on the right is infinite. It may be shown, when $\alpha = 2$, that $1 - \phi(\theta)$ behaves like a negative constant times $\theta^2 \ln |\theta|$ for θ near zero (see problem 6).

In the next example we shall encounter a specific random walk of the type discussed in E2 with $\alpha = 1$. The recurrence of this random walk will be obvious from the probabilistic context in which it arises.

8. RECURRENCE CRITERIA AND EXAMPLES

E3 *Consider simple random walk* $\mathbf{x}_n$ *in the plane*, starting at $\mathbf{x}_0 = 0$. We write $\mathbf{x}_n$ *in complex form* (i.e., for each $n \geq 0$, $\mathbf{x}_n = \mathbf{a}_n + i\mathbf{b}_n$, so that the random variable $\mathbf{a}_n$ is the real part of $\mathbf{x}_n$, and $\mathbf{b}_n$ the imaginary part). Let us now define the stopping time

(1) $$\mathbf{T} = \min\,[k \mid 1 \leq k \leq \infty;\, \mathbf{a}_k = \mathbf{b}_k].$$

Thus $\mathbf{T}$ is the first time the simple random walk in the plane visits the diagonal $\mathrm{Re}\,(x) = \mathrm{Im}\,(x)$. According to T1 the simple random walk is recurrent and by P3.3 this means that $\mathbf{T} < \infty$ with probability one.

We shall be interested in the *hitting place* $\mathbf{x_T}$ rather than in the hitting time $\mathbf{T}$ of the diagonal and define

(2) $$Q(0,n) = \mathbf{P}_0[\mathbf{x_T} = n(1+i)], \qquad n = 0, \pm 1, \pm 2, \ldots.$$

The foregoing remarks concerning recurrence show that

$$\sum_{n=-\infty}^{\infty} Q(0,n) = 1,$$

or, expressing it differently, $Q(m,n) = Q(0, n-m)$, $m,n = 0, \pm 1, \ldots$, is the transition function of a one-dimensional random walk. In fact, we can even deduce from T1 that $Q(m,n)$ is the transition function of a *recurrent* one-dimensional random walk (which is just *the original simple random walk* in the plane, *observed only at those times when it visits the diagonal* $\mathrm{Re}\,(x) = \mathrm{Im}\,(x)$).

It is of interest to determine $Q(m,n)$ explicitly, and we do so, showing that its characteristic function $\psi(\theta)$ is given by

(3) $$\psi(\theta) = \sum_{n=-\infty}^{\infty} Q(0,n)e^{in\theta} = 1 - \sin\left|\frac{\theta}{2}\right|, \qquad -\pi \leq \theta \leq \pi.$$

Using (3) a simple calculation (which we omit) will show that

(4) $$Q(0,0) = 1 - \frac{2}{\pi}, \qquad Q(0,n) = \frac{2}{\pi}\frac{1}{4n^2 - 1} \text{ for } n \neq 0.$$

In view of E2, equation (4) leads to the already known conclusion that the random walk defined by $Q(m,n)$ is recurrent.

Our proof of (3) depends on a trick. Remembering that $\mathbf{x}_n = \mathbf{a}_n + i\mathbf{b}_n$, $\mathbf{x}_0 = 0$, we define

(5) $$\mathbf{u}_0 = \mathbf{v}_0 = 0, \qquad \mathbf{u}_n = \mathbf{a}_n + \mathbf{b}_n, \qquad \mathbf{v}_n = \mathbf{a}_n - \mathbf{b}_n, \qquad n \geq 1.$$

Then we may write

$$\mathbf{T} = \min\,[k \mid k \geq 1,\, \mathbf{v}_k = 0], \qquad \mathbf{x_T} = \frac{\mathbf{u_T}}{2}(1+i),$$

and conclude that

(6) $$\psi(\theta) = \sum_{k=1}^{\infty} \mathbf{E}_0[e^{i\frac{\theta}{2}\mathbf{u}_k};\, \mathbf{T} = k].$$

The trick we mentioned consists in observing that the sequence of random variables $\mathbf{u}_n$ is *independent* of the sequence $\mathbf{v}_n$. One simply checks that for every pair of positive integers m and n

(7) $\quad \mathbf{P}_0[\mathbf{u}_m - \mathbf{u}_{m-1} = r; \mathbf{v}_n - \mathbf{v}_{n-1} = s]$
$\quad\quad = \mathbf{P}_0[\mathbf{u}_m - \mathbf{u}_{m-1} = r]\mathbf{P}_0[\mathbf{v}_n - \mathbf{v}_{n-1} = s].$

Moreover, in so doing, one makes the pleasant discovery that the probability in (7) is zero unless $|r| = 1$ and $|s| = 1$, in which case it is one fourth. Thus we have observed that $\mathbf{u}_n$ and $\mathbf{v}_n$ *are a pair of independent simple one-dimensional random walks.*

The rest is easy. Since $\mathbf{T}$ depends only on the sequence $\mathbf{v}_n$, but not on the random walk $\mathbf{u}_n$, we may apply P3.1 to (6), to obtain

$$\psi(\theta) = \sum_{k=1}^{\infty} \mathbf{E}_0[e^{i\frac{\theta}{2}\mathbf{u}_k}]\mathbf{P}_0[\mathbf{T} = k],$$

and the observation that $\mathbf{u}_n$ is a simple random walk gives

(8) $\quad \psi(\theta) = \sum_{k=1}^{\infty} \left(\cos\frac{\theta}{2}\right)^k \mathbf{P}_0[\mathbf{T} = k] = \mathbf{E}_0\left[\left(\cos\frac{\theta}{2}\right)^{\mathbf{T}}\right].$

It remains only to evaluate $\mathbf{E}_0(s^{\mathbf{T}})$ for arbitrary s in $[-1,1]$. $\mathbf{T}$ is the first time of return to zero for the simple random walk $\mathbf{v}_n$, and as shown in equation (5) of E1.2,

(9) $\quad \mathbf{E}_0(s^{\mathbf{T}}) = \sum_{n=1}^{\infty} s^n F_n(0,0) = 1 - \sqrt{1 - s^2}, \quad |s| \leq 1.$

We conclude from (8) and (9) that

(10) $\quad \psi(\theta) = 1 - \sqrt{1 - \left(\cos\frac{\theta}{2}\right)^2} = 1 - \sin\left|\frac{\theta}{2}\right|,$

which demonstrates (3) and completes this example.

Remark: A very similar problem concerns *two independent simple random walks* $\mathbf{u}_n$ and $\mathbf{v}_n$, both starting at the origin. Let

$$\mathbf{T} = \min[k \mid k \geq 1, \mathbf{u}_k = \mathbf{v}_k]$$

denote the first time $\mathbf{u}_n$ and $\mathbf{v}_n$ meet, so that $\mathbf{u}_\mathbf{T} = \mathbf{v}_\mathbf{T}$ is the *meeting place*. It may be shown that, just as in equations (9) and (10), $\mathbf{T}$ and $\mathbf{u}_\mathbf{T}$ have infinite expectation. It may also be shown (see problem 9) that

(11) $\quad \psi(\theta) = \mathbf{E}[e^{i\theta\mathbf{u}_\mathbf{T}}] = 1 - \frac{1}{2}\sqrt{(1 - \cos\theta)(3 - \cos\theta)}.$

8. RECURRENCE CRITERIA AND EXAMPLES

Whereas E1, E2, and E3 were designed to establish or to disprove recurrence for specific random walks, the last example is intended to illuminate general principles. We shall take a brief look at *abstract harmonic analysis* on Abelian groups G and exhibit a class of functions $\chi_\lambda(g)$, $g \in G$, called the *characters* of G, which exhibit exactly the same behavior as the exponential functions $e^{i\lambda x}$ on R. Actually we shall discuss only one specific group G, but one whose structure differs considerably from that of the group of lattice points R.

E4 We take for G the following countably infinite Abelian group. The elements of G are infinite sequences

$$g = (\epsilon_1, \epsilon_2, \epsilon_3, \ldots)$$

where each $\epsilon_k = \epsilon_k(g)$ is either 0 or 1, and only a finite number of 1's occur in each g. Addition is defined modulo 2; when $g \in G$ and $h \in G$, $g + h$ is defined by

$$\epsilon_k(g + h) = \begin{cases} 0 & \text{if } \epsilon_k(g) = \epsilon_k(h), \\ 1 & \text{otherwise}. \end{cases}$$

Each g in G can be expressed in a unique way as a finite sum of generators $g_k \in G$, g_k being defined by

$$\epsilon_n(g_k) = \delta(n,k), \qquad k \geq 1, \quad n \geq 1.$$

The identity element of G will be

$$e = (0,0,\ldots).$$

A complex valued function $\chi(g)$ on G will be called a *character* of G, if

(1) $\qquad |\chi(g)| = 1, \qquad \chi(g + h) = \chi(g)\chi(h)$ for $g, h \in G$.

It follows from (1) that $\chi(e) = 1$ (using none of the special properties of G) and that $\chi(g)$ is either $+1$ or -1 for each g in G (because every element of G is of order 2).

A collection of characters (sufficiently large to be useful) will now be constructed as follows. Let I denote the interval $[-1,1]$. Each $\lambda \in I$ can be represented in binary form (uniquely if one adopts a suitable convention concerning repeating decimals) as

(2) $\qquad \lambda = \sum\limits_{k=1}^{\infty} \dfrac{\lambda_k}{2^k}$, where each $\lambda_k = \lambda_k(\lambda) = +1$ or -1.

Now we define, for each $\lambda \in I$,

(3) $\qquad \chi_\lambda(g_k) = \lambda_k, \qquad k \geq 1,$
$\qquad \chi_\lambda(e) = 1,$
$\qquad \chi_\lambda(g) = \chi_\lambda(g_{i_1})\chi_\lambda(g_{i_2}) \cdots \chi_\lambda(g_{i_n}),$

for every $g = g_{i_1} + g_{i_2} + \cdots + g_{i_n} \in G$. Clearly (3) implies that $\chi_\lambda(g)$ is a character.

Classical harmonic analysis is based on the orthogonality of the exponential functions in P6.1. The analogue of this proposition in the present setting is the orthogonality relation

(4) $$\frac{1}{2}\int_{-1}^{1} \chi_\lambda(g)\chi_\lambda(h)\,d\lambda = \begin{cases} 1 & \text{if } g = h \\ 0 & \text{if } g \neq h, \end{cases}$$

where $d\lambda$ is ordinary Lebesgue measure on I. Although the proof of (4), based on (1), (2), and (3) is not at all difficult, the observation that such orthogonality relations hold in a very general setting is profound enough to have played a major role in the modern development of probability and parts of analysis.[6] Continuing to imitate the development in section 6, our next step is the definition of a "transition function"

(5) $$P(g,h) = P(e, h - g),$$

satisfying

$$P(e,g) \geq 0, \qquad \sum_{g \in G} P(e,g) = 1$$

and of its "characteristic function"

(6) $$\phi(\lambda) = \sum_{g \in G} P(e,g)\chi_\lambda(g), \qquad \lambda \in I.$$

Now we can generalize parts (a) and (b) of P6.3. If we define the iterates of $P(g,h)$ by

$$P_0(g,h) = \delta(g,h), \qquad P_1(g,h) = P(g,h),$$
$$P_{n+1}(g,h) = \sum_{f \in G} P(g,f)P_n(f,h), \qquad n \geq 0,$$

then P6.3(a) becomes

(7) $$\phi^n(\lambda) = \sum_{g \in G} P_n(e,g)\chi_\lambda(g), \qquad \lambda \in I, \quad n \geq 0,$$

and part (b) of P6.3 turns out to be

(8) $$P_n(e,g) = \frac{1}{2}\int_{-1}^{1} \phi^n(\lambda)\chi_\lambda(g)\,d\lambda, \qquad g \in G, \quad n \geq 0.$$

The proof of (7) and (8) depends on (4) in an obvious manner.

At this point we have all we need to associate a "random walk" on the group G with a transition function $P(g,h)$ on G. Even the measure theoretical considerations of section 3 can of course be introduced without

[6] See Loomis [74]. There are some indications that much of the theory of random walk extends to arbitrary Abelian groups; it is even possible that such an extension might shed new light on purely algebraic problems concerning the structure of Abelian groups.

any difficulty. Given a transition function, we may therefore call the corresponding random walk recurrent, if

$$\text{(9)} \qquad \sum_{n=0}^{\infty} P_n(e,e) = \infty,$$

and ask for criteria for (9) to hold, just as was done in this section, in terms of the characteristic function $\phi(\lambda)$. This is easy, as equation (8) implies that (9) holds if and only if

$$\text{(10)} \qquad \sum_{n=0}^{\infty} \int_{-1}^{1} \phi^n(\lambda)\, d\lambda = \infty.$$

Let us "test our theory" in terms of a very down-to-earth example. We consider *an infinite sequence of light bulbs*, and an infinite sequence of numbers p_k such that

$$p_k \geq 0, \qquad \sum_{k=1}^{\infty} p_k = 1.$$

At time 0 all the light bulbs are "off." At each unit time ($t = 1, 2, 3, \ldots$) thereafter one of the light bulbs is selected (the kth one with probability p_k), and its switch is turned. Thus it goes on if it was off, and off if it was on. What is the probability that all the light bulbs are off at time $t = n$? A moment of thought will show that this is a problem concerning a random walk on the particular group G under discussion, and that the desired probability is $P_n(e,e)$, provided we define the transition function $P(g,h)$ by

$$\text{(11)} \qquad P(e,g) = p_k \text{ if } g = g_k, \quad 0 \text{ otherwise.}$$

(More specifically, the probability that exactly bulbs number 3, 5, 7, and 11 are burning at time n, and no others, is $P_n(e,g)$ where $g = g_3 + g_5 + g_7 + g_{11}$, and so forth.)

In this "applied" setting the recurrence question of equations (9) and (10) has some intuitive interest. Equation (10) is equivalent to the statement that *with probability one the system of light bulbs will infinitely often be in the state where they are all off*, and we obviously want a criterion in terms of the sequence $\{p_k\}$ which specifies the problem. (We shall say that the system is *recurrent* if $\{p_k\}$ is such that (10) holds.)

By equation (11) we get from (6)

$$\phi(\lambda) = \sum_{k=1}^{\infty} p_k \chi_\lambda(g_k)$$

and using (3)

$$\phi(\lambda) = \sum_{k=1}^{\infty} p_k \lambda_k,$$

so that (10) becomes

$$\text{(12)} \qquad \sum_{n=0}^{\infty} \int_{-1}^{1} \left(\sum_{k=1}^{\infty} p_k \lambda_k \right)^n d\lambda = \infty.$$

Just as in the proof of P1 we can write

(13) $$\sum_0^\infty P_n(e,e) = \lim_{t \nearrow 1} \sum_0^\infty t^n P_n(e,e) = \lim_{t \nearrow 1} \frac{1}{2} \sum_0^\infty t^n \int_{-1}^1 \phi^n(\lambda) \, d\lambda$$
$$= \lim_{t \nearrow 1} \frac{1}{2} \int_{-1}^1 \frac{d\lambda}{1 - t\phi(\lambda)}.$$

Here it is possible to interchange limits and integration, since

(14) $$\int_{-1}^1 \frac{d\lambda}{1 - t\phi(\lambda)} = \int_{[\lambda | \phi(\lambda) \leq 0]} \frac{d\lambda}{1 - t\phi(\lambda)} + \int_{[\lambda | \phi(\lambda) > 0]} \frac{d\lambda}{1 - t\phi(\lambda)}.$$

The first of these tends to the integral of $[1 - \phi(\lambda)]^{-1}$ by dominated convergence, and to the second integral one can apply the monotone convergence theorem. Thus (13) and (14) give

(15) $$\sum_0^\infty P_n(e,e) = \frac{1}{2} \int_{-1}^1 \frac{d\lambda}{1 - \phi(\lambda)} \leq \infty.$$

Let us now partition the interval $[-1,1]$ into the sets

$$A_k = [\lambda | \lambda_1 = \lambda_2 = \cdots = \lambda_{k-1} = +1, \lambda_k = -1], \; k \geq 1,$$

noting that $\lambda \in A_k$ implies that

$$1 - 2f_k \leq \phi(\lambda) = \sum p_k \lambda_k \leq 1 - 2p_k,$$

where $f_k = p_k + p_{k+1} + \cdots$. Hence (15) yields

$$\sum \frac{\mu(A_k)}{2f_k} \leq \int_{-1}^1 \frac{d\lambda}{1 - \phi(\lambda)} \leq \sum \frac{\mu(A_k)}{2p_k},$$

where $\mu(A_k) = 2^{-k}$ is the Lebesgue measure of A_k. Therefore we have obtained sufficient conditions for recurrence and transcience of the "light bulb random walk":

(16) $$\text{a sufficient condition for recurrence is } \sum_{k=1}^\infty \frac{1}{2^k f_k} = \infty$$

(17) $$\text{a sufficient condition for transcience is } \sum_{k=1}^\infty \frac{1}{2^k p_k} < \infty.$$

Remark: By direct probabilistic methods Darling and Erdös [S5] have shown that condition (16) is in fact *necessary and sufficient* for recurrence. For a generalization to random walk on more general Abelian groups see [S10].

Remark: The most interesting general result, so far, concerning random walk on groups is a generalization of T1 due to Dudley [26]. He considers countable Abelian additive groups G, and asks what groups admit an aperiodic recurrent random walk. In other words, on what groups G can one define a transition function $P(x,y)$, $x,y \in G$, such that $P(x,y) = P(e, y - x)$, and

(a) no proper subgroup of G contains $[x \mid P(e,x) > 0]$,

(b) $\sum_{n=0}^{\infty} P_n(e,e) = \infty$,

where e is the identity element and $P_n(x,y)$ is defined as in E4? The answer is that *G admits an aperiodic recurrent random walk if and only if it contains no subgroup which is isomorphic to R_3* (the triple direct product of the group of integers). Note that, consequently, it is possible to define an aperiodic recurrent random walk on the additive group of rational numbers!

9. THE RENEWAL THEOREM

In the study of random walk, as in any other area of mathematics, there are some quite indispensable propositions, intimately related to the subject but not actual goals of the theory. Two such propositions will be proved here, the Riemann Lebesgue Lemma (P1) and the renewal theorem (P3). The former belongs in this chapter, being a basic result in Fourier analysis. The latter does not; it properly belongs in Chapter VI where its most general form appears as T24.2. But we shall wish to use P3 long before that and since there is a simple proof of P3 due to Feller ([31], Vol. 2) which is based on P1, we chose to put the renewal theorem in this section.

The *Riemann Lebesgue Lemma* concerns the Fourier coefficients of functions $f(\theta)$, integrable on the cube

$$C = [\theta \mid |\theta_k| \leq \pi, k = 1, \ldots, d] \subset E.$$

All integrals will be over C, as in D6.3, and $d\theta$ is Lebesgue measure (volume). The space R of lattice points is of the same dimension d as E, and for a function $g(x)$ on R

$$\lim_{|x|\to\infty} g(x) = c$$

has the obvious meaning, i.e., g is arbitrarily close to c outside sufficiently large spheres in R.

P1 *If $f(\theta)$ is integrable on C, then*

$$\lim_{|x|\to\infty} \int_C e^{ix\cdot\theta} f(\theta)\, d\theta = 0.$$

Proof: First we derive Bessel's inequality: if a function $g(\theta)$ on C is *square integrable* and has Fourier coefficients

$$a(x) = (2\pi)^{-d} \int e^{ix\cdot\theta} g(\theta)\, d\theta, \qquad x \in R,$$

then

(1) $$\sum_{x\in R} |a(x)|^2 \leq (2\pi)^{-d} \int |g(\theta)|^2\, d\theta.$$

The proof of (1) is elementary, unlike that of Parseval's identity which was mentioned in section 6; for if

$$g_M(\theta) = \sum_{[x|\,|x|\leq M]} a(x) e^{ix\cdot\theta}, \qquad \theta \in E,$$

then P6.1 gives

$$0 \leq (2\pi)^{-d} \int |g(\theta) - g_M(\theta)|^2\, d\theta = (2\pi)^{-d} \int |g(\theta)|^2\, d\theta - \sum_{|x|\leq M} |a(x)|^2$$

for every $M > 0$, which implies (1).

To prove P1 we now decompose $f(\theta)$. For every $A > 0$,

$$f(\theta) = g_A(\theta) + h_A(\theta), \qquad \theta \in E$$

where

$$g_A(\theta) = f(\theta) \text{ when } |f(\theta)| \leq A, \quad 0 \text{ otherwise,}$$
$$h_A(\theta) = f(\theta) - g_A(\theta).$$

If $a_A(x)$ are the Fourier coefficients of $g_A(\theta)$ and $b_A(x)$ those of $h_A(\theta)$, then

(2) $$(2\pi)^{-d} \int e^{ix\cdot\theta} f(\theta)\, d\theta = a_A(x) + b_A(x), \qquad x \in R.$$

9. THE RENEWAL THEOREM

Since $g_A(\theta)$ is bounded, it is square integrable, so that (1) implies that $a_A(x) \to 0$ as $|x| \to \infty$. Also

(3) $\quad b_A(x) \le (2\pi)^{-d} \int |h_A(\theta)| \, d\theta = (2\pi)^{-d} \int_{[\theta \mid |f(\theta)| > A]} f|(\theta)| \, d\theta = \beta_A$

where β_A tends to zero as $A \to \infty$. It follows from (2) that

$$\varlimsup_{|x| \to \infty} (2\pi)^{-d} \int e^{ix \cdot \theta} f(\theta) \, d\theta \le \varlimsup_{|x| \to \infty} a_A(x) + \varlimsup_{|x| \to \infty} b_A(x) \le \beta_A$$

for every $A > 0$. By letting $A \to \infty$ we are therefore able to complete the proof of P1.

Now we shall apply the Riemann Lebesgue Lemma to the discussion of certain aspects of *aperiodic transient random walk*. According to T2.1

$$G(0,x) = \sum_{n=0}^{\infty} P_n(0,x) < \infty$$

for all x in R, if $P(x,y)$ is the transition function of an arbitrary transient random walk. We shall prove (a brief discussion of this result precedes the proof of P3)

P2 *For every aperiodic transient random walk*

$$\lim_{|x| \to \infty} [G(0,x) + G(x,0)] \text{ exists.}$$

Proof: As in the proof of P8.1 we write, for $0 \le t < 1$,

(1) $\quad \displaystyle\sum_{n=0}^{\infty} t^n P_n(0,x) = \sum_{n=0}^{\infty} (2\pi)^{-d} t^n \int e^{-i\theta \cdot x} \phi^n(\theta) \, d\theta,$

where $\phi(\theta)$ is the characteristic function of the random walk. The dimension is $d \ge 1$, and the integration is over the same d-dimensional cube C as in P1. Replacing x by $-x$ in (1), and adding the two equations,

(2) $\quad \displaystyle\sum_{n=0}^{\infty} t^n [P_n(0,x) + P_n(x,0)] = 2(2\pi)^{-d} \int \frac{\cos x \cdot \theta}{1 - t\phi(\theta)} \, d\theta$

$$= 2(2\pi)^{-d} \int \cos x \cdot \theta \, \text{Re} \left[\frac{1}{1 - t\phi(\theta)} \right] d\theta,$$

so that

(3) $\quad G(0,x) + G(x,0) = \displaystyle\lim_{t \nearrow 1} 2(2\pi)^{-d} \int \cos x \cdot \theta \, \text{Re} \left[\frac{1}{1 - t\phi(\theta)} \right] d\theta.$

For convenience, let us call

$$w_t(\theta) = 2(2\pi)^{-d} \operatorname{Re}\left[\frac{1}{1 - t\phi(\theta)}\right], \qquad \theta \in C, \quad 0 \le t < 1.$$

Now we use the aperiodicity of the random walk to conclude, from T7.1, that

$$w(\theta) = \lim_{t \nearrow 1} w_t(\theta) < \infty, \qquad \theta \in C - \{0\}$$

exists for every non-zero value of θ in C. For every real r, $0 < r < \pi$ we define the sphere $S_r = [\theta \mid |\theta| < r]$. Then, again using T7.1, equation (3) may be written

(4) $\quad G(0,x) + G(x,0) = \int_{C-S_r} \cos x \cdot \theta\, w(\theta)\, d\theta + \lim_{t \nearrow 1} \int_{S_r} \cos x \cdot \theta\, w_t(\theta)\, d\theta.$

Now we call

$$\lim_{t \nearrow 1} \int_{S_r} w_t(\theta)\, d\theta = L_r,$$

observing that this limit exists, in view of (4). It is crucial to note, at several points in the proof, that $w_t(\theta) \ge 0$ on C, and hence also $w(\theta) \ge 0$, $w(\theta)$ being the limit of $w_t(\theta)$. Setting $x = 0$ in (4), it follows from the positivity of $w(\theta)$ that

$$0 \le L_r \le 2G(0,0),$$

and that

$$\lim_{r \to 0} L_r = L < \infty$$

exists. Now we shall estimate the second integral in (4) with an arbitrary, fixed, value of x. Given any $\epsilon > 0$ we can choose $\rho > 0$ so that $|1 - \cos x \cdot \theta| < \epsilon$ for all θ in S_ρ. It follows (since $w_t(\theta) \ge 0$) that

$$(1 - \epsilon)L_r \le \lim_{t \nearrow 1} \int_{S_r} \cos x \cdot \theta\, w_t(\theta)\, d\theta \le (1 + \epsilon)L_r$$

when $0 < r < \rho$, and since ϵ is arbitrary

(5) $\quad \lim_{r \to 0} \lim_{t \nearrow 1} \int_{S_r} \cos x \cdot \theta\, w_t(\theta)\, d\theta = L.$

Substituting (5) into (4) we obtain

(6) $\quad G(0,x) + G(x,0) = \lim_{r \to 0} \int_{C-S_r} \cos x \cdot \theta\, w(\theta)\, d\theta + L.$

Now let us set $x = 0$ in (6). Since $w(\theta) \geq 0$ on C we can conclude that

(7) $\qquad w(\theta)$ *is integrable on* C.

Hence (6) takes on the simple form

(8) $\quad G(0,x) + G(x,0) = \int_C \cos x \cdot \theta \, w(\theta) \, d\theta + L, \qquad x \in R,$

and the stage is finally set for the application of the Riemann Lebesgue Lemma (P1). The function $\cos x \cdot \theta$ is the sum of two exponentials, and as $w(\theta)$ is integrable we have

$$\lim_{|x| \to \infty} [G(0,x) + G(x,0)] = L < \infty,$$

which completes the proof of P2.

It is not easy to evaluate the limit in P2 explicitly; nor is it clear, perhaps, why this limit should be of any interest. Therefore we shall discuss a special case of obvious interest, namely so called *positive one-dimensional* random walk with the property that

$$P(0,x) = 0 \text{ for } x \leq 0.$$

Obviously every such random walk is transient so that P2 may be applied. It is more to the point, however, that $G(0,x) = 0$ for $x < 0$, so that P2 reduces to

$$\lim_{x \to +\infty} G(0,x) = L < \infty.$$

The limit L will be evaluated in P3, where it is shown that

$$L = \frac{1}{\mu}, \quad \left(\mu = \sum_{x=1}^{\infty} xP(0,x)\right)$$

when $\mu < \infty$, while $L = 0$ when $\mu = \infty$.

This is the now "classical" renewal theorem in the form in which it was first conjectured and proved by Feller, Erdös, and Pollard[7] (1949). Its name is due to certain of its applications. If **T** is a positive random variable, whether integer valued or not, it may be thought of

---

[7] Cf. [31], Vol. 1, p. 286, where Feller credits Chung with the observation that this theorem is entirely equivalent to a fundamental law governing the ergodic behavior of Markov chains which was discovered by Kolmogorov [67] in 1936. Given any probability measure $p_k = P(0,k)$ defined on the integers $k \geq 1$, it is easy to construct a Markov chain with states $a_1, a_2, \ldots$, so that p_k is the probability that the first return to the state a_1 occurs at time k. That is why our renewal theorem follows if one knows that the probability of being in state a at time n converges to the reciprocal of the mean recurrence time of state a. But this is exactly the theorem of Kolmogorov.

as a random span of time. Frequently it is a lifetime, of an individual in a large population—or of a mass produced article. Suppose now that identical individuals, or articles, have life times $\mathbf{T}_1, \mathbf{T}_2, \ldots$, which are identically distributed independent random variables. If each $\mathbf{T}_k$ is integer valued we define the transition function $P(x,y)$ so that

$$P(0,x) = \mathbf{P}[\mathbf{T}_k = x] \text{ for } x > 0, \qquad k \geq 1,$$

and $P(0,x) = 0$ for $x \leq 0$. Then, evidently,

$$P_n(0,x) = \mathbf{P}[\mathbf{T}_1 + \mathbf{T}_2 + \cdots + \mathbf{T}_n = x], \qquad x > 0, \quad n \geq 1,$$

and

$$G(0,x) = \sum_{n=1}^{\infty} \mathbf{P}[\mathbf{T}_1 + \mathbf{T}_2 + \cdots + \mathbf{T}_n = x].$$

Thus $G(0,x)$ is a sum of probabilities of disjoint events. We can say that $\mathbf{T}_1 + \cdots + \mathbf{T}_n = x$ if the nth lifetime ends at time x (for example, if the nth individual dies, or if a mass produced article has to be "renewed" for the nth time at time x). Thus $G(0,x)$ is the *probability that a renewal takes place at time* x, and according to P3 it *converges to the reciprocal of the expected lifetime*

$$\mu = \sum x P(0,x) = \mathbf{E}[\mathbf{T}].$$

This conclusion is false, when $\mathbf{T}$ is integer valued, unless $P(0,x)$ is aperiodic, or equivalently, unless the greatest common divisor

$$\text{g.c.d. } \{x \mid \mathbf{P}[\mathbf{T} = x] > 0\}$$

has the value one.

P3 *For aperiodic one-dimensional random walk with $P(0,x) = 0$ for $x \leq 0$, $\mu = \sum_{x=1}^{\infty} x P(0,x) \leq \infty$,*

$$\lim_{n \to +\infty} G(0,x) = \frac{1}{\mu} \quad (= 0 \text{ if } \mu = \infty).$$

Proof: If

$$G_n(0,x) = \sum_{k=0}^{n} P_k(0,x)$$

we have

$$G_{n+1}(0,x) = \sum_{t \in R} P(0,t) G_n(t,x) + \delta(0,x),$$

and, letting $n \to \infty$,

(1) $$G(0,x) = \sum_{t \in R} P(0,t) G(t,x) + \delta(0,x), \qquad x \in R.$$

This equation will be used to determine the value of

(2) $$L = \lim_{x \to +\infty} G(0,x)$$

which exists according to P2. We shall sum the variable x in (1) from 0 to n. Observing that $P(0,t) = 0$ for $t \leq 0$ and $G(t,x) = 0$ for $t > x$, one obtains

$$\begin{aligned}
1 &= \sum_{x=0}^{n} G(0,x) - \sum_{x=0}^{n} \sum_{t \in R} P(0,t)G(t,x) \\
&= \sum_{x=0}^{n} G(0,x) - \sum_{x=0}^{n} G(0,x)[P(0,0) + P(0,1) + \cdots + P(0,n-x)] \\
&= \sum_{x=0}^{n} G(0,x)[1 - \sum_{y=1}^{n-x} P(0,y)] \\
&= \sum_{x=0}^{n} G(0,n-x)f(x),
\end{aligned}$$

where

$$f(k) = \sum_{x=k+1}^{\infty} P(0,x).$$

Observe that

(3) $$\sum_{k=0}^{\infty} f(k) = \mu$$

whether μ is finite or not. When $\mu < \infty$, we let $n \to \infty$ in

(4) $$1 = \sum_{x=0}^{n} G(0,n-x)f(x),$$

concluding by a dominated convergence argument from (2) and (3) that $1 = L\mu$ or $L = \mu^{-1}$, which is the desired limit. Finally, when $\mu = \infty$, the value of L must be zero, for otherwise the right-hand side in (4) would tend to infinity as $n \to \infty$. Hence the proof of P3 is complete.

Problems

1. Extend P6.4 by proving that $m_{2k} < \infty$ whenever the characteristic function has a derivative of order $2k$ at the origin.

2. Extend the result of P6.6 to

$$m_3 = \frac{m_1}{m_2} + \frac{3}{2\pi m_2} \int_{-\pi}^{\pi} \operatorname{Re} \frac{\phi(\theta) - 1 + m_2(1 - \cos\theta)}{(1 - \cos\theta)^2} d\theta,$$

m_3 being finite if and only if the integral on the right exists in the sense of Lebesgue.

3. Show that two-dimensional random walk is transient if $m < \infty$ and if the mean vector $\mu \neq 0$.

4. Prove that if $\phi(\theta)$ is the characteristic function of a recurrent random walk, then so is $|\phi(\theta)|^2$. Consequently two identical, independent, recurrent random walks, starting at the same point, will meet infinitely often.

5. For what real values of α is

$$\phi(\theta) = 1 - \left|\sin\frac{\theta}{2}\right|^\alpha$$

the characteristic function of a one-dimensional random walk? For each such value of α, describe the asymptotic behavior of $P(0,x)$ as $|x| \to \infty$.

6. If a one-dimensional random walk satisfies

$$P(0,x) \sim |x|^{-3} \quad \text{as} \quad |x| \to \infty,$$

what can be said about the asymptotic behavior of $1 - \phi(\theta)$ as $|\theta| \to 0$?

7. *Simple random walk in the plane.* In complex notation the random walk is $\mathbf{z}_n$, with $\mathbf{z}_0 = 0$. If T is the time of the first visit of the random walk to the line $z = m + i$, $m = 0, \pm 1, \ldots$, let

$$Q(0,m) = \mathbf{P}_0[\operatorname{Re}(\mathbf{z}_T) = m].$$

Show that

$$\sum_{m=-\infty}^{\infty} Q(0,m)e^{im\theta} = 2 - \cos\theta - \sqrt{(1 - \cos\theta)(3 - \cos\theta)}.$$

8. *Continuation.* Use the remarks preceding the proof of P6.8 to calculate

$$\lim_{n\to\infty} \mathbf{P}_0[\operatorname{Re}(\mathbf{z}_{T_n}) \leq nx] \qquad -\infty < x < \infty,$$

where

$$\mathbf{T}_n = \min[k \mid k \geq 1, \operatorname{Im}(\mathbf{z}_k) = n].$$

9. Let $\mathbf{u}_n$ and $\mathbf{v}_n$ be independent simple random walks on the line, with $\mathbf{u}_0 = \mathbf{v}_0 = 0$. Let T be the first time they meet. Prove that $T < \infty$ with probability one, and use the result of problem 7 to obtain the characteristic function of $\mathbf{u}_T = \mathbf{v}_T$. Is $T < \infty$ also if $\mathbf{u}_n$ and $\mathbf{v}_n$ are identical but independent Bernoulli walks?

10. *Simple random walk in three space.* Let $Q(0,n)$ denote the probability that the first visit to the x^3-axis occurs at the point $x = (x^1, x^2, x^3)$ with $x^1 = x^2 = 0$, $x^3 = n$. (The starting point is the origin.) Show that if

$$\psi(\theta) = \sum_{n=-\infty}^{\infty} Q(0,n)e^{in\theta},$$

then

$$[1 - \psi(\theta)]^{-1} = K\left(\frac{2}{3 - \cos \theta}\right),$$

where K is the *elliptic integral*

$$K(z) = \frac{2}{\pi} \int_0^{\pi/2} \frac{dt}{\sqrt{1 - z^2 \sin^2 t}} = \sum_{n=0}^{\infty} \binom{2n}{n}^2 \left(\frac{z}{4}\right)^{2n}.$$

11. *Continued.* If F is the probability to return to zero, explain why

$$G = (1 - F)^{-1} = \frac{1}{2\pi} \int_{-\pi}^{\pi} K\left(\frac{2}{3 - \cos \theta}\right) d\theta.$$

This integral was evaluated by G. N. Watson [104], who showed that

$$G = 3(18 + 12\sqrt{2} - 10\sqrt{3} - 7\sqrt{6})K^2(2\sqrt{3} + \sqrt{6} - 2\sqrt{2} - 3)$$
$$= 1.5163860591\ldots.$$

Thus *the probability that simple random walk in three-space will ever return to its starting point is* $F = 1 - G^{-1} = 0.340537330\ldots$.

12. In the *light bulb problem* of E8.4, let $\mathbf{N}_n$ denote the number of bulbs burning at time n ($\mathbf{N}_0 = 0$). Show that for $0 \le s < 1$, $0 \le t < 1$

$$\sum_{n=0}^{\infty} \frac{s^n}{n!} \mathbf{E}[t^{\mathbf{N}_n}] = \prod_{k=1}^{\infty} [\cosh (sp_k) + t \sinh (sp_k)],$$

$$e^{-s} \sum_{n=0}^{\infty} \frac{s^n}{n!} \mathbf{E}[\mathbf{N}_n] = \frac{1}{2} \sum_{k=1}^{\infty} [1 - e^{-2sp_k}].$$

13. A die is cast repeatedly, the scores are added and p_n is the probability that the total score ever has the value n. What is the limit of p_n as $n \to \infty$ (Putnam Competition 1960)? Generalize this problem by proving P9.3 for a "generalized loaded die" with m sides which have the probabilities $q_1, q_2, \ldots, q_m$ of coming up. In other words, prove the renewal theorem in P9.3, by a simple method which avoids Fourier analysis, for positive bounded random variables.

14. Is there a recurrent one-dimensional random walk $\mathbf{x}_n$ which has the remarkable property that each random walk $\mathbf{y}_n = \mathbf{x}_n + an$ is also recurrent, for arbitrary $a \in R$?

Hint: One of the random walks in problem 5 will do.

15. A particularly elegant continuous analogue of two-dimensional simple random walk is the *random flight* investigated by Lord Rayleigh (see [103], p. 419). Here

$$\mathbf{x}_n = \mathbf{X}_1 + \mathbf{X}_2 + \cdots + \mathbf{X}_n, \qquad n \geq 1$$

where the $\mathbf{X}_i$ are independent complex random variables with $|\mathbf{X}_i| = 1$, whose arguments are equidistributed between 0 and 2π. Develop the requisite Fourier analysis for spherically symmetric functions (see [5], Ch. 9) to conclude that

$$\tfrac{1}{2}\mathbf{P}[|\mathbf{x}_n| < r] + \tfrac{1}{2}\mathbf{P}[|\mathbf{x}_n| \leq r] = r \int_0^\infty J_1(rt) J_0^n(t)\, dt$$

for $n \geq 1$, $r > 0$. When $n \geq 2$ the right-hand side is a continuous function of r, and it follows that

$$\mathbf{P}[|\mathbf{x}_n| < 1] = \mathbf{P}[|\mathbf{x}_n| \leq 1] = 1/(n+1).$$

Remark: $J_k(x)$ is the Bessel function of order k, which enters the picture because, for $k = 0, 1, 2, \ldots$

$$J_k(x) = \frac{1}{2\pi} \int_{-\pi}^{\pi} e^{-ik\theta + ix\sin\theta}\, d\theta.$$

Chapter III

TWO-DIMENSIONAL RECURRENT RANDOM WALK

10. GENERALITIES

Just about all worthwhile known results concerning random walk (or concerning any stochastic process for that matter) are closely related to some stopping time $\mathbf{T}$, as defined in definition D3.3. Thus we plan to investigate stopping times. Given a stopping time $\mathbf{T}$ we shall usually be concerned with the random variable $\mathbf{x}_T$, the *position of the random walk at a random time which depends only on the past* of the process. There can be no doubt that problems concerning $\mathbf{x}_T$ represent a natural generalization of the theory in Chapters I and II; for in those chapters our interest was confined to the iterates $P_n(0,x)$ of the transition function—in other words, to the probability law governing $\mathbf{x}_n$ at an arbitrary but nonrandom time.

Unfortunately it must be admitted that the study of arbitrary stopping times is far too ambitious a program. As an example of the formidable difficulties that arise even in apparently simple problems consider the famous problem of the *self-avoiding random walk*. Let $\mathbf{x}_n$ be simple random walk in the plane with $\mathbf{x}_0 = 0$, and let $\mathbf{T}$ denote the first time that $\mathbf{x}_n$ visits a point in R which was occupied at some time less than n. Thus

$$\mathbf{T} = \min\,[k \mid k \geq 1;\, \mathbf{x}_k \in \{\mathbf{x}_0, \mathbf{x}_1, \ldots, \mathbf{x}_{k-1}\}].$$

This is certainly a stopping time, but one whose dependence on the past turns out to be of much too elaborate a character to be susceptible to a simple treatment. Although it is of course possible to calculate the probability that $\mathbf{T} = n$ for any fixed n, the distribution of $\mathbf{x}_T$ presents a problem that seems far out of reach; for each x in R, the event "$\mathbf{x_T} = x$" depends on the entire history of the random walk up to time $\mathbf{T}$, and $\mathbf{T}$ may be arbitrarily large. Even a seemingly more modest problem, namely that of finding

$$\lim_{n \to \infty} \{\mathbf{P}[\mathbf{T} > n]\}^{1/n}$$

lies beyond the horizon of the present state of the art. (The above limit, which may be shown to exist without any trouble, holds the key to a number of interesting physical problems. Its recent calculation to four presumably significant digits [33], offers no significant clue to the mathematical complexities of the problem. We shall conclude this chapter, in example E16.2, by discussing a related but much easier problem that will nevertheless use in an essential way the results of this entire chapter.)

That is one of the reasons why our field of inquiry has to be narrowed. Another reason, of greater depth, was mentioned in the Preface. We shall study exactly those stopping times which render the "stopped" random walk Markovian. These will all be of the same simple type. We shall take a subset $A \subset R$ of the state space. It will always be a *proper nonempty subset*, and its cardinality will be denoted by $|A|$, a positive integer if A is finite, and $|A| = \infty$ otherwise. And our stopping time will always be of the form

$$\mathbf{T} = \mathbf{T}_A = \min\,[k \mid 1 \leq k \leq \infty;\, \mathbf{x}_k \in A],$$

in other words, $\mathbf{T}$ *will be the time of the first visit of the random walk to a set A.*

Now we give (in D1 below) the principal definitions using measure theoretical language, in the terminology of D3.2. This is undoubtedly the most intuitive way—but nevertheless we repeat the process once more afterward. In equations (1) through (3), following D1, we shall give some of these definitions in purely analytical form, in other words, directly in terms of $P(x,y)$ without the intermediate construction of a measure space $(\Omega, \mathscr{F}, \mathbf{P})$.

One notational flaw should not be passed over in silence. We shall define, in D1, a function $Q(x,y)$ and its iterates $Q_n(x,y)$. These will depend on the set A and should therefore logically be called $Q_A(x,y)$ and $Q_{A,n}(x,y)$—but no confusion will result from the omission of A as a subscript.

First we indicate the domain of the functions to be defined. They will all be real valued; so, if we say that $f(x,y)$ is defined on $R \times (R - A)$, this means that $f(x,y)$ is defined for every x in R and every y in R but not in A (f maps $R \times (R - A)$ into the real numbers). We shall define

D1 $\quad Q_n(x,y) \quad$ for $n \geq 0$ on $(R - A) \times (R - A)$,
$\quad\quad H_A^{(n)}(x,y)$ for $n \geq 0$ on $R \times A$,
$\quad\quad H_A(x,y) \quad\quad\quad$ on $R \times A$,
$\quad\quad \Pi_A(x,y) \quad\quad\quad$ on $A \times A$,
$\quad\quad g_A(x,y) \quad\quad\quad$ on $R \times R$.

The definitions are, in terms of the time $\mathbf{T} = \mathbf{T}_A$ of the first visit to A,

$$Q_n(x,y) = \mathbf{P}_x[\mathbf{x}_n = y; \mathbf{T} > n];$$
$$H_A^{(n)}(x,y) = \mathbf{P}_x[\mathbf{x}_\mathbf{T} = y; \mathbf{T} = n] \text{ for } x \in R - A,$$
$$= 0 \text{ for } x \in A, n \geq 1,$$
$$= \delta(x,y) \text{ for } x \in A, n = 0;$$
$$H_A(x,y) = \mathbf{P}_x[\mathbf{x}_\mathbf{T} = y; \mathbf{T} < \infty] \text{ for } x \in R - A,$$
$$= \delta(x,y) \text{ for } x \in A;$$
$$\Pi_A(x,y) = \mathbf{P}_x[\mathbf{x}_\mathbf{T} = y; \mathbf{T} < \infty];$$
$$g_A(x,y) = \sum_{n=0}^{\infty} Q_n(x,y) \text{ for } x \in R - A, y \in R - A,$$
$$= 0 \text{ otherwise.}$$

It should be clear how to cast these definitions in a purely analytic form. For $Q_n(x,y)$ we define $Q_0(x,y) = \delta(x,y)$; then $Q(x,y) = Q_1(x,y) = P(x,y)$ when (x,y) is in $(R - A) \times (R - A)$. Finally, just as in P1.1

$$Q_{n+1}(x,y) = \sum_{t \in R - A} Q_n(x,t) Q(t,y).$$

As for $H_A^{(n)}(x,y)$, with $x \in R - A$, $n \geq 1$, we could now simply take

(1) $$H_A^{(n)}(x,y) = \sum_{t \in R - A} Q_{n-1}(x,t) P(t,y).$$

Alternatively, we could write, when $x \in R - A$, $n \geq 1$,

(2) $$H_A^{(n)}(x,y) = \sum_{x_1 \in R - A} \cdots \sum_{x_{n-1} \in R - A} P(x,x_1) \cdots P(x_{n-1}, y)$$

and then prove the equivalence of (1) and (2). It should also be clear that $H_A(x,y)$, in the case when $x \in R - A$, is simply

(3) $$H_A(x,y) = \sum_{n=1}^{\infty} H_A^{(n)}(x,y).$$

Now it is time to employ these definitions to obtain some simple identities that will be extremely useful in the sequel. One of them

(P1(b) below) concerns the convergence of the sum defining $g_A(x,y)$—which perhaps is not quite obvious from D1; others concern relations between H_A, Π_A, and g_A. In later sections such relations will be shown to be familiar ones in the mathematical formulation of potential theory—but that is irrelevant at this juncture.

P1 *For arbitrary random walk (recurrent or transient)*

(a) $$\sum_{t \in R} P(x,t)H_A(t,y) - H_A(x,y)$$
$$= \begin{cases} \Pi_A(x,y) - \delta(x,y) & \text{for } x \in A, \ y \in A, \\ 0 & \text{for } x \in R - A, \ y \in A. \end{cases}$$

(b) $0 \le g_A(x,y) \le g_A(y,y)$ *for all* $x \in R, \ y \in R$

and if, in addition, the random walk is aperiodic, then

$$g_A(x,x) < \infty, \quad x \in R.$$

(c) *For* $x \in R - A$, $y \in A$,

$$H_A(x,y) = \sum_{t \in R} g_A(x,t) P(t,y).$$

(d) *For* $x \in R - A$, $y \in A$,

$$G(x,y) = \sum_{t \in A} H_A(x,t) G(t,y).$$

Proof: When $x \in A$ and $y \in A$, the left-hand side in P1(a) is

$$\sum_{t \in R} P(x,t)H_A(t,y) - H_A(x,y)$$
$$= P(x,y) + \sum_{t \in R - A} P(x,t)H_A(t,y) - \delta(x,y).$$

Thus we have to show that

$$\Pi_A(x,y) = P(x,y) + \sum_{t \in R - A} P(x,t) H_A(t,y), \quad x \in A, \ y \in A.$$

Decomposing $\Pi_A(x,y)$ according to the value of **T**,

$$\Pi_A(x,y) = \mathbf{P}_x[\mathbf{x}_T = y; \mathbf{T} < \infty] = \sum_{k=1}^{\infty} \mathbf{P}_x[\mathbf{x}_T = y; \mathbf{T} = k]$$

$$= \mathbf{P}_x[\mathbf{x}_T = y; \mathbf{T} = 1] + \sum_{k=2}^{\infty} \mathbf{P}_x[\mathbf{x}_T = y; \mathbf{T} = k].$$

Here
$$\mathbf{P}_x[\mathbf{x_T} = y; \mathbf{T} = 1] = P(x,y),$$
and when $k \geq 2$,
$$\mathbf{P}_x[\mathbf{x_T} = y; \mathbf{T} = k] = \sum_{t \in R-A} \mathbf{P}_x[\mathbf{x}_1 = t; \mathbf{x_T} = y; \mathbf{T} = k]$$
$$= \sum_{t \in R-A} P(x,t) \mathbf{P}_t[\mathbf{x_T} = y; \mathbf{T} = k-1].$$
Hence
$$\Pi_A(x,y) = P(x,y) + \sum_{t \in R-A} P(x,t) \sum_{k=2}^{\infty} \mathbf{P}_t[\mathbf{x_T} = y; \mathbf{T} = k-1]$$
$$= P(x,y) + \sum_{t \in R-A} P(x,t) \mathbf{P}_t[\mathbf{x_T} = y; 1 \leq \mathbf{T} < \infty]$$
$$= P(x,y) + \sum_{t \in R-A} P(x,t) H_A(t,y).$$

When $x \in R - A$ and $y \in A$ we have to show that
$$H_A(x,y) = \sum_{t \in R} P(x,t) H_A(t,y),$$
the right-hand side in P1(a) being zero. This is done, just as in the first half of the proof, by decomposing $H_A(x,y)$ according to the possible values of $\mathbf{T}$.

The proof of (b) depends on the interpretation of $g_A(x,y)$ as the expected number of visits of the random walk $\mathbf{x}_n$ with $\mathbf{x}_0 = x$ to the point y before the time $\mathbf{T}$ of the first visit to the set A. Since there is nothing to prove when either x or y lies in A, we assume the contrary. If in addition $x \neq y$, then
$$g_A(x,y) = \sum_{n=0}^{\infty} Q_n(x,y) = \sum_{n=1}^{\infty} \mathbf{P}_x[\mathbf{x}_n = y; \mathbf{T} > n],$$
and letting $\mathbf{T}_y = \min[k \mid 1 \leq k \leq \infty, \mathbf{x}_k = y]$,
$$g_A(x,y) = \sum_{k=1}^{\infty} \sum_{n=1}^{\infty} \mathbf{P}_x[\mathbf{x}_n = y; \mathbf{T}_y = k; \mathbf{T} > n]$$
$$= \sum_{k=1}^{\infty} \sum_{n=k}^{\infty} \mathbf{P}_x[\mathbf{T}_y = k < \mathbf{T}] \mathbf{P}_y[\mathbf{x}_{n-k} = y; \mathbf{T} > n-k]$$
$$= \sum_{k=1}^{\infty} \mathbf{P}_x[\mathbf{T}_y = k < \mathbf{T}] \sum_{j=0}^{\infty} \mathbf{P}_y[\mathbf{x}_j = y; \mathbf{T} > j]$$
$$= \mathbf{P}_x[\mathbf{T}_y < \mathbf{T}; \mathbf{T}_y < \infty] g_A(y,y) \leq g_A(y,y).$$

It follows that for all x, y in R

$$g_A(x,y) \leq g_A(y,y).$$

The problem of showing that $g_A(y,y) < \infty$ for aperiodic random walk can be reduced to a slightly simpler one by the observation that $g_A(y,y) \leq g_B(y,y)$ when B is a subset of A. Thus it suffices to let $B = \{0\}$, $g_B(x,y) = g(x,y)$, and to show that

$$g(x,x) < \infty$$

when $x \neq 0$. This is true for transient, as well as for recurrent random walk, but for different reasons. In the transient case clearly

$$g(x,x) < G(x,x) < \infty.$$

In the recurrent case

$$g(x,x) = 1 + \sum_{n=1}^{\infty} \mathbf{P}_x[\mathbf{x}_n = x; \mathbf{T}_0 > n]$$

where $\mathbf{T}_y = \min [k \mid 1 \leq k < \infty, \mathbf{x}_k = y]$. Since the random walk is aperiodic, $\mathbf{P}_x[\mathbf{T}_0 < \infty] = F(x,0) = 1$ by T2.1. In addition it is easy to see that

$$\mathbf{P}_x[\mathbf{T}_0 < \mathbf{T}_x] = \Pi_{\{0,x\}}(x,0) > 0.$$

But $g(x,x)$ is the expected value of the number of visits to x (counting the visit at time 0) before the first visit to 0. It is quite simple to calculate this expectation (it is the mean of a random variable whose distribution is geometric) and the result of the calculation is

$$g(x,x) = [\Pi_{\{0,x\}}(x,0)]^{-1} < \infty.$$

The reader who is reluctant to fill in the details will find that P3 below yields a much shorter proof that $g_{\{0\}}(x,x) = g(x,x) < \infty$.

To prove (c) we decompose $H_A(x,y)$ as follows. For $x \in R - A$, $y \in A$,

$$H_A(x,y) = \sum_{n=1}^{\infty} \sum_{t \in R-A} \mathbf{P}_x[\mathbf{x}_{n-1} = t; \mathbf{T} = n; \mathbf{x}_\mathbf{T} = y]$$

$$= \sum_{n=1}^{\infty} \sum_{t \in R-A} \mathbf{P}_x[\mathbf{x}_{n-1} = t; \mathbf{T} \geq n] P(t,y)$$

$$= \sum_{t \in R-A} g_A(x,t) P(t,y) = \sum_{t \in R} g_A(x,t) P(t,y),$$

which proves (c).

Part (d) is of absolutely no interest when the random walk is recurrent, for then both sides are infinite (or possibly both are zero in the periodic case). However, the proof which follows is quite general. We write, for $x \in R - A$, $y \in A$,

$$\begin{aligned}
G(x,y) &= \sum_{n=0}^{\infty} P_n(x,y) = \sum_{n=0}^{\infty} \mathbf{P}_x[\mathbf{x}_n = y] \\
&= \sum_{n=0}^{\infty} \sum_{k=1}^{n} \mathbf{P}_x[\mathbf{x}_n = y; \mathbf{T} = k] \\
&= \sum_{t \in A} \sum_{n=0}^{\infty} \sum_{k=1}^{n} \mathbf{P}_x[\mathbf{x}_n = y; \mathbf{T} = k; \mathbf{x}_\mathbf{T} = t] \\
&= \sum_{t \in A} \sum_{n=0}^{\infty} \sum_{k=1}^{n} \mathbf{P}_x[\mathbf{T} = k, \mathbf{x}_\mathbf{T} = t] \mathbf{P}_t[\mathbf{x}_{n-k} = y] \\
&= \sum_{t \in A} \left\{ \sum_{k=1}^{\infty} \mathbf{P}_x[\mathbf{T} = k, \mathbf{x}_\mathbf{T} = t] \right\} \left\{ \sum_{n=0}^{\infty} \mathbf{P}_t[\mathbf{x}_n = y] \right\} \\
&= \sum_{t \in A} H_A(x,t) G(t,y).
\end{aligned}$$

That completes the proof of P1.

A very simple but powerful technique, successful because the random variables $\mathbf{X}_k = \mathbf{x}_{k+1} - \mathbf{x}_k$ are identically distributed and independent, consists of *reversing a random walk*.

D2 *If $P(x,y)$ is a random walk with state space R, then the random walk with transition function $P^*(x,y) = P(y,x)$ and with the same state space is called the reversed random walk. We shall denote by $G^*(x,y)$, $Q_n^*(x,y)$, $H_A^*(x,y)$, $\Pi_A^*(x,y)$, $g_A^*(x,y)$, etc., ... the functions defined in D1 for the reversed random walk.*

As obvious consequences of this definition

P2 $\qquad G^*(x,y) = G(y,x), \qquad Q_n^*(x,y) = Q_n(y,x),$
$\qquad\qquad \Pi_A^*(x,y) = \Pi_A(y,x), \qquad g_A^*(x,y) = g_A(y,x),$

each identity being valid on the entire domain of definition of the function in question. Note that, for good reasons, no identity for $H_A^*(x,y)$ is present in P2.

As the first application of reversal of the random walk we derive an identity concerning the number of visits $\mathbf{N}_x$ of a random walk to the point $x \in R$, before its return to the starting point. We shall set

$$\mathbf{T} = \min [k \mid 1 \le k \le \infty; \mathbf{x}_k = 0], \qquad \mathbf{N}_x = \sum_{n=0}^{\mathbf{T}} \delta(x, \mathbf{x}_n).$$

Then the expected value of $\mathbf{N}_x$, when $\mathbf{x}_0 = 0$, is

$$E_0[\mathbf{N}_x] = E_0[\sum_{n=0}^{T} \delta(x,\mathbf{x}_n)] = \sum_{n=1}^{\infty} \mathbf{P}_0[\mathbf{x}_n = x; \mathbf{T} > n].$$

But

$$\mathbf{P}_0[\mathbf{x}_n = x; \mathbf{T} > n] = \sum_{t \neq 0} P(0,t) Q_{n-1}(t,x),$$

where $Q_n(x,y)$ is defined according to D1 for the set $A = \{0\}$. Using P2, for $n \geq 1$

$$\mathbf{P}_0[\mathbf{x}_n = x; \mathbf{T} > n] = \sum_{t \neq 0} P^*(t,0) Q_{n-1}^*(x,t) = \sum_{t \neq 0} Q_{n-1}^*(x,t) P^*(t,0).$$

But this is the probability that the reversed random walk, starting at x, visits 0 for the first time at time n. Thus

$$\mathbf{P}_0[\mathbf{x}_n = x; \mathbf{T} > n] = F_n^*(x,0), \qquad E_0[\mathbf{N}_x] = \sum_{n=1}^{\infty} F_n^*(x,0) = F^*(x,0).$$

This enables us to prove

P3 *For recurrent aperiodic random walk*

$$E_0[\mathbf{N}_x] = 1 \text{ for all } x \neq 0.\text{[1]}$$

Proof: All that remains is to observe that

$F^*(x,0) = F(0,x)$, and by T2.1 $F(0,x) = 1$ for all x in R.

As a corollary of P3 we get a simple proof that $g_A(x,y) < \infty$ for all x and y if the random walk is aperiodic. The reduction to the case when $A = \{0\}$ and $x = y$ was accomplished in the proof of P1, and it is easy to see that

$$1 = E_0[\mathbf{N}_x] > Q_n(0,x) g_{\{0\}}(x,x)$$

for every $n \geq 1$. By choosing n so that $Q_n(0,x) > 0$ we see that $g_{\{0\}}(x,x) < \infty$.

As a second application one can prove

P4 *For an arbitrary random walk*

$$\sum_{y \in A} \Pi_A(x,y) \leq 1 \text{ for all } x \in A$$

and

$$\sum_{x \in A} \Pi_A(x,y) \leq 1 \text{ for all } y \in A.$$

[1] This answers problem 6 in Chapter I. In the theory of recurrent Markov chains $f(x) = E_0[\mathbf{N}_x]$ was shown by Derman [21] to be the unique nonnegative solution of the equation $f(x) = Pf(x)$—which happens to be constant for recurrent random walk. We shall prove uniqueness in P13.1.

Proof: According to D1

$$\sum_{y \in A} \Pi_A(x,y) = \mathbf{P}_x[\mathbf{T} < \infty] \le 1,$$

where $\mathbf{T} = \min[k \mid 1 \le k \le \infty, \mathbf{x}_k \in A]$. The second part follows from the observation that $\Pi_A^*(x,y) = \Pi_A(y,x)$, so that

$$\sum_{x \in A} \Pi_A(x,y) = \sum_{x \in A} \Pi_A^*(y,x) \le 1$$

by the first part of P4. For recurrent aperiodic random walk P4 will be strengthened considerably in the beginning of the next section (P11.2).

Remark: The terminology and notation introduced in this section will be used throughout the remainder of the book. The rest of this chapter (sections 11 through 16) will be devoted to two-dimensional recurrent random walk. For this case a rather complete theory of stopping times $\mathbf{T} = \mathbf{T}_A$ for finite sets $A \subset R$ may be developed using quite elementary methods. These methods provide a good start, but turn out to be insufficient in other cases—namely for transient random walk as well as in the one-dimensional recurrent case. Therefore the theory in this chapter will have to perform the service of motivating and illustrating many later developments.

11. THE HITTING PROBABILITIES OF A FINITE SET

Through the remainder of this chapter we assume that we are dealing with *aperiodic recurrent random walk in two dimensions*.[2] Further, all probabilities of interest, i.e., all the functions in D10.1 will be those associated with a nonempty *finite* subset B of R. The methods and results will be independent of the cardinality $|B|$ except in a few places where the case $|B| = 1$ requires special consideration.

The primary purpose of this section is to show how one can calculate $H_B(x,y)$ which we shall call the *hitting probability* measure of the set B. To begin with, we base the calculation on part (a) of P10.1 from which one obtains

---

[2] This will be tacitly assumed. Only the statements of the principal theorems will again explicitly contain these hypotheses.

P1 *For $x \in R$, $y \in B$, and $n \geq 0$*

$$\sum_{t \in R} P_{n+1}(x,t) H_B(t,y) = H_B(x,y) + \sum_{t \in B} G_n(x,t)[\Pi_B(t,y) - \delta(t,y)].$$

Proof: Operating on part (a) of P10.1 by the transition function $P(x,y)$ on the left gives

$$\sum_{t \in R} P_2(x,t) H_A(t,y) - \sum_{t \in R} P(x,t) H_A(t,y) = \sum_{t \in B} P(x,t)[\Pi_B(t,y) - \delta(t,y)].$$

Now we do this again $n - 1$ times, and the resulting $n + 1$ equations, counting the original one, may be expressed in the abbreviated form

$$PH = H + \Pi - I, \quad P_2 H = PH + P(\Pi - I), \ldots,$$
$$P_{n+1} H = P_n H + P_n(\Pi - I).$$

Adding these equations, using I to denote the identity operator, one finds

$$P_{n+1} H = H + (I + P + \cdots + P_n)(\Pi - I).$$

This shows that P1 holds, if we remember that $I + P + \cdots + P_n$ stands for

$$G_n(x,y) = \sum_{k=0}^{n} P_k(x,y),$$

in view of D1.3.

The next lemma uses the recurrence of the random walk in an essential way. It is

P2 $\sum_{t \in B} \Pi_B(t,y) = \sum_{t \in B} \Pi_B(y,t) = 1$ *for every y in B.*

Proof: Consider the reversed random walk with transition function $P^*(x,y) = P(y,x)$. If we call $\Pi_B^*(x,y)$ what one gets if one forms Π_B for the transition function P^* in accordance with D10.1, then by P10.2

$$\Pi_B^*(x,y) = \Pi_B(y,x), \text{ for } x,y \text{ in } B.$$

This fact was already exploited in the proof of P10.4. Since $P_n^*(0,0) = P_n(0,0)$, it is obvious that the random walk P^* is recurrent whenever the original random walk is recurrent. Therefore P2 will be proved if we verify that

$$\sum_{y \in B} \Pi_B(x,y) = 1, \qquad x \in B$$

for every aperiodic recurrent random walk. Here it is convenient to use the measure theoretical definition of Π_B, giving

$$\sum_{y \in B} \Pi_B(x,y) = \mathbf{P}_x[\mathbf{T}_B < \infty] \geq \mathbf{P}_x[\mathbf{T}_x < \infty],$$

since $\mathbf{T}_x = \min [k \mid 1 \leq k \leq \infty; \mathbf{x}_k = x] \geq \mathbf{T}_B$. But

$$\mathbf{P}_x[\mathbf{T}_x < \infty] = \sum_{k=1}^{\infty} F_k(x,x) = \sum_{k=1}^{\infty} F_k(0,0) = F$$

in the terminology of section 1, and $F = 1$ since the process is recurrent. That completes the proof.

Now P2 opens up an important possibility. It enables us to transform P1 into

$$\sum_{t \in R} P_{n+1}(x,t) H_B(t,y)$$
$$= H_B(x,y) + \sum_{t \in B} [c_n + G_n(x,t)][\Pi_B(t,y) - \delta(t,y)],$$

where the c_n are arbitrary constants (independent of t). We choose to let $c_n = -G_n(0,0)$ and define

D1 $\qquad A_n(x,y) = G_n(0,0) - G_n(x,y), \qquad x,y$ in R,
$\qquad\qquad a_n(x) = A_n(x,0), \qquad x$ in R.

At this point P1 has beome

P3 $\qquad \sum_{t \in R} P_{n+1}(x,t) H_B(t,y)$
$\qquad\qquad = H_B(x,y) - \sum_{t \in B} A_n(x,t)[\Pi_B(t,y) - \delta(t,y)],$

for $x \in R$, $y \in B$, $n \geq 0$.

The further development is now impeded by two major problems. We wish to let $n \to \infty$ in P3 and naturally ask

(i) *Does* $\lim_{n \to \infty} A_n(x,t)$ *exist?*

(ii) *If the answer to* (i) *is affirmative, what can be said about the other limit in* P3, *namely*

$$\lim_{n \to \infty} \sum_{t \in R} P_{n+1}(x,t) H_B(t,y)?$$

In order not to interrupt the continuity of the presentation we shall answer these two questions here, and then devote the next two sections to proving that the answers are correct. In P12.1 of section 12 we shall prove in answer to (i) that the limit of $A_n(x,y)$ exists for all x,y in R, which we record as

(i)' $$\lim_{n\to\infty} A_n(x,y) = A(x,y), \qquad \lim_{n\to\infty} a_n(x) = a(x).$$

The proof of this assertion in section 12 will use in an essential way two of our assumptions. The first is that the random walk is aperiodic, which is important, because the limits in question need not exist in the periodic case. The second assumption is that the dimension of the process is two. This is essential only for the proof and in fact we shall present a different proof, in Chapter VII, which shows that

$$\lim_{n\to\infty} A_n(x,y) = A(x,y)$$

exists for every aperiodic random walk.

Concerning question (ii), let us try to simplify the problem. If the answer to (i) is affirmative, then it follows from P3 that

$$\lim_{n\to\infty} \sum_{t\in R} P_{n+1}(x,t) H_B(t,y)$$

exists for every pair $x \in R$, $y \in B$. Let us fix y and call the above sum $f_{n+1}(x)$. Iterated operation by the transition function $P(x,y)$ shows that

$$f_{n+1}(x) = \sum_{t\in R} P(x,t) f_n(t).$$

Here we have used the obvious fact that $0 \leq f_n(x) \leq 1$, to justify an interchange of summation. As we know that the limit of $f_n(x)$ exists, we may call it $f(x)$, and conclude, again by dominated convergence, that

$$f(x) = \sum_{y\in R} P(x,y) f(y), \qquad x \in R.$$

In section 13, P13.1, we shall show, again using the aperiodicity and recurrence of the random walk, that the only non-negative solutions of this equation are the constant functions. This means that the limit we inquired about in (ii) above exists and is independent of x, so we may call it $\mu_B(y)$ (as it may well depend on y and on the set B) and record the answer to question (ii) as

(ii)' $$\lim_{n\to\infty} \sum_{t\in R} P_{n+1}(x,t) H_B(t,y) = \mu_B(y), \qquad x \in R, \quad y \in B.$$

In the remainder of this section we proceed under the assumption that (i)' and (ii)' are correct and this reservation will be explicitly stated in the conclusions we obtain in propositions P4 through P8 and in the main theorem T1 at the end. The reader to whom this departure from a strictly logical development is abhorrent is invited to turn to the proof of propositions P12.1 and P13.1 at this point.

By applying (i)' and (ii)' to P3 one can immediately conclude

P4 $\qquad H_B(x,y) = \mu_B(y) + \sum_{t \in B} A(x,t)[\Pi_B(t,y) - \delta(t,y)],$

for $x \in R$, $y \in B$ (subject to (i)' and (ii)').

Our next move is to consider a very special case of P4, by restricting attention to sets B of cardinality two. Nothing is lost, and some simplification gained by letting the origin belong to B, so that we let $B = \{0,b\}$ where b is any point in R except 0. Consider now the two by two matrix Π_B. In view of P2 it is doubly stochastic. This means that if we call $\Pi_B(0,b) = \Pi$, then $\Pi_B(b,0) = \Pi$,

$$\Pi_B(0,0) = \Pi_B(b,b) = 1 - \Pi.$$

To gain further insight it is of interest to restrict both x and y in P4 to the set B. P4 then consists of four equations, and keeping in mind that $H_B(x,y) = \delta(x,y)$, we represent them in matrix form as follows:

$$\begin{pmatrix} 1 & 0 \\ 0 & 1 \end{pmatrix} = \begin{pmatrix} \mu_B(0) & \mu_B(b) \\ \mu_B(0) & \mu_B(b) \end{pmatrix} + \Pi \begin{pmatrix} a(-b) & -a(-b) \\ -a(b) & a(b) \end{pmatrix}.$$

The above matrix identity shows that Π cannot be zero, and it readily yields the following solution (in the sense that we think of $a(x)$ as a known function, and of Π, $\mu_B(0)$, $\mu_B(b)$ as unknowns).

P5 *For $B = \{0,b\}$, $b \neq 0$,* $\quad \Pi_B(0,b) = \dfrac{1}{a(b) + a(-b)} > 0,$

$$\mu_B(0) = \frac{a(b)}{a(b) + a(-b)}, \qquad \mu_B(b) = \frac{a(-b)}{a(b) + a(-b)},$$

(subject to (i)' and (ii)').

By substitution of the result of P5 into P4 for values of x not in B one can now obtain explicit formulas for $H_B(x,0)$ and $H_B(x,b)$ in the case when $B = \{0,b\}$. For example

(1) $\qquad H_B(x,b) = \dfrac{A(x,0) + A(0,b) - A(x,b)}{A(0,b) + A(b,0)}, \qquad x \in R, \quad b \neq 0.$

A result that will be of more lasting interest, however, concerns the function $g_B(x,y)$, in an important special case.

D2 *If $B = \{0\}$, i.e., the set consisting of the origin, $g_B(x,y)$ is denoted by $g_{\{0\}}(x,y) = g(x,y)$.*

Thus $g(x,y)$ is the expected number of visits of the random walk, starting at x, to the point y, before the first visit to 0. We shall show that

P6 $g(x,y) = A(x,0) + A(0,y) - A(x,y), \qquad x \in R, \quad y \in R,$
(*subject to* (i)′ *and* (ii)′).

Proof: When either $x = 0$ or $y = 0$ or both we get $g(x,y) = A(0,0) = 0$, as we should, according to D10.1. When $x = y$, $g(x,y) = g(x,x)$ is the expected number of returns to x before visiting 0, or

(2) $$g(x,x) = 1 + \sum_{k=1}^{\infty} (1 - \Pi)^k = \Pi^{-1},$$

where $\Pi = \Pi_{\{0,x\}}(x,0)$. Therefore the result of P6 is verified in this case by consulting the formula for Π in P5. Finally, when $x \neq y$, and neither is 0,

(3) $$g(x,y) = H_{\{0,y\}}(x,y)g(y,y)$$

by an obvious probabilistic argument. But now the substitution of the formula (1) for $H_B(x,y)$ and of equation (2) for $g(y,y)$ into equation (3) completes the proof.

We need one last auxiliary result before proceeding to draw conclusions from P4 for finite sets B of arbitrary size. It is a subtler result than perhaps appears at first sight—in fact there are one-dimensional aperiodic random walks for which it is false, as we shall discover in Chapter VII.

P7 *For $x \neq 0$, $a(x) > 0$; (subject to* (i)′ *and* (ii)′*).*

Proof: Remember first of all that

$$a(x) = \lim_{n \to \infty} [G_n(0,0) - G_n(x,0)] \geq 0$$

by P1.3, so that only the strict positivity of $a(x)$ is in doubt. Suppose now that there is a point $b \neq 0$ in R where $a(x)$ vanishes. Using P6 we then have $g(b,-b) = 2a(b) - a(2b) = 0$. Since $a(b) = 0$ and $a(2b) \geq 0$, it follows that $g(b,-b) = 0$. This is impossible *in two dimensions* as it says that there is no path (of positive probability) from b to

$-b$ which does not go through the origin. Being precise, it states that whenever

$$P(b,x_1)P(x_1,x_2) \cdots P(x_n,-b) > 0$$

then at least one of the points $x_1, x_2, \ldots, x_n$ is the origin. A simple combinatorial argument settles the matter. There is a path of positive probability from any point to any other. Therefore there is a positive product of the type above, such that not all of the differences

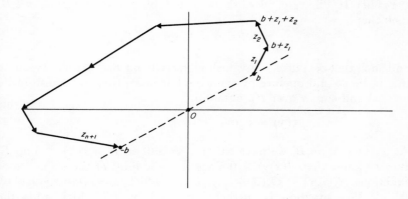

$x_1 - b, x_2 - x_1, \ldots, -b - x_n$ are multiples of b (parallel to the vector b). Call these differences $y_1, y_2, \ldots, y_{n+1}$ and reorder them so that the arguments $\arg(y_k)$ are in nondecreasing order. We can do this so that, calling the newly ordered set $z_1, z_2, \ldots, z_{n+1}$, we have $\arg b \leq \arg z_1 \leq \arg z_2 \leq \ldots \leq \arg z_{n+1} \leq \arg b + 2\pi$. Then we clearly have a path with probability (see diagram)

$$P(b,b+z_1)P(b+z_1,b+z_1+z_2) \cdots P(b+\cdots+z_n,-b) > 0$$

which goes from b to $-b$ without going through 0. That proves P7 by contradiction.

We are nearing our goal of expressing H_B, Π_B, and μ_B in terms of the function $a(x)$, for arbitrary finite sets B. Since this is a trivial matter when $|B| = 1$, but one which requires annoying modifications in the analysis, we assume from now on that $|B| \geq 2$. The key to the necessary calculations is the inversion of the operator $A(x,y)$, with x and y restricted to B. We shall say that *the operator $A(x,y)$ restricted* to the set B *has an inverse if* there exists a function $K_B(x,y)$, with x and y in B, such that

D3 $\qquad \sum_{t \in B} A(x,t) K_B(t,y) = \delta(x,y), \quad \text{for } x,y \text{ in } B.$

This is just the definition of a matrix-inverse, and according to a standard theorem of algebra, this inverse is unique, if it exists.

P8 *The operator $A(x,y)$, restricted to any finite subset B of R with $|B| \geq 2$ has an inverse K_B (subject to* (i)' *and* (ii)'*).*

Proof: If P8 were false, there would be a function $v(x)$, $x \in B$, such that

$$\sum_{s \in B} v(s) A(s,y) = 0 \quad \text{for } y \in B,$$

and such that $v(x)$ does not vanish identically on B. We shall assume this to be so and operate by v on the left in P4 (i.e., we multiply P4 by $v(x)$ and sum x over B), obtaining

$$v(y) = \mu_B(y) \sum_{x \in B} v(x), \qquad y \in B.$$

As $\mu_B(y) \geq 0$ on B, we have either $v(y) \geq 0$ on B or $v(y) \leq 0$ on B. But we know that $A(x,y) \geq 0$ since it is the limit of the sequence of functions $A_n(x,y) = G_n(0,0) - G_n(x,y)$, which are non-negative by P1.3. The situation is further improved by P7 which adds the information that $A(x,y) > 0$ unless $x = y$. Therefore the operator A, with x, y restricted to the set B, represents a matrix, with zeros on the diagonal, and with all other elements positive. Now this matrix A has the property that $vA = 0$ where v is a nonzero vector, all of whose components are of the same sign. This is obviously impossible, and the contradiction proves P8.

Next we introduce the notation

D4 *For $B \subset R$, $2 \leq |B| < \infty$,*

$$K_B(\cdot x) = \sum_{y \in B} K_B(y,x), \qquad K_B(x \cdot) = \sum_{y \in B} K_B(x,y), \qquad x \in R,$$

$$K_B(\cdot \cdot) = \sum_{x \in B} \sum_{y \in B} K_B(x,y),$$

and prove the main result of this section.[3]

---

[3] The first proof of T1, for symmetric random walk, is in [93], and a complete treatment in [94], which contains the main results of this chapter, with the exception of section 16. Partial results may also be found in [43]. Recently Kemeny and Snell [55] have extended T1 to a very large class of recurrent Markov chains.

T1 *For aperiodic recurrent random walk in two dimensions, with* $B \subset R$, $2 \le |B| < \infty$,

$$H_B(x,y) = \mu_B(y) + \sum_{t \in B} A(x,t)[\Pi_B(t,y) - \delta(t,y)], \qquad x \in R, \quad y \in B,$$

where

$$K_B(\cdot\cdot) > 0, \qquad \mu_B(y) = \frac{K_B(\cdot y)}{K_B(\cdot\cdot)} \text{ for } y \in B,$$

$$\Pi_B(x,y) - \delta(x,y) = K_B(x,y) - \frac{K_B(x\cdot)K_B(\cdot y)}{K_B(\cdot\cdot)} \text{ for } x, y \in B.$$

T1 *is subject to the truth of* (i)′ *and* (ii)′.

Proof: Restrict x to B in P4 and operate by K_B on the left in P4. This yields

(1) $$K_B(x,y) = K_B(x\cdot)\mu_B(y) + \Pi_B(x,y) - \delta(x,y)$$

for $x, y \in B$. Summing x over B one has, in view of P2,

(2) $$K_B(\cdot y) = K_B(\cdot\cdot)\mu_B(y).$$

If we had $K_B(\cdot\cdot) = 0$ here, it would follow that $K_B(\cdot y) = 0$ for all y in B. But then K_B, regarded as a matrix operator, would be singular, which is impossible since K_B has as its inverse the operator $A(x,y)$ restricted to B. Therefore $K_B(\cdot\cdot) \ne 0$.

Now $K_B(\cdot y) = K_B(\cdot\cdot)\mu_B(y)$ shows that either $K_B(\cdot y) \ge 0$ on B or $K_B(\cdot y) \le 0$ on B. Since $\sum_{t \in B} K_B(\cdot t)A(t,x) = 1$ for $x \in B$ and $A(x,y) \ge 0$, $K_B(\cdot y)$ must be non-negative on B so that $K_B(\cdot\cdot) > 0$.

Equations (1) and (2) yield the proper formulas for μ_B and Π_B. Finally the proof of T1 is completed by using P4 to express H_B in terms of μ_B and Π_B.

12. THE POTENTIAL KERNEL $A(x,y)$

First we shall prove that

P1 $$\lim_{n \to \infty} A_n(x,y) = \lim_{n \to \infty} \sum_{k=0}^{n} [P_k(0,0) - P_k(x,y)] = A(x,y)$$

exists for aperiodic two-dimensional random walk (and is given by (3) *below).*

Proof: Of course we are interested only in the recurrent case, the truth of P1 in the transient case being not only obvious, but also quite irrelevant to the work in the preceding section. The proof will be Fourier analytical, and we adopt the notation of Chapter II, letting $\phi(\theta)$ denote the *characteristic function* of the random walk. Using P6.3 we have

(1) $$a_n(x) = (2\pi)^{-2} \int \frac{1 - e^{ix\cdot\theta}}{1 - \phi(\theta)} [1 - \phi^{n+1}(\theta)]\, d\theta.$$

(Equation (1) is the result of a straightforward calculation based on

$$\sum_{k=0}^{n} P_k(x,y) = (2\pi)^{-2} \int e^{i\theta\cdot(x-y)} [1 + \phi(\theta) + \cdots + \phi^n(\theta)]\, d\theta.)$$

The integration is over the usual square $C = [\theta \mid |\theta_i| \leq \pi; i = 1, 2]$ and it is clearest and simplest to use Lebesgue integration for the following reasons. First of all the integrand in (1) is undefined at $\theta = 0$ when $1 - \phi(\theta) = 0$. Secondly, we then have the dominated convergence theorem to work with. It will be useful since $|1 - \phi^n(\theta)| \leq 2$ and $1 - \phi^n(\theta)$ tends to zero, as $n \to \infty$, almost everywhere on C (indeed at all but finitely many points). Hence the existence of the limit in (1) will be assured if we can prove that

(2) $\dfrac{1 - e^{ix\cdot\theta}}{1 - \phi(\theta)}$ is Lebesgue integrable on the square C.

If (2) is true, then we shall know that the limit in P1 exists and has the representation

(3) $$a(x) = \lim_{n \to \infty} a_n(x) = (2\pi)^{-2} \int \frac{1 - e^{ix\cdot\theta}}{1 - \phi(\theta)}\, d\theta, \qquad x \in R.$$

To prove (2) we use two elementary inequalities,

$$|1 - e^{ix\cdot\theta}| \leq |x||\theta|,$$

and

$$|1 - \phi(\theta)| \geq \operatorname{Re}[1 - \phi(\theta)],$$

which together imply

$$\left|\frac{1 - e^{ix\cdot\theta}}{1 - \phi(\theta)}\right| \leq |x| \frac{|\theta|}{\operatorname{Re}[1 - \phi(\theta)]}.$$

Thus it suffices to show that the function on the right is Lebesgue integrable on the square C.

This last step uses P7.5 which asserts that for *aperiodic* random walk one can find a constant $\lambda > 0$ such that $\text{Re}\,[1 - \phi(\theta)] \geq \lambda|\theta|^2$ when $\theta \in C$. Thus

$$\frac{|\theta|}{\text{Re}\,[1 - \phi(\theta)]} \leq \frac{1}{\lambda|\theta|}, \qquad \theta \in C,$$

and $|\theta|^{-1}$ is indeed Lebesgue integrable since we are in dimension $d = 2$, where

$$\int_C \frac{d\theta}{|\theta|} \leq \int_{|\theta| \leq \pi\sqrt{2}} \frac{d\theta}{|\theta|} = 2^{3/2}\pi^2 < \infty.$$

Observe that this completes the proof—but that the same proof would break down at the very last step in one dimension, due to the fact that there

$$\int_C \frac{d\theta}{|\theta|} = \int_{-\pi}^{\pi} \frac{dx}{|x|} = \infty.$$

Much more sophisticated methods are required to show (in Chapter VII) that P1 is true even for one-dimensional random walk.

We proceed to two more results which have a natural proof by Fourier analysis, and which will be useful in continuing the work of section 11. The first proposition (P2) is a general law governing the asymptotic behavior of $A(x,y)$ for two-dimensional aperiodic recurrent random walk. The second result (P3) is a sharp version of P2, valid only for a very restricted class of random walks. Incidentally, for reasons which will appear later, we provide the function $A(x,y)$ with the name of *potential kernel*.

P2 *Let $A(x,y)$ be the potential kernel of an aperiodic two-dimensional recurrent random walk. For every fixed pair of points y_1 and y_2 in R,*

$$\lim_{|x| \to \infty} [A(x,y_1) - A(x,y_2)] = 0.$$

Proof: The statement of P2 is clearly equivalent to stating that for an arbitrary fixed y in R

$$\lim_{|x| \to \infty} [a(x + y) - a(x)] = 0,$$

i.e., that given any $\epsilon > 0$, there is a positive M (which may depend on ϵ and y) such that $|a(x + y) - a(x)| < \epsilon$ when $|x| > M$. The proof is based on (3) in the proof of P1, which gives

$$a(x + y) - a(x) = (2\pi)^{-2} \int e^{ix\cdot\theta} \psi(\theta)\,d\theta,$$

where

$$\psi(\theta) = \frac{1 - e^{iy\cdot\theta}}{1 - \phi(\theta)}.$$

However, we saw in the proof of P1 that $\psi(\theta)$ is Lebesgue integrable on the square C. Therefore the Riemann Lebesgue Lemma (P9.1) gives

$$\lim_{|x|\to\infty} \int e^{ix\cdot\theta}\psi(\theta)\,d\theta = 0,$$

which completes the proof of P2.

It is of course of considerable interest to know more about the asymptotic behavior of $a(x)$ for the *simple random walk* with $P(0,x) = \frac{1}{4}$ when $|x| = 1$. But it turns out to be hardly more work to discuss a somewhat larger class of random walks, which satisfies the following conditions

(a) $P(x,y)$ is two dimensional, and aperiodic,
(b) $\qquad \mu = \sum xP(0,x) = 0,$
(c) $\qquad Q(\theta) = \sum (x\cdot\theta)^2 P(0,x) = \mathbf{E}[(\mathbf{X}\cdot\theta)^2] = \sigma^2|\theta|^2 < \infty,$
(d) $\qquad \mathbf{E}[|\mathbf{X}|^{2+\delta}] = \sum |x|^{2+\delta}P(0,x) < \infty$ for some $\delta > 0$.

Of course simple random walk satisfies these conditions, and we also know (T8.1) that (b) and (c) imply that the random walk is recurrent. In addition (c) introduces the isotropy requirement that $Q(\theta)$ be proportional to $|\theta|^2$ rather than to any positive definite quadratic form. That is essential and so is (d) in order to obtain a result as sharp and simple as the following.

P3 *A random walk satisfying* (a) *through* (d) *has the property that*

$$\lim_{|x|\to\infty} \left[a(x) - \frac{1}{\pi\sigma^2} \ln|x| \right] = c + \frac{1}{\pi\sigma^2} d.$$

The definition of the constants c and d is:

$$0 < c = (2\pi)^{-2} \int \left[\frac{1}{1-\phi(\theta)} - \frac{2}{Q(\theta)} \right] d\theta < \infty,$$

$$d = \gamma + \ln\pi - \frac{2}{\pi}\lambda,$$

where

$\gamma = 0.5572\ldots$ *is Euler's constant,*

$$\lambda = \sum_{n=0}^{\infty} \frac{(-1)^n}{(2n+1)^2} \text{ is Catalan's constant.}$$

For the simple random walk, $\sigma^2 = 1/2$ and

$$\lim_{|x|\to\infty} \left[a(x) - \frac{2}{\pi} \ln |x|\right] = \frac{1}{\pi} \ln 8 + \frac{2\gamma}{\pi}.$$

Proof: We begin by checking that

$$\psi(\theta) = \frac{1}{1 - \phi(\theta)} - \frac{2}{Q(\theta)}$$

is Lebesgue integrable on the square C. Let us write, as we may in view of (c),

$$\psi(\theta) = \frac{2}{\sigma^2} \cdot \frac{|\theta|^2}{1 - \phi(\theta)} \chi(\theta),$$

$$\chi(\theta) = |\theta|^{-4} \left[\sigma^2 \frac{|\theta|^2}{2} - 1 + \phi(\theta)\right].$$

By P6.7

$$\lim_{|\theta|\to 0} \frac{|\theta|^2}{1 - \phi(\theta)} = \frac{2}{\sigma^2} > 0,$$

and according to T7.1, $1 - \phi(\theta) = 0$ only at $\theta = 0$ in the square C. Therefore it suffices to prove that $\chi(\theta)$ is integrable on C.

Now

$$\sigma^2 \frac{|\theta|^2}{2} - 1 + \phi(\theta) = \mathbf{E}[e^{i\mathbf{X}\cdot\theta} - 1 - i\mathbf{X}\cdot\theta - \tfrac{1}{2}(i\mathbf{X}\cdot\theta)^2],$$

if $\mathbf{X}$ is a random variable with values in R such that

$$\mathbf{P}[\mathbf{X} = x] = P(0,x), \qquad x \in R.$$

Here we used condition (b) in inserting the useful term $i\mathbf{X}\cdot\theta$. Now we use the positive number δ in condition (d) by making the obvious assertion that there is some $h > 0$ such that

$$\left|e^z - \left(1 + z + \frac{z^2}{2}\right)\right| \le h|z|^{2+\delta}$$

for all complex z in the half plane Re $z \le 0$. That gives (setting $z = i\theta \cdot \mathbf{X}$)

$$|\chi(\theta)| \le h|\theta|^{-4} \mathbf{E}[|\theta \cdot \mathbf{X}|^{2+\delta}].$$

Using the Schwarz inequality and condition (d)

$$|\chi(\theta)| \le h|\theta|^{-4} \mathbf{E}[(|\theta|^2|\mathbf{X}|^2)^{1+\delta/2}] = h|\theta|^{\delta-2} \mathbf{E}[|\mathbf{X}|^{2+\delta}] \le M|\theta|^{\delta-2},$$

for some $M < \infty$. Thus $\chi(\theta)$, and hence $\psi(\theta)$, is integrable, for

$$\int_C |\theta|^{\delta-2}\,d\theta \leq \int_{|\theta|\leq \pi\sqrt{2}} |\theta|^{\delta-2}\,d\theta = 2\pi \int_0^{\pi\sqrt{2}} \frac{dt}{t^{1-\delta}} < \infty.$$

The next step is to decompose

$$a(x) = (2\pi)^{-2} \int \frac{1 - e^{ix\cdot\theta}}{1 - \phi(\theta)}\,d\theta$$

$$= \frac{2}{\sigma^2}(2\pi)^{-2} \int \frac{1 - \cos x\cdot\theta}{|\theta|^2}\,d\theta + (2\pi)^{-2} \int (1 - e^{ix\cdot\theta})\psi(\theta)\,d\theta.$$

The last integral tends to the constant c in P3, by the Riemann Lebesgue Lemma in P9.1. Therefore the proof of P3, apart from the calculations for simple random walk, will be complete if we show that

$$\lim_{|x|\to\infty} \left[\frac{1}{2\pi} \int \frac{1 - \cos x\cdot\theta}{|\theta|^2}\,d\theta - \ln|x|\right] = \gamma - \frac{2}{\pi}\lambda + \ln\pi.$$

To accomplish this we decompose

$$\frac{1}{2\pi} \int \frac{1 - \cos x\cdot\theta}{|\theta|^2}\,d\theta = I_1(x) + I_2(x).$$

Here $I_1(x)$ is the integral over the circular disc $|\theta| \leq \pi$ and $I_2(x)$ is the contribution from the rest of the square. The proof will be completed by showing that

$$\lim_{|x|\to\infty} [I_1(x) - \ln|x|] = \gamma + \ln\frac{\pi}{2}$$

and

$$\lim_{|x|\to\infty} I_2(x) = \ln 2 - \frac{2}{\pi}\lambda.$$

Introducing polar coordinates it is seen that $I_1(x)$ depends only on $|x|$. Specifically

$$I_1(x) = \frac{1}{2\pi} \int_{|\theta|\leq\pi} \frac{1 - \cos x\cdot\theta}{|\theta|^2}\,d\theta = \frac{1}{2\pi} \int_0^{2\pi} dt \int_0^\pi \frac{1 - \cos(|x|r\sin t)}{r}\,dr$$

$$= \frac{2}{\pi} \int_0^{\pi/2} dt \int_0^{\pi|x|\sin t} \frac{1 - \cos u}{u}\,du.$$

12. THE POTENTIAL KERNEL $A(x,y)$

Now we shall use as our definition of Euler's constant[4]

$$\gamma = \int_0^1 \frac{1 - \cos u}{u} du - \int_1^\infty \frac{\cos u}{u} du.$$

Substitution into the formula for $I_1(x)$ yields

$$I_1(x) = \frac{2}{\pi} \int_0^{\pi/2} \left[\gamma + \ln |x| + \ln (\sin t) + \ln \pi + \int_{\pi|x| \sin t}^\infty \frac{\cos u}{u} du \right] dt.$$

Taking account of

$$\frac{2}{\pi} \int_0^{\pi/2} \ln (\sin t) \, dt = -\ln 2,$$

and of

$$\lim_{|x| \to \infty} \frac{2}{\pi} \int_0^{\pi/2} dt \int_{\pi|x| \sin t}^\infty \frac{\cos u}{u} du = 0,$$

one has

$$\lim_{|x| \to \infty} [I_1(x) - \ln |x|] = \gamma - \ln 2 + \ln \pi,$$

as was to be shown.

By use of the Riemann Lebesgue Lemma

$$\lim_{|x| \to \infty} I_2(x) = \lim_{|x| \to \infty} \frac{1}{2\pi} \int_{\substack{|\theta_1| \leq \pi, |\theta_2| \leq \pi \\ |\theta| \geq \pi}} \frac{1 - \cos x \cdot \theta}{|\theta|^2} d\theta$$

$$= \frac{1}{2\pi} \int_{\substack{|\theta_1| \leq \pi, |\theta_2| \leq \pi \\ |\theta| \geq \pi}} \frac{d\theta}{|\theta|^2}.$$

Again introducing polar coordinates, one obtains

$$\lim_{|x| \to \infty} I_2(x) = \frac{4}{\pi} \int_0^{\pi/4} dt \int_\pi^{\pi/\cos t} \frac{dr}{r} = -\frac{4}{\pi} \int_0^{\pi/4} \ln (\cos t) \, dt.$$

This definite integral may be shown to have the value $\ln 2 - 2\lambda/\pi$.

---

[4] This happens to agree (the proof is not quite trivial) with the usual definition of

$$\gamma = \lim_{n \to \infty} [1 + 1/2 + \cdots + 1/n - \ln n].$$

We omit the explicit evaluation of the limit in P3 for simple random walk, as this will be done more easily in the beginning of section 15. The result of P3 for simple random walk has incidentally been obtained by several authors in a number of quite unrelated contexts, cf. [76], [96], and Appendix II by Van der Pol in [51].

13. SOME POTENTIAL THEORY

The equation

$$\sum_{y \in R} P(x,y)f(y) = f(x), \quad x \in R, \qquad \text{or briefly } Pf = f,$$

is Laplace's equation in a very thin disguise indeed. When P is the transition operator of simple random walk in two dimensions, we have

$$Pf(x) - f(x) = \tfrac{1}{4}[f(x+1) + f(x-1) + f(x+i) + f(x-i) - 4f(x)],$$

where we have adopted the complex notation $x = (x^1, x^2) = x^1 + ix^2$. But then $Pf - f = (P - I)f$ where $P - I$ is nothing but the two-dimensional *second difference operator*. Hence the equation $Pf = f$ is the discrete analogue of Laplace's equation

$$\Delta f = \frac{\partial^2 f}{\partial x_1^2} + \frac{\partial^2 f}{\partial x_2^2} = 0,$$

where $f(x) = f(x_1, x_2)$ is a twice differentiable function, and therefore we may expect the solutions of $Pf = f$ to have some properties reminiscent of those of entire harmonic functions.

Although our results will automatically lead to the far from obvious conclusion that such analogies can be both useful and far-reaching, the purpose of this section is quite modest. We shall give a very superficial description of three of the *key problems in classical two-dimensional potential theory*. The solution of these three problems in the classical context is quite easy and well known. Admittedly the problems we shall consider will not explain adequately the basic principles of potential theory. But it is merely our purpose to use these three problems to guide us in our treatment of analogous problems we shall encounter in the study of recurrent two-dimensional random walk.

One last point deserves considerable emphasis: Our three problems A, B, C below concern classical two-dimensional potential theory, which is also known as *logarithmic potential theory*. The same problems make sense in three dimensions, in the context of *Newtonian potential theory*, but there they have somewhat different answers. Only in Chapters VI and VII will it become quite clear that the structure of logarithmic potential theory is that encountered in recurrent random walk, whereas Newtonian potentials arise in the context of transient random walk.

The remainder of this section is divided into three parts, where the three problems A, B, C are first discussed in their classical setting, and then stated and solved in the context of two-dimensional random walk.

Problem A. Characterize the real valued harmonic functions, i.e., functions $u(x,y)$ satisfying $\Delta u(x,y) = 0$ at every point in the plane. In particular, *what are all the non-negative harmonic functions?*

The well-known answer is that $u(x,y)$ must be the real part of an analytic function. In particular, suppose that $u(x,y) \geq 0$ and harmonic in the whole plane. Then there exists an entire analytic function of the form

$$f(z) = u(x,y) + iv(x,y).$$

The function

$$g(z) = e^{-f(z)}$$

is also analytic in the whole plane, its absolute value is

$$|g(z)| = e^{-u(x,y)},$$

and since we assume that $u(x,y) \geq 0$ everywhere it follows that $|g(z)| \leq 1$ for all z. But by Liouville's theorem a bounded entire function is constant. Therefore $u(x,y)$ *is a constant function.*

There is another simple proof ([7], p. 146) based on the mean value property of harmonic functions, which works for every dimension $d \geq 1$. Suppose that u is a non-negative harmonic function, and that $u(x) < u(y)$, where x and y are two points of d-dimensional space. To arrive at a contradiction, we integrate u over two (solid) spheres: one of radius r_1 with center at x, and another of radius r_2 with center at y. Calling $I(r_k)$ the two integrals, and $V(r_k)$ the volumes of the two spheres, we have

$$u(x)V(r_1) = I(r_1), \qquad u(y)V(r_2) = I(r_2),$$

so that

$$\frac{u(x)}{u(y)} = \frac{I(r_1)}{I(r_2)} \left(\frac{r_2}{r_1}\right)^d.$$

Since u is non-negative, it is evidently possible to let r_1 and r_2 tend to infinity in such a way that the first sphere contains the second, which implies $I(r_1) \ge I(r_2)$, and so that the ratio r_2/r_1 tends to one. (Simply give r_2 the largest possible value so that the sphere around y is contained in the sphere about x.) Therefore

$$\frac{u(x)}{u(y)} \ge \left(\frac{r_2}{r_1}\right)^d \to 1$$

which shows that $u(x) \ge u(y)$. This contradiction completes the proof that non-negative harmonic functions are constant.

To formulate the analogous random walk problem, we make the definition

D1 *If $P(x,y)$ is the transition function of a recurrent aperiodic random walk, then the non-negative solutions of*

$$\sum_{y \in R} P(x,y)f(y) = f(x), \qquad x \in R$$

are called regular functions.

In the random walk case, the difficulties of problem A (the problem of characterizing regular functions) will now be illustrated with three examples.

E1 Let $P(x,y)$ be the transition function of periodic random walk. We don't care whether it is recurrent or not, nor what the dimension is. Then $\overline{R}$ is a proper subgroup of R. Define

$$f(x) = 1 \text{ for } x \in \overline{R},$$
$$ = 0 \text{ for } x \in R - \overline{R}.$$

Then

$$Pf(x) = \sum_{y \in R} P(x,y)f(y) = \sum_{y \in R} P(0, y-x)f(y) = \sum_{t \in R} P(0,t)f(t+x).$$

But $P(0,t) = 0$ unless t is in $\overline{R}$, so that

$$Pf(x) = \sum_{t \in \overline{R}} P(0,t)f(t+x).$$

When $t \in \overline{R}$, then $t + x$ is in $\overline{R}$ if and only if x is in $\overline{R}$. Thus $Pf(x) = 0$ when x is in $R - \overline{R}$ and $Pf(x) = 1$ when x is in $\overline{R}$. Hence $f(x)$ is a non-constant function. This example in fact shows that aperiodicity is necessary in order that P have the property that all regular functions are constant.

E2 Consider *Bernoulli random walk* in one dimension, with $P(0,1) = p$, $P(0,-1) = q$, $p + q = 1$, $p > q$. As was shown in E1.2 there are two linearly independent regular functions, when $p \neq q$, namely

$$f(x) \equiv 1 \quad \text{and} \quad f(x) = \left(\frac{p}{q}\right)^x.$$

This shows that we cannot hope to prove that all regular functions are constant for transient random walk.

E3 Dropping the restriction of non-negativity, and calling solutions of $Pf = f$ harmonic, let us look at simple symmetric random walk in the plane. There are then so many harmonic functions that it is interesting to look for *harmonic polynomials*,

$$p(z) = \sum_{k,m} a_{k,m} x^k y^m, \quad a_{k,m} \text{ real},$$

where $z = x + iy$, and $\sum$ a sum over a finite number of pairs (k,m) of non-negative integers. The *degree* of $p(z)$ is the largest value of $k + m$ occurring in any term of the sum. In the classical theory of harmonic functions every harmonic polynomial of degree n can be expressed uniquely as a linear combination of

$$a_k \operatorname{Re}(z^k) + b_k \operatorname{Im}(z^k), \quad a_k, b_k \text{ real}, \quad 0 \leq k \leq n.$$

In the random walk case one might expect a similar result and so we make two conjectures, each reflecting certain essential features of the classical result.

(1) For each $n \geq 1$ there are exactly $2n + 1$ linearly independent harmonic polynomials of degree n.

(2) Among the polynomials in (1) there are two which are homogeneous of degree n, i.e., polynomials $p(z)$ such that $p(tz) = |t|^n p(z)$.

Following an elegant approach of Stöhr [96], we show that (1) is correct, but that (2) is false. Two simple lemmas are required. Let ∇ be the operator $P - I$, so that

$$\nabla f(z) = \tfrac{1}{4}[f(z+1) + f(z-1) + f(z+i) + f(z-i)] - f(z).$$

Lemma 1. If $f(z)$ is a polynomial of degree n or less, then $\nabla f(z)$ is a polynomial of degree at most $n - 2$ (or zero if $n = 0$ or 1).

Proof: Since ∇ is linear it suffices to apply ∇ to simple polynomials of the form $x^j y^k$ and to check that the result has no terms of degree $j + k - 1$ or higher.

Lemma 2. Let $p(z)$ be a polynomial of degree $n - 2$ or less ($p(z) = 0$ if $n = 0$ or 1). Then there exists a polynomial $q(z)$ of degree n or less such that $\nabla q(z) = p(z)$.

Proof: We may assume $n > 2$; otherwise the result is trivial. Let us write $p(z)$ lexicographically, as

$$p(z) = a_{n-2,0}x^{n-2}$$
$$+ a_{n-3,1}x^{n-3}y + a_{n-3,0}x^{n-3}$$
$$+ \cdots$$
$$+ a_{0,n-2}y^{n-2} + a_{0,n-3}y^{n-3} + \cdots + a_{0,1}y + a_{0,0}.$$

The first nonvanishing term in this array is called the *leading term*. Suppose it is bx^jy^k. We make the induction hypothesis that Lemma 2 is true for all polynomials with leading term $b'x^{j'}y^{k'}$ "lexicographically later" than bx^jy^k, i.e., satisfying $j' < j$ or $j' = j$ and $k' < k$. Now

$$h(z) = \frac{b}{(k+1)(k+2)} x^j y^{k+2}$$

is a polynomial of at most degree n, and $\nabla h(z)$, according to Lemma 1 is at most of degree $n - 2$. In addition, it is easy to see that $\nabla h(z)$ has the same leading term as $p(z)$. In the equation

$$\nabla g(z) = p(z) - \nabla h(z)$$

the right-hand side is therefore a polynomial of degree at most $n - 2$ which is either identically zero, or has a leading term which is lexicographically later than that of $p(z)$. Using the induction hypothesis, the above equation has as a solution a polynomial $g(z)$ of degree n or less. But if $g(z)$ is such a solution, then

$$\nabla[g + h](z) = p(z),$$

so that $q(z) = g(z) + h(z)$ is a solution of $\nabla q = p$, which is of degree n or less. That proves Lemma 2, since the induction can be started by verifying the induction hypothesis for constants or linear functions.

To get the conclusion (1) let us now look at the collection of all harmonic polynomials of degree n or less as a vector space V_n over the real numbers. We don't know its dimension; in fact, what we want to show is that

$$\dim V_n = 1 + \underbrace{2 + \cdots + 2}_{n \text{ times}} = 2n + 1.$$

(if $n = 0$, V_0 consists of the constant function only, so that "1" forms a basis; V_1 has a basis consisting of the three polynomials 1, x, and y). To get a proof for arbitrary $n \geq 0$ we consider the larger vector space $W_n \supset V_n$ of *all* (not necessarily harmonic) polynomials of degree less than or equal to n. Thus W_1 has the same basis as V_1, but W_2 has the basis consisting of $1, x, y, x^2, y^2$, and xy. It is easy to see that

$$\dim W_n = \frac{(n+1)(n+2)}{2}.$$

According to Lemma 1 the linear operator ∇ maps W_n *into* W_{n-2}, and according to Lemma 2 the mapping is *onto*. But

$$V_n = [p(z) \mid p(z) \in W_n, \nabla p(z) = 0]$$

and by a familiar result from linear algebra, the dimension of the null space (also called the rank) of the transformation ∇ of W_n onto W_{n-2} is

$$\dim V_n = \dim W_n - \dim W_{n-2} = \frac{(n+1)(n+2)}{2} - \frac{(n-1)n}{2} = 2n+1.$$

Hence (1) is proved.

For each n one can exhibit a basis by calculation. Thus we get, for

$n = 0$: $\quad 1$
$n = 1$: $\quad x, y$
$n = 2$: $\quad x^2 - y^2, xy,$
$n = 3$: $\quad x^3 - 3xy^2, 3x^2y - y^3$. So far (2) is correct, *but*, going to degree
$n = 4$: $\quad x^4 - 6x^2y^2 + y^4 - x^2 - y^2, x^3y - xy^3$.

Only one of these is homogeneous of degree 4 as required by (2). Finally, increasing n to

$n = 5$: $\quad x^5 - 10x^3y^2 + 5xy^4 - 10xy^2, 5x^4y - 10x^2y^3 + y^5 - 10x^2y.$

It is therefore impossible, even by forming linear combinations, to produce a single homogeneous harmonic polynomial of degree 5.

After this excursion we return to regular (non-negative harmonic) functions and derive the answer to problem A as stated following D1.

P1 *If P is the transition function of an aperiodic recurrent random walk, then the only non-negative functions regular relative to P are the constant functions.*

Proof:[5] Suppose that $f(x)$ is regular and not constant. If $f(0) = \alpha$ then there is a point z in R where $f(z) = \beta \neq \alpha$. We may obviously assume that $\beta > \alpha$, for otherwise we make a translation such that z becomes the origin, and again obtain the situation $\beta > \alpha$.

We define the random walk as

$$\mathbf{x}_n = x + \mathbf{S}_n = x + \mathbf{X}_1 + \cdots + \mathbf{X}_n,$$

starting at the point x. Then $\mathbf{T} = \min [k \mid k \geq 1; \mathbf{x}_k = z]$ is a stopping time, and since the random walk is aperiodic and recurrent, $\mathbf{T} < \infty$ with probability one.

---

[5] Suggested by J. Wolfowitz. For an alternative proof, see problem 1 in Chapter VI.

Since f is a regular function
$$\mathbf{E}_0[f(\mathbf{x}_n)] = \sum_{x \in R} P_n(0,x) f(x) = f(0) = \alpha$$
for every $n \geq 0$. Similarly
$$\mathbf{E}_x[f(\mathbf{x}_n)] = f(x) \text{ for all } x \text{ in } R.$$
Therefore
$$\alpha = \mathbf{E}_0[f(\mathbf{x}_n)] \geq \mathbf{E}_0[f(\mathbf{x}_n); \mathbf{T} \leq n]$$
$$= \sum_{k=1}^{n} \mathbf{E}_0[f(\mathbf{x}_n); \mathbf{T} = k]$$
$$= \sum_{k=1}^{n} \mathbf{P}_0[\mathbf{T} = k] \mathbf{E}_z[f(\mathbf{x}_{n-k}) = \sum_{k=1}^{n} \mathbf{P}_0[\mathbf{T} = k] f(z)$$
$$= \beta \mathbf{P}_0[\mathbf{T} \leq n],$$
for each $n \geq 1$.

Hence
$$\alpha \geq \beta \lim_{n \to \infty} \mathbf{P}_0[\mathbf{T} \leq n] = \beta,$$
and this contradicts the assumption that $\beta > \alpha$.

In Chapter VI we will encounter several other attacks on this problem—in particular an elementary proof of T24.1 to the effect that the only *bounded* regular functions for any aperiodic random walk are the constant functions. This weaker result holds in the transient case as well.

The reader is now asked to agree that problems (i) and (ii) raised in section 11 have been completely answered. Problem (i) was disposed of through P12.1 and (ii) by P1 above. *Therefore we have now completed the proof of every statement in section 11, in particular of the important* T11.1. Now we are free to use T11.1 in proceeding further. We shall actually do so in connection with problem C, a little later on.

Problem B. The classical problem in question is the *Exterior Dirichlet Problem*. Given a bounded domain (simply connected open set) in the plane whose boundary ∂D is a sufficiently regular simple closed curve we seek a function $u(x)$ such that
 (i) $\Delta u = 0$ on the complement of $D + \partial D$
 (ii) $u(x) = \varphi(x)$ on ∂D
 (iii) $u(x)$ is continuous and bounded on the complement of D.

This problem is known to have a unique solution.[6] This is also the case in the discrete problem as we shall show in P2. The particularly simple representation of the solution in P2 below has the following classical counterpart. Associated with the domain D is an exterior Green function $g_D(x,y)$ defined for x and y outside D. It is a harmonic function in each variable, except at the point $x = y$, where it has a logarithmic singularity. When y is a point on the boundary we call the exterior normal derivative of $g_D(x,y)$

$$H_D(x,y) = \frac{\partial}{\partial n} g_D(x,y), \qquad x \in R - (D + \partial D), \quad y \in \partial D.$$

Then the solution of the exterior Dirichlet problem has the representation as the line integral

$$u(x) = \int_{\partial D} H_D(x,y)\varphi(y)\,dy.$$

As the notation was intended to suggest, H_D is the correct continuous analogue of the *hitting probability measure* $H_D(x,y)$ defined in D10.1, and the Green function g_D is also the counterpart of g_D in D10.1. (Unfortunately one has to look quite hard for the discrete analogue of the above relation between H_D and g_D by means of a normal derivative. It can be shown to be the very unassuming identity (c) in P10.1.)

Here is now the random walk counterpart of problem B. The dimension of the random walk again is irrelevant to the conclusion (the reason being that we treat only recurrent random walk, so that $d \leq 2$ by T8.1).

P2 *If P is the transition function of an aperiodic recurrent random walk, then the exterior Dirichlet problem, defined by*

(i) $Pf(x) = f(x)$ *for x in $R - B$,*
(ii) $f(x) = \varphi(x)$ *for x in B,*
(iii) $f(x)$ *bounded for all x in R,*

has the unique solution

$$f(x) = \sum_{y \in B} H_B(x,y)\varphi(y) \qquad x \in R.$$

Here B is a proper subset of R and if $|B| = \infty$, $\varphi(x)$ is of course assumed to be bounded.

---

[6] For a concise treatment of the two-dimensional case, see [79], Chapter V. The result is false when in dimension $d \geq 3$.

Proof: First we check that the solution given in P2 really satisfies (i), (ii), and (iii). First of all the series representing $f(x)$ converges, since $H_B(x,y)$ is a probability measure on B for each x in R. Condition (i) is satisfied in view of part (a) of P10.1, (ii) is true since $H_B(x,y) = \delta(x,y)$ for x and y both in B, and (iii) is also true since H_B is a probability measure, so that $f(x) \le \sup_{y \in B} |\varphi(y)|$.

To prove uniqueness, suppose that f_1 and f_2 are two solutions and let $h = f_1 - f_2$. Then h is a solution of the exterior Dirichlet problem with boundary value $\varphi(x) \equiv 0$ on B. If we define $Q(x,y)$ and its iterates $Q_n(x,y)$ for x and y in $R - B$ according to D10.1 (Q is simply P restricted to $R - B$), then

$$h(x) = \sum_{y \in R-B} Q_n(x,y) h(y) \text{ for } x \in R - B.$$

Now $h(x)$ is bounded, being the difference of two bounded functions, so if $|h(x)| \le M$ on R, then

$$|h(x)| \le M \sum_{y \in R-B} Q_n(x,y) = M\mathbf{P}_x[\mathbf{T}_B > n]$$

for every x in $R - B$ and every $n \ge 0$. As the random walk is recurrent and aperiodic

$$\lim_{n \to \infty} \mathbf{P}_x[\mathbf{T}_B > n] \le \lim_{n \to \infty} \mathbf{P}_x[\mathbf{T}_{\{y\}} > n] = 0,$$

where y is an arbitrary point in B. Therefore $h(x) \equiv 0$, which proves P2.

Problem C. This problem concerns *Poisson's equation*. On the domain D in problem B, or rather on its compact closure $\bar{D} = D + \partial D$ a non-negative function $\rho(x)$ is given. The problem is to find a function $f(x)$ on $R - \bar{D}$ such that
   (i) $\Delta f(x) = 0$ on $R - \bar{D}$
   (ii) $\Delta f(x) = -\rho(x)$ on D,
and finally a third condition is needed in the setup of logarithmic potential theory to ensure uniqueness. A very strong but convenient one is

   (iii) $f(x) + \dfrac{1}{2\pi}\left[\displaystyle\int_D \rho(y)\,dy\right] \ln|x| \to 0$ as $|x| \to \infty$.

In the physical interpretation of this problem $\rho(x)$ is a given charge density on the body $\bar{D}$, and $f(x)$ is the potential due to this charge in the plane surrounding $\bar{D}$. The volume integral in condition (iii)

13. SOME POTENTIAL THEORY

represents the total charge. It is well known [79] that this problem has a unique solution, given by

$$f(x) = \frac{1}{2\pi} \int_D \rho(y) \ln |x - y|^{-1} \, dy, \qquad x \in R.$$

The solution as given here turns out to be negative for large $|x|$. In the discrete case it will be more convenient to have it positive, in particular since there is no counterpart in the discrete theory for the singularity of $\ln |x - y|^{-1}$ at $x = y$. So there will be a change of sign in the discrete formulation of problem C in condition (ii). That having been done, as the reader may have guessed, we shall find that the kernel $\ln |x - y|$ corresponds to $A(x,y)$ in the discrete theory. (The asymptotic evaluation of $A(x,y)$ in P12.3 is highly suggestive in this direction. Quite naturally it is the simple random walk in the plane which should most closely approximate the continuous analogue, because the second difference operator gives in a sense a better approximation to Laplace's operator than does any other difference operator. That is the reason why $A(x,y)$ was called the *potential kernel* in P12.2.)

Our discrete version of problem C will be further modified by replacing (iii) by the condition that the solution of Poisson's equation be non-negative.

D2 *When $P(x,y)$ is the transition function of aperiodic recurrent two-dimensional random walk, and $B \subset R$, $1 \leq |B| < \infty$, we call a non-negative function $\rho(x)$ such that $\rho(x) = 0$ on $R - B$ a charge distribution on B. A function $f(x)$ on R is called a potential due to the charge ρ on B if*

(i) $Pf(x) - f(x) = 0$ for $x \in R - B$,
(ii) $Pf(x) - f(x) = \rho(x)$ for $x \in B$,
(iii) $f(x) \geq 0$ for $x \in R$.

Problem C consists in finding the potentials due to a charge ρ on B. A partial solution to this problem will be provided in

P3 *The potential kernel $A(x,y) = a(x - y)$ of a recurrent aperiodic random walk of dimension $d = 2$ has the property that*

(a) $$\sum_{y \in R} P(x,y)a(y) - a(x) = \delta(x,0), \qquad x \in R.$$

A solution of problem C is therefore given by

(b) $$f(x) = \sum_{y \in B} A(x,y)\rho(y), \qquad x \in R.$$

Proof: Since the set B is finite, equation (b) is an immediate consequence of (a). The proof of (a) happens to be quite lengthy in general, but a very quick proof is available for symmetric random walk (see problem 3). Going back to fundamentals

$$a_n(x) = \sum_{k=0}^{n} [P_k(0,0) - P_k(x,0)], \qquad x \in R,$$

and one finds

$$\sum_{y \in R} P(x,y) a_n(y) = \sum_{k=0}^{n} [P_k(0,0) - P_{k+1}(x,0)]$$
$$= a_n(x) + \delta(x,0) - P_{n+1}(x,0).$$

Hence, using for instance P7.6

$$\lim_{n \to \infty} \sum_{y \in R} P(x,y) a_n(y) = a(x) + \delta(x,0).$$

Unfortunately this does not enable us to conclude, as we would like, that one can interchange limits to obtain P3 right away. However, one can conclude by an obvious truncation argument (Fatou's lemma for sums instead of integrals) that

$$\sum_{y \in R} P(x,y) a(y) \leq a(x) + \delta(x,0)$$

for all x in R. Calling the difference $f(x)$ we may write

$$\sum_{y \in R} P(x,y) a(y) = a(x) + \delta(x,0) - f(x),$$

where $f(x) \geq 0$ on R.

Now we resort to a complicated probabilistic argument, based on P11.4. It will yield the useful conclusion that $f(x)$ is constant (independent of x). We shall need only the special case of P11.4 where $B = \{0,b\}$, $b \neq 0$. Abbreviating the notation to

$$H(x) = H_B(x,0), \qquad \mu = \mu_B(0), \qquad \Pi = \Pi_B(b,0),$$

one obtains, setting $y = 0$ in P11.4,

$$H(x) = \mu - [a(x) - a(x-b)] \Pi.$$

Now the transition operator $P(x,y)$ is applied to the last equation on the left, i.e., x is replaced by t, we multiply the equation through by $P(x,t)$ and sum over t. Using part (a) of P10.1 for the left-hand side, and our conclusion that $Pa - a = \delta - f$ for the right-hand side, one gets

$H(x) + \Pi_B(x,0) - \delta(x,0)$
$= \mu - \Pi[a(x) + \delta(x,0) - f(x) - a(x - b) - \delta(x,+b) + f(x - b)].$

Here $\Pi_B(x,0)$ should be interpreted as zero when x is not in B. In the above equation we now replace $H(x)$ by $\mu - [a(x) - a(x - b)]\Pi$, and then set $x = 0$. The result turns out to be

$$\Pi = \Pi[1 - f(0) + f(-b)].$$

Since Π is not zero, $f(0) = f(-b)$, and since b is an arbitrary point in $R - \{0\}$, $f(x)$ is independent of x!

Now call $f(x) = f$ and return to the equation

$$\sum_{y \in R} P(x,y)a(y) = a(x) + \delta(x,0) - f.$$

Operating on this equation by the transition operator P on the left, and repeating this process $n - 1$ times gives $n + 1$ equations (counting the original equation). When these $n + 1$ equations are added some cancellation takes place, and the final result is

$$\sum_{y \in R} P_{n+1}(x,y)a(y) = a(x) + G_n(x,0) - nf.$$

This calls for division by n, and when we let $n \to \infty$, noting that

$$\lim_{n \to \infty} \frac{a(x)}{n} = \lim_{n \to \infty} \frac{G_n(x,0)}{n} = 0,$$

we obtain

$$0 \le \lim_{n \to \infty} \frac{1}{n} \sum_{y \in R} P_{n+1}(x,y)a(y) = -f.$$

Hence $f \le 0$, but since we know that $f \ge 0$, we conclude that $f = 0$, completing the proof of P3.

Remark: The proof of P11.4 goes through for any recurrent aperiodic random walk with the property that

$$a(x) = \lim_{n \to \infty} a_n(x) = \lim_{n \to \infty} [G_n(0,0) - G_n(x,0)]$$

exists. And the proof we just gave of P3 used only P11.4. Hence the equation

$$\sum_{y \in R} P(x,y)a(y) - a(x) = \delta(x,0), \qquad x \in R,$$

holds for every aperiodic recurrent random walk which has a potential kernel $A(x,y)$ in the sense that $\lim_{n\to\infty} a_n(x) = a(x)$ exists.

In Chapter VII we shall find that *the potential kernel $A(x,y)$ does exist for every aperiodic recurrent random walk*. In addition the result of P3 will be strengthened considerably. The stronger version of P3 for aperiodic two-dimensional recurrent random walk will read:

If $Pf - f = 0$ on $R - B$, $Pf - f = \rho \geq 0$ on B, and $f \geq 0$ on R then

$$f(x) = \text{constant} + \sum_{t \in B} A(x,t)\rho(t), \qquad x \in R.$$

In particular the equation

$$\sum_{y \in R} P(x,y)g(y) - g(x) = \delta(x,0)$$

has only the non-negative solutions $g(x) = \text{constant} + a(x)$. To illustrate why the one-dimensional case is going to offer greater difficulties, it is sufficient to consider simple random walk in one dimension with $P(0,1) = P(0,-1) = 1/2$. This process is recurrent, and it can easily be shown (see E29.1) that

$$a(x) = \lim_{n \to \infty} a_n(x) = \frac{1}{2\pi} \int_{-\pi}^{\pi} \frac{1 - \cos x\theta}{1 - \cos \theta} d\theta = |x|.$$

However there are now other non-negative solutions of $Pg - g = \delta$, besides $g(x) = a(x) = |x|$. In fact, it is easy to check that the most general non-negative solution is

$$a + |x| + bx \quad \text{with} \quad a \geq 0, \quad -1 \leq b \leq 1.$$

14. THE GREEN FUNCTION OF A FINITE SET

This section will complete the potential theory of two-dimensional recurrent random walk. We resume where we left off in section 11, namely with P11.4 or T.11.1:

(1) $\quad H_B(x,y)$
$$= \mu_B(y) + \sum_{t \in B} A(x,t)[\Pi_B(t,y) - \delta(t,y)], \qquad x \in R, \quad y \in B,$$

where $2 \leq |B| < \infty$. When $|B| = 1$, the situation is of course uninteresting since $H_B(x,y) = 1$.

In addition to the results which were available when equation (1) was derived, we now have P12.2 which says that for fixed y_1 and y_2 in R,

(2) $$\lim_{|x| \to \infty} [A(x,y_1) - A(x,y_2)] = 0.$$

By combining (1) and (2) we shall prove in a few lines that the hitting probabilities $H_B(x,y)$ have the following remarkable property. *As $|x| \to \infty$, $H_B(x,y)$ tends to a limit for each y in B.* We shall call this limit $H_B(\infty,y)$ and observe that $H_B(\infty,y)$ is again, just as is each $H_B(x,y)$, a probability measure on the set B. In other words,

$$H_B(\infty,y) \geq 0 \quad \text{and} \quad \sum_{y \in B} H_B(\infty,y) = 1.$$

The intuitive probability meaning of $H_B(\infty,y)$ is quite obvious. It is still the probability of first hitting the set B at the point y, the initial condition being that the random walk starts "at infinity." The existence of the limit of $H_B(x,y)$ as $|x| \to \infty$ thus implies that the phrase "at infinity" has an unambiguous meaning at least for certain purposes. The direction of x (say the angle between the segment 0 to x and the horizontal axis) has no influence on the value of $H_B(x,y)$ when $|x|$ is very large. In anticipation of certain results in the one-dimensional case in Chapter VII (where the analogous theorem may be false) the following explanation for the existence of the limit is particularly appealing. The space R of dimension 2 is rather "large." The random walk starting at x with $|x|$ large will traverse a path which with high probability winds around the set B many times before hitting it. Its final approach to B will therefore be from a direction which is stochastically independent of the direction of x. A still more picturesque way of saying the same thing is that the particle is almost sure to have forgotten which direction it originally came from—provided it came from sufficiently far away. Thus we shall prove

T1 *For aperiodic recurrent two-dimensional random walk, with $1 \leq |B| < \infty$ the limit*

$$H_B(\infty,y) = \lim_{|x| \to \infty} H_B(x,y), \quad y \in B$$

exists and determines a probability measure on B. If $|B| = 1$, $H_B(\infty,y) = 1$ and if $|B| \geq 2$, then

$$H_B(\infty,y) = \frac{K_B(\cdot y)}{K_B(\cdot \cdot)} = \mu_B(y).$$

K_B is the inverse of the potential kernel $A(x,y)$ restricted to B, as defined in D11.3 and D11.4.

Proof: When $|B| = 1$, there is nothing to prove. When $|B| \geq 2$, we consult equation (1). The function $\mu_B(y)$ in T11.1 is precisely the desired limit. Therefore we have to show that the sum in equation (1) tends to zero as $|x| \to \infty$. Now we fix y, and call

$$c(t) = \Pi_B(t,y) - \delta(t,y), \qquad t \in B.$$

In view of P11.2, $c(t)$ has the property that $\sum_{t \in B} c(t) = 0$. Therefore we may express

$$\sum_{t \in B} A(x,t)c(t) = \sum_{t \in B} [A(x,t) - A(x,0)]c(t)$$

as a finite sum of terms which according to equation (2) tend to zero as $|x| \to \infty$. This observation yields

$$\lim_{|x| \to \infty} \sum_{t \in B} A(x,t)[\Pi_B(t,y) - \delta(t,y)] = 0$$

and completes the proof of the theorem.

E1 Consider *symmetric aperiodic recurrent two-dimensional random walk*, i.e., we are adding the requirement $P(x,y) = P(y,x)$. When $|B| = 2$, what about $H_B(\infty, y)$? We saw that $H_B(\infty, y) = \mu_B(y)$ in equation (1) above and in P11.5 $\mu_B(y)$ was calculated: If $B = \{0,b\}$

$$\mu_B(0) = \frac{a(b)}{a(b) + a(-b)}, \qquad \mu_B(b) = \frac{a(-b)}{a(b) + a(-b)}.$$

But symmetric random walk has the property that $a(x) = a(-x)$ for all x in R (from the definition of $a(x)$ as the limit of $a_n(x)$). Therefore we have

$$H_B(\infty, y) = 1/2 \text{ at each of the two points of } B.$$

Next we turn to the Green function $g_B(x,y)$ of the finite set B, defined in D10.1. When the set B consists of a single point, $g_B(x,y)$ was calculated explicitly in P11.6. This leaves only the case $|B| \geq 2$, for which we shall derive

14. THE GREEN FUNCTION OF A FINITE SET

T2 *For aperiodic recurrent two-dimensional random walk, and* $2 \le |B| < \infty$

$$g_B(x,y) = -A(x,y) - \frac{1}{K_B(\cdot\cdot)} + \sum_{s\in B} \mu(s)A(s,y) + \sum_{t\in B} A(x,t)\mu^*(t)$$
$$+ \sum_{s\in B}\sum_{t\in B} A(x,t)[\Pi_B(t,s) - \delta(t,s)]A(s,y), \quad x,y \in R.$$

The matrix Π_B *is given by* T11.1 *and so are*

$$\mu(s) = \mu_B(s) = \frac{K_B(\cdot s)}{K_B(\cdot\cdot)},$$

$$\mu^*(t) = \mu_B^*(t) = \frac{K_B(t\cdot)}{K_B(\cdot\cdot)}.$$

Proof: First we check that the formula in T2 gives $g_B(x,y) = 0$ when either x or y or both are in B. Supposing that x is in B,

$$\sum_{t\in B} A(x,t)[\Pi_B(t,s) - \delta(t,s)] = H_B(x,s) - \mu_B(s).$$

The right-hand side in T2 therefore becomes

$$-A(x,y) - \frac{1}{K_B(\cdot\cdot)} + \sum_{s\in B} \mu(s)A(s,y) + \sum_{t\in B} A(x,t)\mu^*(t)$$
$$+ \sum_{s\in B} H_B(x,s)A(s,y) - \sum_{s\in B} \mu(s)A(s,y)$$
$$= -A(x,y) - \frac{1}{K_B(\cdot\cdot)} + \sum_{t\in B} A(x,t)\mu^*(t) + \sum_{s\in B} \delta(x,s)A(s,y)$$
$$= -\frac{1}{K_B(\cdot\cdot)} + \sum_{t\in B} A(x,t)\mu^*(t) = 0.$$

When y is in B the verification is the same, using in this case

$$\sum_{s\in B} [\Pi_B(t,s) - \delta(t,s)]A(s,y) = H_B^*(y,t) - \mu_B^*(t) = \delta(y,t) - \mu_B^*(t).$$

Here $H_B^*(y,s)$ is the hitting probability measure of the reversed random walk (D10.2) with $P^*(x,y) = P(y,x)$. That is clearly the correct result in view of

$$\Pi_B^*(x,y) = \Pi_B(y,x), \qquad A^*(x,y) = A(y,x)$$

and

$$\sum_{s\in B} [\Pi_B(t,s) - \delta(t,s)]A(s,y) = \sum_{s\in B} A^*(y,s)[\Pi_B^*(s,t) - \delta(s,t)]$$
$$= H_B^*(y,t) - \mu_B^*(t).$$

For the rest of the proof we may assume that neither x nor y is in B. We shall use a method which is often useful in classical potential theory, i.e., we find a "partial differential equation" satisfied by $g_B(x,y)$ and solve it. The theory of the Poisson equation in P13.3 will turn out to provide the appropriate existence theorem.

The equation we start with is

(1) $$H_B(x,y) - \delta(x,y) = \sum_{t \in R} g_B(x,t) P(t,y) - g_B(x,y),$$

for all x,y in R. This is an obvious extension of part (c) of P10.1, to the full product space of all pairs (x,y) in $R \times R$. Here, and in the rest of the proof, we define $H_B(x,y) = 0$ for $x \in R$, $y \in R - B$.

Now we fix a value of x in $R - B$ and let

(2) $$u(t) = g_B(x,t) + A(x,t), \qquad t \in R.$$

Substitution into equation (1), using P13.3, gives

(3) $$\sum_{t \in R} u(t) P(t,y) - u(y) = H_B(x,y), \qquad y \in R.$$

The next step consists in solving (3) and we shall show that every solution of (3) is of the form

(4) $$u(y) = \text{constant} + \sum_{t \in B} H_B(x,t) A(t,y), \qquad y \in R,$$

where of course the constant may depend on x.

First of all we rewrite equation (3) in the form

(5) $$\sum_{t \in R} P^*(y,t) u(t) - u(y) = 0 \text{ for } y \in R - B$$
$$= \rho(y) \geq 0 \text{ for } y \in B,$$

where $\rho(y) = H_B(x,y)$. Thus $u(t)$ is a solution of Poisson's equation discussed in D13.2, and by P13.3 equation (4) indeed is a solution of (3). But since P13.3 did not guarantee uniqueness we have to resort to other devices to show that (4) holds.

From equation (2) we have about $u(t)$ the a priori information that

$$A(x,t) \leq u(t) \leq A(x,t) + M(x),$$

where

$$M(x) = \sup_{t \in R} g_B(x,t).$$

14. THE GREEN FUNCTION OF A FINITE SET

Introducing the reversed random walk, we have

$$M(x) = \sup_{t \in R} g_B^*(t,x) = \sup_{t \in R} H_{B \cup \{x\}}^*(t,x) g_B^*(x,x)$$

$$\leq g_B^*(x,x) = g_B(x,x) < \infty.$$

Thus the function $u(t)$ has the property that $u(t) - A(x,t)$ is bounded, and, in view of P12.2,[7] this is the same as saying that $u(t) - a(-t)$ is a bounded function on R.

Now let

$$w(y) = \sum_{t \in B} H_B(x,t) A(t,y).$$

Then

$$h(y) = u(y) - w(y) = u(y) - a(-y) + \sum_{t \in B} H_B(x,t)[A(0,y) - A(t,y)]$$

is a bounded function of y (since $A(0,y) - A(t,y)$ is bounded for each t according to P12.2). Furthermore $u(y)$ satisfies (5), so that

$$\sum_{s \in R} P^*(y,s) h(s) = h(y) \quad \text{for } y \in R.$$

By P13.1, $h(y) = c(x)$, a constant depending on x of course—and therefore we have proved that $u(y) = c(x) + w(y)$ is given by equation (4).

The rest is easy. We now have

$$g_B(x,y) = -A(x,y) + c(x) + \sum_{t \in B} H_B(x,t) A(t,y)$$

$$= -A(x,y) + c(x) + \sum_{s \in B} \mu(s) A(s,y)$$

$$+ \sum_{s \in B} \sum_{t \in B} A(x,s)[\Pi_B(s,t) - \delta(s,t)] A(t,y).$$

To complete the proof of T2, we now have only to show that

$$c(x) = -\frac{1}{K_B(\cdot\cdot)} + \sum_{t \in B} A(x,t) \mu^*(t),$$

but this was done at the very start of this proof, when we verified that T2 was correct whenever y is in B.

---

[7] Observe that we are not using the full strength of P12.2 which states that $A(x,t) - A(0,t) \to 0$ as $|t| \to \infty$. This seemingly academic point will be crucial in section 30, T30.2, where we shall imitate the present proof for one-dimensional random walk. Even then $A(x,t)$ exists and $A(x,t) - A(0,t)$ is a bounded function of t, but it may not tend to zero as $|t| \to \infty$. We shall have occasion to refer to this remark in section 30.

The result of T2 can be used to develop further the theory along the lines of classical logarithmic potential theory. There are no great surprises in store as the entire development exactly parallels the classical theory. We shall outline the principal facts—delegating the proofs to the problem section at the end of the chapter.

We fix a set B, with cardinality $1 \leq |B| < \infty$ and omit the subscript B in H_B, μ_B, K_B, Π_B, g_B, as this will cause no confusion at all. Remember that a *charge* on B is a non-negative function vanishing off the set B. And let us define the *potential due to the charge* ψ on B as

$$(1) \qquad A\psi(x) = \sum_{t \in B} A(x,t)\psi(t), \qquad x \in R.$$

It may be shown that potentials satisfy the *minimum principle*; i.e., every potential $A\psi(x)$ assumes its minimum on the set B, or rather on the subset of B where the charge ψ is positive (for a proof see T31.2).

The minimum principle can be shown to imply that among all charges ψ on B with total charge $\sum_{x \in B} \psi(x) = 1$, there is exactly one with the property that its potential is constant on B. This particular unit charge is called the *equilibrium charge* $\mu^*(x)$ of the set B. It is easily shown that

$$(2) \qquad \mu^*(x) = 1 \text{ if } |B| = 1$$

$$\mu^*(x) = \frac{K(x \cdot)}{K(\cdot \cdot)} \text{ when } |B| \geq 2,$$

and of course,

$$(3) \qquad A\mu^*(x) = \frac{1}{K(\cdot \cdot)} \text{ when } x \in B$$

$$\geq \frac{1}{K(\cdot \cdot)} \text{ when } x \in R - B.$$

To make the last and subsequent statements meaningful when $|B| = 1$, we define

$$(4) \qquad K(\cdot \cdot) = \infty, \qquad \frac{1}{K(\cdot \cdot)} = 0 \text{ when } |B| = 1.$$

The potential $A\mu^*$ is called the *equilibrium potential* of the set B. The constant $[K(\cdot \cdot)]^{-1}$, the "boundary value" of the equilibrium potential on B, is called the *capacity*

$$(5) \qquad C(B) = \frac{1}{K(\cdot \cdot)}$$

of the set B.

There are two other equivalent definitions of capacity. It is easily shown that there is one and only one constant C such that for every unit charge ψ on B

(6) $$\min_{x \in B} A\psi(x) \leq C \leq \max_{x \in B} A\psi(x).$$

This constant C is defined to be the capacity of B, and it is true that $C = C(B)$.

Another useful definition of capacity is in terms of the Green function $g(x,y)$ of the set B. It is

(7) $$C' = \sum_{t \in B} A(x,t)\mu^*(t) - \lim_{|y| \to \infty} g(x,y).$$

It will presently be shown in T3 below that

$$\lim_{|y| \to \infty} g(x,y) = g(x,\infty)$$

exists for all x in R, that C' is indeed independent of x, and finally that $C' = [K(\cdot\cdot)]^{-1}$. This then is a third definition of capacity. Of all the three definitions it is the most useful one in uncovering the properties of $C(B)$ as a set function on the finite subsets of R. The two most important laws governing $C(B)$ are:

(8) *If $B_1 \subset B_2 \subset R$, then $C(B_1) \leq C(B_2)$.*

(9) *If B_1 and B_2 have a nonempty intersection, then*

$$C(B_1 \cup B_2) + C(B_1 \cap B_2) \leq C(B_1) + C(B_2).$$

The reader who investigates the analogy with classical logarithmic theory in [79] will find the name of capacity for $C(B)$ to be a misnomer. $C(B)$ is really the discrete analogue of the so-called *Robin's constant*, which is the logarithm of the logarithmic capacity.

Because of its importance in the study of time dependent phenomena in section 16, we single out the asymptotic behavior of $g_B(x,y)$ for further study.

T3 *For recurrent aperiodic two-dimensional random walk*

$$\lim_{|y| \to \infty} g_B(x,y) = g_B(x,\infty), \quad x \in R$$

exists for subsets $B \subset R$ with $1 \leq |B| < \infty$. When $B = \{b\}$,

$$g_B(x,\infty) = a(x - b).$$

When $|B| \geq 2$,

$$g_B(x,\infty) = \sum_{t \in B} A(x,t)\mu_B^*(t) - \frac{1}{K_B(\cdot\cdot)},$$

where $\mu_B^*(t) = \dfrac{K_B(t\cdot)}{K_B(\cdot\cdot)}$, $\quad t \in B$.

Proof: When $|B| = 1$, it suffices to assume $b = 0$. Then

$$g_B(x,y) = A(x,0) + A(0,y) - A(x,y),$$

by the remarkable result of P11.6 in section 11. By P12.2

$$\lim_{|y| \to \infty} g_B(x,y) = a(x) + \lim_{|y| \to \infty} [A(0,y) - A(x,y)] = a(x).$$

The proof when $|B| \geq 2$ is in fact just as straightforward. One simply uses the explicit representation of $g_B(x,y)$ in T2 and then applies P12.2 to it. The verification is left to the reader.

15. SIMPLE RANDOM WALK IN THE PLANE

The potential kernel for simple random walk can be calculated numerically by a very simple iterative scheme. We have

(1) $$a(x) = (2\pi)^{-2} \int \frac{1 - \cos x \cdot \theta}{1 - \phi(\theta)} \, d\theta,$$

where

$$\phi(\theta) = \tfrac{1}{2}[\cos \theta_1 + \cos \theta_2],$$

but it will turn out to be unnecessary to evaluate this integral except for points x on the diagonal $x^1 = x^2$. Using P13.3 one has, first of all,

(2) $\quad \tfrac{1}{4}[a(z + 1) + a(z - 1) + a(z + i) + a(z - i)] - a(z) = \delta(z,0),$

using complex notation for the sake of convenience. The obvious symmetry properties of the integral in (1) show that, if $z = x + iy$, $\bar{z} = x - iy$,

$$a(z) = a(-z) = a(\bar{z}) = a(-\bar{z}) = a(iz) = a(-iz) = a(i\bar{z}) = a(-i\bar{z}).$$

Hence it suffices to calculate $a(z)$ for z in the half-quadrant $0 \leq y \leq x$. Setting $z = 0$ in (2) gives, since $a(0) = 0$,

$$a(1) = a(-1) = a(i) = a(-i) = 1.$$

15. SIMPLE RANDOM WALK IN THE PLANE

Next we evaluate the integral in (1) for $z = n(1 + i)$, $n \geq 1$.

$$a(n + ni) = \frac{1}{4\pi^2} \int_{-\pi}^{\pi} \int_{-\pi}^{\pi} \frac{1 - \cos n(\theta_1 + \theta_2)}{1 - \frac{1}{2}(\cos \theta_1 + \cos \theta_2)} d\theta_1 d\theta_2$$

$$= \frac{1}{4\pi^2} \int_{-\pi}^{\pi} \int_{-\pi}^{\pi} \frac{1 - \cos n(\theta_1 + \theta_2)}{1 - \cos\left(\frac{\theta_1 + \theta_2}{2}\right) \cos\left(\frac{\theta_1 - \theta_2}{2}\right)} d\theta_1 d\theta_2.$$

The transformation $\alpha = (\theta_1 + \theta_2)/2$, $\beta = (\theta_1 - \theta_2)/2$ has Jacobian $\frac{1}{2}$ and transforms the region of integration into the set $|\alpha| + |\beta| \leq \pi$. Obvious symmetry properties of the integrand enable us to write

$$a(n + ni) = \frac{1}{4\pi^2} \int_{-\pi}^{\pi} \int_{-\pi}^{\pi} \frac{1 - \cos 2n\alpha}{1 - \cos \alpha \cos \beta} d\alpha\, d\beta$$

$$= \frac{2}{\pi} \int_{-\pi}^{\pi} \frac{1 - \cos 2n\alpha}{|\sin \alpha|} d\alpha = \frac{1}{\pi} \int_{-\pi}^{\pi} \frac{\sin^2 n\alpha}{|\sin \alpha|} d\alpha$$

$$= \frac{2}{\pi} \int_0^{\pi} \left[\sum_{k=1}^{n} \sin(2k - 1)\alpha\right] d\alpha$$

$$= \frac{4}{\pi} \left[1 + \frac{1}{3} + \frac{1}{5} + \cdots + \frac{1}{2n - 1}\right].$$

The values of $a(z)$ computed so far suffice, by proper use of (2), to construct a table for arbitrary values of z. Here is a sample.

	$y = 0$	1	2	3	4
$x = 0$	0				
1	1	$\dfrac{4}{\pi}$			
2	$4 - \dfrac{8}{\pi}$	$-1 + \dfrac{8}{\pi}$	$\dfrac{16}{3\pi}$		
3	$17 - \dfrac{48}{\pi}$	$\dfrac{92}{3\pi} - 8$	$\dfrac{8}{3\pi} + 1$	$\dfrac{92}{15\pi}$	

The first table of this kind was constructed in 1940 by McCrea and Whipple [76]. As they pointed out, once one row is filled in it is very easy to obtain the entries in the next one by the following iterative scheme.

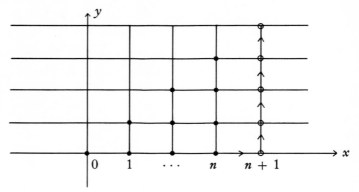

Suppose the numbers corresponding to the black dots are known, i.e., we know the values of $a(k + im)$ for $0 \le m \le k \le n$. Then one can get $a(n + 1)$ since $a(n)$ is the average of $a(n + 1)$, $a(n - 1)$, $a(n + i)$, and $a(n - i) = a(n + i)$. Next $a(n + 1 + i)$ is found, $n + 1 + i$ being the only neighbor of $n + i$ where the value of $a(z)$ is unknown. In this way the values of $a(z)$ in the $(n + 1)$st column indicated above by circles, can be determined in the order indicated by the arrows.

As $|z| \to \infty$ along the diagonal $z = n(1 + i)$ one obtains

$$\lim_{|z| \to \infty} \left[a(z) - \frac{2}{\pi} \ln |z| \right]$$

$$= \lim_{n \to \infty} \left[\frac{4}{\pi} \left(1 + \frac{1}{3} + \cdots + \frac{1}{2n - 1} \right) - \frac{2}{\pi} \ln (n\sqrt{2}) \right]$$

$$= \lim_{n \to \infty} \left[\frac{4}{\pi} \sum_{k=1}^{2n} \frac{1}{k} - \frac{4}{\pi} \ln 2n \right] - \lim_{n \to \infty} \left[\frac{2}{\pi} \sum_{k=1}^{n} \frac{1}{k} - \frac{2}{\pi} \ln n \right] + \frac{1}{\pi} \ln 8$$

$$= \frac{4}{\pi} \gamma - \frac{2}{\pi} \gamma + \frac{1}{\pi} \ln 8 = \frac{2}{\pi} \gamma + \frac{1}{\pi} \ln 8.$$

But in view of P12.3 this limit must exist as $|z| \to \infty$, quite independently of the direction of z, so that we have confirmed the limit given in P12.3 for simple random walk.

Furthermore, as pointed out by McCrea and Whipple, the approximation of $a(z)$ by

$$\frac{1}{\pi} [2 \ln |z| + \ln 8 + 2\gamma]$$

is remarkably good even for rather small values of z. Part of their table of the approximate values for $0 \leq \text{Re}(z) < \text{Im}(z) \leq 5$ is

Im (z)	0	1	2	3	4	5
Re (z)						
0	---		1.0294	1.4706	1.7288	1.9119	2.0540
1		1.2500	1.5417	1.7623	1.9312	2.0665
2			1.6913	1.8458	1.9829	2.1012
3				1.9494	2.0540	2.1518

which should be compared to the table of $a(z)$.

Im (z)	0	1	2	3	4	5
Re (z)						
0	0	1	1.4535	1.7211	1.9080	2.0516
1		1.2732	1.5465	1.7615	1.9296	2.0650
2			1.6977	1.8488	1.9839	2.1012
3				1.9523	2.0558	2.1528

As an application of the preceding result let us calculate the hitting probabilities $H_B(x,y)$ for some finite set B. By T11.1 and T14.1

$$H_B(x,y) = H_B(\infty,y) + \sum_{t \in B} A(x,t) \left[K_B(t,y) - \frac{K_B(t,\cdot)K_B(\cdot y)}{K_B(\cdot,\cdot)} \right]$$

for all $y \in B$, $x \in R$. Suppose that the largest distance between any two points in B is $m > 0$. It follows then that $K_B(t,y)$ for $t \in B$, $y \in B$ as well as

$$H_B(\infty,y) = \frac{K_B(\cdot y)}{K_B(\cdot,\cdot)}, \quad y \in B,$$

can be expressed in terms of known values of the function $a(x)$ with $|x| \leq m$. Furthermore our remarks about the accuracy of approximation of $a(x)$ by $(\pi)^{-1}[2 \ln |x| + \ln 8 + 2\gamma]$ imply that, when the point x is reasonably far away from the nearest point of B (say at a distance 3 or more), then

$$H_B(\infty,y) + \sum_{t \in B} \frac{2}{\pi} \ln |x - t| \left[K_B(t,y) - \frac{K_B(t,\cdot)K_B(\cdot y)}{K_B(\cdot,\cdot)} \right]$$

will yield an excellent approximation to $H_B(x,y)$. Observe that the term $\ln 8 + 2\gamma$ could be dropped in view of

$$\sum_{t \in B}' \left[K_B(t,y) - \frac{K_B(t \cdot) K_B(\cdot y)}{K_B(\cdot \cdot)} \right] = 0, \qquad y \in B.$$

Finally, when x is *very* far away from the set B, $H_B(x,y)$ will of course be well approximated by $H_B(\infty,y)$. For the sake of specific illustration we let B be the three point set

$$B = \{z_1, z_2, z_3\}, \qquad z_1 = i, \quad z_2 = 0, \quad z_3 = 1.$$

Then the operator $A(x,y)$, restricted to the set B, has the matrix representation

$$A = (A(z_i, z_j)) = \begin{pmatrix} 0 & 1 & \frac{4}{\pi} \\ 1 & 0 & 1 \\ \frac{4}{\pi} & 1 & 0 \end{pmatrix}, \qquad i,j = 1, 2, 3.$$

Its inverse is

$$K_B = \frac{\pi}{8} \begin{pmatrix} -1 & \frac{4}{\pi} & 1 \\ \frac{4}{\pi} & -\left(\frac{4}{\pi}\right)^2 & \frac{4}{\pi} \\ 1 & \frac{4}{\pi} & -1 \end{pmatrix},$$

$$K_B(\cdot y) = \left\{ \frac{1}{2}, 1 - \frac{2}{\pi}, \frac{1}{2} \right\},$$

so that by T14.1, $H_B(\infty, y)$ is the vector

$$\left\{ \frac{\pi}{4(\pi - 1)}, \frac{\pi - 2}{2(\pi - 1)}, \frac{\pi}{4(\pi - 1)} \right\}.$$

The irrationality of the answer reflects its dependence upon the *entire history of the process*, as the probability of any event determined by a finite number of steps (say n steps) of the simple random walk would necessarily have to be rational (in fact an integer times 4^{-n}). There are curious exceptions to this phenomenon, however, as the following example shows.

15. SIMPLE RANDOM WALK IN THE PLANE 153

The hitting probabilities $H_B(\infty, y)$ of the set $B = \{-1 - i, 0, 1 + i\}$ "should be" $\{3/8, 2/8, 3/8\}$ as "shown" by the following "heuristic" argument. The points in B each have four neighbors, not in B, from which they may be reached.

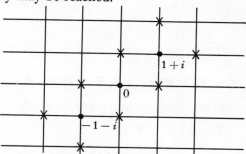

The point $1 + i$ shares two of its neighbors with the origin. Counting each of the shared neighbors as $1/2$, this gives the point $1 + i$ three neighbors, the origin two and the point $-1 - i$ three. Assuming the chance of being hit proportional to the number of neighbors from which this event can take place (and having divided the neighbors up fairly!) the hitting probabilities should be those given above. A short calculation, just like the one carried out before, *confirms* that $\{3/8, 2/8, 3/8\}$ is the correct answer!

Now we embark on some calculations for infinite subsets B of R which make use of a special property of simple random walk: the continuity of its paths, i.e., the fact that $P(x,y) = 0$ unless x and y are neighbors.

Let B be the lower half plane, i.e.,

$$B = [z \mid \operatorname{Im} z \le 0].$$

We define $g_B(x,y)$ as usual, to wit $g_B(x,y) = 0$ unless both x and y are in the upper half plane $R - B$, and in this case D10.1 gives

$$g_B(x,y) = \sum_{n=0}^{\infty} Q_n(x,y),$$

where $Q(x,y)$ is the transition function restricted to $R - B$ and Q_n its iterates. We shall show

(3) $\qquad g_B(x,y) = A(x,\bar{y}) - A(x,y)$ for x,y in $R - B$.

Here $\bar{y} = y^1 - iy^2$.

The proof is an example of the so-called *reflection principle*.[8] Letting $T = T_B$ denote the time of the first visit of the random walk to B, one can write, for x and y in $R - B$ and $n \geq 1$,

$$P_n(x,y) = \mathbf{P}_x[\mathbf{x}_n = y] = \mathbf{P}_x[\mathbf{x}_n = y; T > n] + \mathbf{P}_x[\mathbf{x}_n = y; T \leq n]$$
$$= Q_n(x,y) + \sum_{k=1}^{n} \mathbf{P}_x[\mathbf{x}_n = y; T = k].$$

But

$$\mathbf{P}_x[\mathbf{x}_n = y; T = k] = \mathbf{P}_x[\mathbf{x}_n = \bar{y}; T = k] \text{ for } k \leq n,$$

which is quite obvious although a formal proof proceeds via

$$\mathbf{P}_x[\mathbf{x}_n = y; T = k] = E_x\{\mathbf{P}_{\mathbf{x}_T}[\mathbf{x}_{n-k} = y]; T = k\}$$
$$= E_x\{\mathbf{P}_{\mathbf{x}_T}[\mathbf{x}_{n-k} = \bar{y}]; T = k\}$$
$$= \mathbf{P}_x[\mathbf{x}_n = \bar{y}; T = k],$$

the idea being simply that Im $(\mathbf{x}_T) = 0$ so that transitions from $\mathbf{x}_T$ to y in time $n - k$ have the same probability as those from $\mathbf{x}_T$ to $\bar{y}$ in time $n - k$. Hence

$$P_n(x,y) = Q_n(x,y) + \sum_{k=1}^{n} \mathbf{P}_x[\mathbf{x}_n = \bar{y}; T = k]$$
$$= Q_n(x,y) + \mathbf{P}_x[\mathbf{x}_n = \bar{y}; T \leq n].$$

The event $T \leq n$ being implied by the event $\mathbf{x}_n = \bar{y}$ (as $x \in R - B$ and $\bar{y} \in B$), we conclude

$$P_n(x,y) = Q_n(x,y) + P_n(x,\bar{y}), \quad n \geq 1.$$

Also $P_0(x,y) = Q_0(x,y) = \delta(x,y)$ and $P_0(x,\bar{y}) = 0$, so that

$$g_B(x,y) = \sum_{n=0}^{\infty} Q_n(x,y) = \sum_{n=0}^{\infty} [P_n(x,y) - P_n(x,\bar{y})]$$
$$= \sum_{n=0}^{\infty} [P_n(0,0) - P_n(x,\bar{y})] - \sum_{n=0}^{\infty} [P_n(0,0) - P_n(x,y)]$$
$$= A(x,\bar{y}) - A(x,y),$$

which was promised in equation (3).

[8] The proof of (3) which follows was invented by McCrea and Whipple [76], who lamented the fact that it lacked rigor at one point. Their difficulty lay in having defined $A(x,0) = a(x)$ by the integral in (1), fully aware, but without being able to prove that $a(x) = \sum_{n=0}^{\infty} [P_n(0,0) - P_n(x,0)]$.

As for the reflection principle, its roots go back much farther. In physical applications it often appears under the name of the *method of images*. See pp. 70 and 335 in [31], Vol. 1, for further comments.

To calculate the hitting probabilities $H_B(x,y)$ of the lower half plane B the theorems of sections 11 and 14 are not available, since B is an infinite set. Instead we can use the completely general identity

(4) $\qquad H_A(x,y) = \sum_{t \in R} g_A(x,t) P(t,y), \qquad x \in R - A, \quad y \in A$

which was part (c) of P10.1. Due to the symmetry of the present problem, nothing of interest is lost if we take the point $x = in, n \geq 1$ on the imaginary axis, and the point $y = k$, $-\infty < k < \infty$ on the real axis. Because of the continuity of the paths, it is only for y on the real axis, the boundary of B, that the probability in (4) will be positive. In this case (4) gives

(5) $\qquad\qquad H_B(in, k) = \tfrac{1}{4} g_B(in, k + i),$

and in view of (3)

(6) $\qquad H_B(in, k) = \tfrac{1}{4} a[(n+1)i - k] - \tfrac{1}{4} a[(n-1)i - k].$

First we let $n = 1$ and calculate, using (1)

$$H_B(i,k) = \tfrac{1}{4} a(2i + k) - \tfrac{1}{4} a(k)$$
$$= \frac{1}{16\pi^2} \int_{-\pi}^{\pi} \int_{-\pi}^{\pi} \frac{\cos k\theta_1 - \cos(k\theta_1 + 2\theta_2)}{1 - \tfrac{1}{2}(\cos \theta_1 + \cos \theta_2)} d\theta_1 \, d\theta_2.$$

The evaluation of this integral seems hopelessly tedious, so that a different method is indicated (see problem 7 in Chapter II). We let

$$\phi(\theta) = \sum_{k=-\infty}^{\infty} H_B(i,k) e^{ik\theta}, \qquad \theta \text{ real},$$

noting that this Fourier series is a characteristic function, since

$$H_B(i,k) \geq 0, \qquad \sum_{k=-\infty}^{\infty} H_B(i,k) = 1.$$

In addition,

$$\phi^n(\theta) = \sum_{k=-\infty}^{\infty} H_B(ni,k) e^{ik\theta}.$$

A formal proof of the last identity, based on (5) or (6) would be lengthy. The easiest way of getting it is to regard the hitting place $\mathbf{x}_T$ of the process $\mathbf{x}_n$ with $\mathbf{x}_0 = in$, as the sum of n identically distributed independent random variables with the distribution $H_B(i,k)$ (the process having to hit the lines Im $z = n - 1, n - 2, \ldots, 0$ in succession). Then one gets the difference equation (part (a) of P10.1)

$$H_B(i,k) = \tfrac{1}{4}H_B(i,k-1) + \tfrac{1}{4}H_B(i,k+1) + \tfrac{1}{4}H_B(2i,k) + \tfrac{1}{4}\delta(k,0).$$

Taking the Fourier series on each side

$$\phi(\theta) = \tfrac{1}{4}e^{i\theta}\phi(\theta) + \tfrac{1}{4}e^{-i\theta}\phi(\theta) + \tfrac{1}{4}\phi^2(\theta) + \tfrac{1}{4},$$

and this quadratic equation has two solutions, the only possible one, in view of $|\phi(\theta)| \le 1$, being

(7) $\qquad \phi(\theta) = 2 - \cos\theta - \sqrt{(1 - \cos\theta)(3 - \cos\theta)}.$

A very simple calculation shows that

(8) $$\lim_{n\to\infty} \phi^n\left(\frac{\theta}{n}\right) = e^{-|\theta|}.$$

It so happens that

$$e^{-|\theta|} = \frac{1}{\pi} \int_{-\infty}^{\infty} e^{i\theta x} \frac{dx}{1 + x^2},$$

is the characteristic function of the Cauchy distribution. Using the Lévy continuity theorem, mentioned in connection with the Central limit theorem in P6.8, one can conclude that

(9) $$\lim_{n\to\infty} \sum_{k=-\infty}^{[nx]} H(in,k) = \frac{1}{\pi} \int_{-\infty}^{x} \frac{dt}{1 + t^2} = \frac{1}{2} + \frac{1}{\pi} \tan^{-1} x.$$

To obtain an amusing application of this result, let us consider an apparently more complicated problem. Let C be the whole plane with the exception of the first quadrant, i.e.,

$$C = R - [z \mid \text{Re } z > 0 \text{ and Im } z > 0].$$

What is $g_C(x,y)$ when x and y are in $R - C$? Using the reflection principle this question is just as easy to answer as the corresponding one answered for the half plane B in equation (3). One simply performs two reflections, once in the real axis and once in the imaginary one and gets

(10) $\qquad g_C(x,y) = A(x,\bar{y}) - A(x,y) + A(x,-\bar{y}) - A(x,-y).$

Now we take for the starting point of the random walk $\mathbf{x}_0 = z = r + is$, a point in the first quadrant $R - C$ and ask for the *probability*

$\rho(z)$ that absorption occurs on the real axis rather than on the imaginary one. This probability is

$$\rho(z) = \sum_{k=1}^{\infty} H_C(z,k) = \frac{1}{4} \sum_{k=1}^{\infty} g_C(z, i + k).$$

Using (10) this is

$$\rho(z) = \frac{1}{4} \sum_{k=1}^{\infty} [a(z - k + i) - a(z - k - i) + a(z + k - i) - a(z + k + i)].$$

Using (6) one finally gets

$$\rho(z) = \sum_{k=1}^{\infty} H_B(is, r - k) - \sum_{k=1}^{\infty} H_B(is, r + k),$$

or

(11) $$\rho(z) = \sum_{k=-\infty}^{r-1} H_B(is,k) - \sum_{k=r+1}^{\infty} H_B(is,k).$$

Finally we shall let $|z| \to \infty$ in such a way that

$$\lim_{|z| \to \infty} \tan^{-1} \frac{r}{s} = \alpha, \qquad 0 \le \alpha \le \frac{\pi}{2}.$$

In other words, $|z| \to \infty$, z making an angle α with the imaginary axis. Now we apply (9) to (11), writing $r = s \tan \alpha + \epsilon s$ where $\epsilon = \epsilon(z) \to 0$. The two error terms (a sum of the hitting probabilities from $k = s \tan \alpha$ to $s \tan \alpha + \epsilon s$ and another term like it) go to zero in view of (9). Hence

(12) $$\lim_{|z| \to \infty} \rho(z) = \lim_{s \to \infty} \sum_{k=-\infty}^{s \tan \alpha} H_B(is,k) - \lim_{s \to \infty} \sum_{k=s \tan \alpha}^{\infty} H_B(is,k)$$

$$= 2 \lim_{s \to \infty} \sum_{k=0}^{s \tan \alpha} H_B(is,k) = \frac{2}{\pi} \tan^{-1} (\tan \alpha) = \frac{2}{\pi} \alpha.$$

Any other answer would of course have been disconcerting (see problem 11).

16. THE TIME DEPENDENT BEHAVIOR

Consider the following simple case of the Exterior Dirichlet problem (problem B in section 13). D is a given domain in the plane, and we seek a function $u(x)$ such that
 (i) $\Delta u(x) = 0$ on the complement of $D + \partial D$,
 (ii) $u(x) \equiv 1$ on ∂D,
 (iii) $u(x)$ is continuous and bounded on the complement of D.

This problem has the very unspectacular unique solution, $u(x) \equiv 1$. The situation becomes interesting only when we compare $u(x)$ to the solution of a nonstationary problem, which has the above Exterior Dirichlet problem as its limiting, or steady-state case. This is the following

Diffusion problem: Find the solutions $u(x,t)$ of
(i) $\partial u/\partial t = \tfrac{1}{2}\Delta u$, for $t > 0$, x outside D,
(ii) $u(x,t) = 1$ for $t > 0$, x on ∂D
(iii) $\lim_{t \to 0+} u(x,t) = 0$ for all x outside D.

Such problems arise in the theory of heat conduction. The solution $u(x,t)$ represents the temperature at the point x at time t in the region exterior to the body D. The boundary of D is kept at temperature one (condition (ii)) and the initial temperature is zero (condition (iii)). It is known that this problem has a unique solution $u(x,t)$ and that

$$\lim_{t \to \infty} u(x,t) = u(x) \equiv 1 \text{ off } D.$$

In other words, the temperature of the surrounding medium rises, approaching at every point the boundary temperature.

If this seems obvious, consider as a precautionary measure the analogous Dirichlet and Diffusion problems in three dimensions. To make everything as simple as possible, let D be the unit sphere. Then the exterior Dirichlet problem has *at least two* solutions

$$u(x) = 1 \quad \text{and} \quad u(x) = |x|^{-1}.$$

The solution to the diffusion problem is again unique, but its limit behavior is now different, as one finds that

$$\lim_{t \to \infty} u(x,t) = |x|^{-1}.$$

Thus a compact set in three-space is apparently not "large" enough to "warm up the surrounding space" to its own temperature. In the next few pages we shall find further hints concerning this phenomenon, but the complete probabilistic explanation will not be clear until Chapter VII. It is intimately connected with the difference between recurrent random walk (such as simple random walk in the plane) and transient random walk (every genuinely three-dimensional random walk is transient).

A far more delicate question, and one which is indigenous to the recurrent case, concerns the *rate of convergence* of $u(x,t)$ to the stationary state solution $u(x)$. As conjectured by Kac, and proved by Hunt [44], 1956, the answer is

(1) $$\lim_{t \to \infty} [1 - u(x,t)] \ln t = 2\pi g_D(x,\infty).$$

Here $g_D(x,y)$ is the exterior Green function of the domain D mentioned in the discussion of problem B in section 13, and $g_D(x,\infty)$ in equation (1) is of course the limit of $g_D(x,y)$ as $|y| \to \infty$.

The plan of work for the remainder of this chapter is now very simple. We shall formulate the discrete analogue of the diffusion problem of arbitrary aperiodic recurrent two-dimensional random walk, and then try to prove a theorem which corresponds to equation (1) above. The content of this theorem will be probabilistically interesting. It will concern the probability law governing the time $\mathbf{T}_B$ when the random walk first visits a given set B. This will be the first time in this chapter that we are able to obtain a result concerning the *time dependent behavior* of the random walk. All the previous results concerned the hitting place $\mathbf{x}_{\mathbf{T}_B}$ rather than the hitting time $\mathbf{T}_B$ itself. For example

$$H_B(x,y) = \mathbf{P}_x[\mathbf{x}_{\mathbf{T}_B} = y] \text{ when } x \in R - B,$$

and even our explicit formula for $g_B(x,y)$ gave no information at all concerning the distribution of $\mathbf{T}_B$.

We shall use the notation

D1 $\mathbf{T}_B = \min [k \mid k \geq 1; \mathbf{x}_k \in B]$, *and call*
$\mathbf{T}_B = \mathbf{T}$ *when* $B = \{0\}$, *and* $R_n = \mathbf{P}_0[\mathbf{T} > n]$, *for* $n \geq 1$,
$F_0 = 0$, $F_n = \mathbf{P}_0[\mathbf{T} = n]$ *for* $n \geq 1$, $U_n = \mathbf{P}_0[\mathbf{x}_n = 0]$
$= P_n(0,0)$ *for* $n \geq 0$.

Since the random walk is recurrent, $\mathbf{T}_B < \infty$ with probability one whenever B is a nonempty finite set.

The diffusion problem is formulated as follows. The function $u(x,t)$ becomes a sequence of functions $f_n(x)$ with $n \geq 0$, $x \in R$, since the time parameter must be discrete. The time derivative $\partial/\partial t[u(x,t)]$ then should correspond to $f_{n+1}(x) - f_n(x)$, and since the Laplacian $\frac{1}{2}\Delta$ corresponds to the operator $P - I$, we arrive at the equation

$$Pf_n - f_n = f_{n+1} - f_n$$

or more simply $Pf_n = f_{n+1}$. Thus the discrete version of the diffusion problem consists of the three conditions

(i) $\sum_{y \in R} P(x,y) f_n(y) = f_{n+1}(x)$, $x \in R - B$, $n \geq 0$,
(ii) $f_n(x) = 1$ for all $n \geq 0$, $x \in B$,
(iii) $f_0(x) = 0$ for all $x \in R - B$.

It is a trivial matter to verify that this system of equations has a unique solution. The function $f_0(x)$ is specified completely for all x by (ii) and (iii). Substituting its values into (i) determines $f_1(x)$ on $R - B$. But on B, $f_1(x)$ is given by (ii) so that $f_1(x)$ is uniquely determined. And so on, by induction: once f_n is determined everywhere, equation (i) determines f_{n+1} on $R - B$, and then (ii) gives f_{n+1} on B.

To avoid time consuming iteration, we shall simply write down the solution, and then verify that (i), (ii), and (iii) hold. Let $f_0(x) = 0$ on $R - B$, $f_0(x) = 1$ on B, and for $n \geq 1$ let

$$f_n(x) = \mathbf{P}_x[\mathbf{T}_B \leq n] \quad \text{for } x \in R - B,$$
$$= 1 \quad \text{for } x \in B.$$

Conditions (ii) and (iii) hold trivially, and (i) follows from

$$\mathbf{P}_x[\mathbf{T} \leq n + 1] = \sum_{y \in B} P(x,y) + \sum_{y \in R-B} P(x,y) \mathbf{P}_y[\mathbf{T} \leq n],$$

for $x \in R - B$, $n \geq 1$.

Therefore the random walk analogue of equation (1) should concern the rate of convergence of the sequence

$$1 - f_n(x) = \mathbf{P}_x[\mathbf{T}_B > n]$$

to zero as n tends to infinity. We shall show that the correct analogue of $\ln t$ in (1) is the reciprocal of

$$R_n = \mathbf{P}_0[\mathbf{T}_{\{0\}} > n] = \mathbf{P}_0[\mathbf{T} > n].$$

In fact, simple random walk in the plane has the property that R_n is asymptotically $\pi(\ln n)^{-1}$, as will be shown later, in E1.

T1 *For aperiodic recurrent two-dimensional random walk, let $B = \{0\}$. Then*

$$\lim_{n \to \infty} \frac{\mathbf{P}_x[\mathbf{T}_B > n]}{\mathbf{P}_0[\mathbf{T} > n]} = g_B(x,\infty) = a(x) \text{ for } x \text{ in } R - B$$

where $g_B(x,\infty)$ is defined and shown equal to $a(x)$ in T14.3.

Proof: The proof of this theorem depends entirely on the following lemma, which will be proved as soon as we have explained why it implies the truth of T1.

P1 *For aperiodic recurrent two-dimensional random walk*
$$\lim_{n\to\infty} \frac{R_n}{R_{n+1}} = 1.$$

We may write
$$\mathbf{P}_x[T > n] = \sum_{k=n}^{\infty} \mathbf{P}_x[T = k+1] = \sum_{k=n}^{\infty} \sum_{y \neq 0} Q_k(x,y) P(y,0),$$
where, of course, $Q(x,y)$ is the transition function $P(x,y)$ restricted to $R - \{0\}$, and $Q_n(x,y)$ are its iterates, i.e.,
$$Q_k(x,y) = \mathbf{P}_x[\mathbf{x}_k = y; T > k] \text{ for } k \geq 1, \quad Q_0(x,y) = \delta(x,y).$$
Letting
$$g(x,y) = g_{\{0\}}(x,y) = \sum_{n=0}^{\infty} Q_n(x,y)$$
one obtains
$$\mathbf{P}_x[T > n] = \sum_{k=0}^{\infty} \sum_{y \neq 0} Q_{n+k}(x,y) P(y,0) = \sum_{t \neq 0} g(x,t) \sum_{y \neq 0} Q_n(t,y) P(y,0).$$
Setting
$$v_n(t) = \frac{1}{R_{n+1}} \sum_{y \neq 0} Q_n(t,y) P(y,0), \quad t \in R - \{0\},$$
one has

(1) $$\frac{\mathbf{P}_x[T > n]}{R_n} = \frac{R_{n+1}}{R_n} \sum_{t \neq 0} g(x,t) v_n(t), \quad x \neq 0.$$

To complete the proof of T1 it suffices to show that the right-hand side in (1) tends to $a(x)$ as $n \to \infty$. In view of P1 it will suffice to prove

(2) $$\lim_{n\to\infty} \sum_{t \neq 0} g(x,t) v_n(t) = a(x), \quad x \neq 0.$$

To reduce the problem still further, observe that (2) could easily be deduced from

(3) $$\lim_{|t|\to\infty} g(x,t) = a(x),$$

(4) $$v_n(t) \geq 0, \quad \sum_{t \neq 0} v_n(t) = 1,$$

(5) $$\lim_{n\to\infty} v_n(t) = 0 \text{ for each } t \neq 0.$$

But we know that (3) holds from T14.3. To establish (4) and (5) we go back to the definition of $v_n(t)$ which implies

$$v_{n-1}(t) = \frac{1}{R_n} \mathbf{P}_t[\mathbf{T} = n].$$

Introducing the reversed random walk with probability measure $\mathbf{P}_0^*[\ ]$ defined by $P^*(x,y) = P(y,x)$, one finds that

$$v_{n-1}(t) = \frac{1}{R_n} \mathbf{P}_0^*[\mathbf{x}_n = t; \mathbf{T} > n]$$
$$= \frac{\mathbf{P}_0^*[\mathbf{x}_n = t; \mathbf{T} > n]}{\mathbf{P}_0[\mathbf{T} > n]} = \frac{\mathbf{P}_0^*[\mathbf{x}_n = t; \mathbf{T} > n]}{\mathbf{P}_0^*[\mathbf{T} > n]}.$$

Hence $v_{n-1}(t)$ is a conditional probability measure on $R - \{0\}$. Summing on t over $R - \{0\}$ gives (4).

To prove (5) we have to show that for each t

(6) $$\lim_{n \to \infty} \frac{\mathbf{P}_0[\mathbf{x}_n = t; \mathbf{T} > n]}{\mathbf{P}_0[\mathbf{T} > n]} = 0.$$

Here we have dropped the "stars" as we shall prove (6) for an arbitrary aperiodic two-dimensional recurrent random walk. We can select the integer $m > 0$ so that $P_m(t,0) = \alpha > 0$,

$$\frac{\mathbf{P}_0[\mathbf{x}_n = t; \mathbf{T} > n]}{\mathbf{P}_0[\mathbf{T} > n]} P_m(t,0) \leq \frac{\mathbf{P}_0[n < \mathbf{T} \leq n + m]}{\mathbf{P}_0[\mathbf{T} > n]}$$

or

$$\frac{\mathbf{P}_0[\mathbf{x}_n = t; \mathbf{T} > n]}{\mathbf{P}_0[\mathbf{T} > n]} \leq \frac{1}{\alpha} \frac{R_n - R_{n+m}}{R_n},$$

and the right-hand side goes to zero as $n \to \infty$ in view of P1.

Now the proof of T1 is complete, except for the proof of P1. But P1 is equivalent to

$$\lim_{n \to \infty} \frac{F_n}{R_n} = 0,$$

and as $F_n \leq U_n$ for $n \geq 0$ it will suffice to show that

(7) $$\lim_{n \to \infty} \frac{U_n}{R_n} = 0.$$

We shall make use of P7.6 to the effect that

(8) $$U_n \leq \frac{A}{n}, \qquad n \geq 1$$

for some constant $A > 0$. (This requires some care, as (8) is valid only when condition (3) in P7.6 is satisfied. But since our random walk is recurrent one can find three distinct non-zero points x, not colinear, such that $P(0,x) > 0$, and that suffices to satisfy condition (3) in P7.6.) In addition, we only require the identity

$$(9) \qquad \sum_{k=0}^{n} U_k R_{n-k} = 1, \qquad n \geq 0,$$

which follows by summing on n in

$$U_n = \sum_{k=1}^{n} F_k U_{n-k}, \qquad n \geq 1,$$

which is nothing but P1.2. For each integer $m \leq n$ we have

$$1 = \sum_{k=0}^{n} U_k R_{n-k}$$
$$\leq (U_0 + U_1 + \cdots + U_m) R_{n-m} + U_{m+1} + \cdots + U_n,$$

or

$$R_k \geq \frac{1 - (U_{m+1} + \cdots + U_{m+k})}{U_0 + \cdots + U_m} \geq \frac{1 - A\left(\dfrac{1}{m+1} + \cdots + \dfrac{1}{m+k}\right)}{1 + \dfrac{A}{1} + \dfrac{A}{2} + \cdots + \dfrac{A}{m}},$$

for each choice of positive integers k and m. Letting $m = [ck]$ (the greatest integer in ck)

$$1 - A\left(\frac{1}{m+1} + \cdots + \frac{1}{m+k}\right) \sim 1 - A \ln\left(1 + \frac{1}{c}\right)$$

as $k \to \infty$. Choosing $c > 0$ so that $A \ln(1 + 1/c) < 1/2$,

$$R_k \geq \frac{1}{2A \ln k}$$

for sufficiently large k. Hence, using (8) for the second time,

$$\frac{U_k}{R_k} \leq \frac{2 \ln k}{k}$$

when k is sufficiently large. This upper bound tends to zero, proving (7), completing the proof of P1 and hence of T1.

Theorem T1 was worded in a somewhat peculiar manner in order to suggest that perhaps

(1) $$\lim_{n\to\infty} \frac{\mathbf{P}_x[\mathbf{T}_B > n]}{\mathbf{P}_0[\mathbf{T} > n]} = g_B(x,\infty), \qquad x \in R - B,$$

for every set B of cardinality $1 \le |B| < \infty$. Unfortunately the step up from $|B| = 1$ to $|B| > 1$ causes considerable difficulty, so that we refer to the recent (1963) literature [60] for the proof of the above assertion. As an instructive way of countering the difficulty we offer the following attempt at proving (1) for arbitrary finite sets B by mathematical induction on the size $|B|$ of B. It has been proved for $|B| = 1$, so that we assume it proved for sets of cardinality p and proceed to $|B| = p + 1$. When $|B| = p + 1$ we write $B = B' \cup z$, where $|B'| = p$, and z is the new point. Then the following equation is probabilistically obvious. For $x \in R - B$

$$\mathbf{P}_x[\mathbf{T}_{B'} > n] = \mathbf{P}_x[\mathbf{T}_B > n] + \sum_{k=1}^{n} \mathbf{P}_x[\mathbf{T}_B = k; \mathbf{T}_{B'} > n]$$
$$= \mathbf{P}_x[\mathbf{T}_B > n] + \sum_{k=1}^{n} \mathbf{P}_x[\mathbf{T}_B = k; \mathbf{x}_{\mathbf{T}_B} = z]\mathbf{P}_z[\mathbf{T}_{B'} > n - k].$$

We simplify the notation by writing

(2) $$\frac{a_n}{b_n} = \frac{\mathbf{P}_x[\mathbf{T}_B > n]}{\mathbf{P}_0[\mathbf{T} > n]} + \frac{1}{b_n} \sum_{k=1}^{n} c_k d_{n-k}$$

where

$$a_n = \mathbf{P}_x[\mathbf{T}_{B'} > n], \qquad b_n = \mathbf{P}_0[\mathbf{T} > n],$$
$$c_n = \mathbf{P}_x[\mathbf{T}_B = n; \mathbf{x}_{\mathbf{T}_B} = z], \qquad d_n = \mathbf{P}_z[\mathbf{T}_{B'} > n].$$

According to the induction hypothesis we know that

(3) $$\lim_{n\to\infty} \frac{a_n}{b_n} = g_{B'}(x,\infty), \qquad x \in R - B'.$$

Also, using P1 together with the induction hypothesis,

(4) $$\lim_{n\to\infty} \frac{d_{n-k}}{b_n} = g_{B'}(z,\infty), \qquad z \in R - B', \quad k \ge 0.$$

Therefore it seems *reasonable to conjecture that*

(5) $$\lim_{n\to\infty} \frac{1}{b_n} \sum_{k=1}^{n} c_k d_{n-k} = g_{B'}(z,\infty) \sum_{k=1}^{\infty} c_k$$
$$= g_{B'}(z,\infty) H_B(x,z) \text{ for } x, z \in R - B'.$$

But if (5) holds, then substitution of (3), (4), and (5) into (2) shows that

$$\lim_{n \to \infty} \frac{\mathbf{P}_x[\mathbf{T}_B > n]}{\mathbf{P}_0[\mathbf{T} > n]} = g_{B'}(x,\infty) - g_{B'}(z,\infty)H_B(x,z)$$

for all x in $R - B$. Observe that

(6) $\quad g_{B'}(x,y) - g_{B'}(z,y)H_B(x,z) = g_B(x,y), \quad y \in R - B,$

which is obtained by decomposing $g_{B'}(x,y)$ into the sum of the expected number of visits to y before hitting B, and those visits to y occurring after the first visit to the point z in B. It is obvious how to construct a formal proof, using P3.2. In view of T14.3 we can let $|y| \to \infty$ in (6). The result is

$$\lim_{n \to \infty} \frac{\mathbf{P}_x[\mathbf{T}_B > n]}{\mathbf{P}_0[\mathbf{T} > n]} = g_B(x,\infty), \quad x \in R - B,$$

which completes the induction and therefore proves (1), assuming that (5) is true.

Finally, we shall derive a *sufficient condition* for (5), and hence (1) to hold. It is quite simply that

(7) $\quad \overline{\lim_{n \to \infty}} \dfrac{R_n}{R_{2n}} < \infty.$

(Before reducing (5) to (7) it should be pointed out that Kesten has proved (7) for arbitrary recurrent random walk [60]. And in the two-dimensional case even the limit in (7) exists and equals one. For simple random walk this will be verified in E1 below.)

Now we shall assume (7) and for a fixed point $x \neq 0$ in R and some integer $M > 0$, we consider the sum

$$I(n,M,x) = \frac{1}{R_n} \sum_{k=M}^{n-M} \mathbf{P}_x[\mathbf{T} = k] R_{n-k}.$$

Decomposing $I(n,M,x)$ into a sum from $k = M$ to $k = [n/2]$ and another one from $[n/2] + 1$ to $n - M$ one gets

$$I(n,M,x) \leq \frac{R_{n-[n/2]}}{R_n} \mathbf{P}_x[M \leq \mathbf{T} \leq [n/2]]$$

$$+ \frac{R_M}{R_n} \mathbf{P}_x[[n/2] + 1 \leq \mathbf{T} \leq n - M]$$

$$\leq \frac{R_{[n/2]}}{R_n} \mathbf{P}_x[M \leq \mathbf{T}] + \frac{R_M}{R_n} \mathbf{P}_x[[n/2] \leq \mathbf{T}].$$

Using T1 we get
$$I(n,M,x) \le ka(x) R_M \frac{R_{[n/2]}}{R_n}$$
for some $k > 0$, and finally, using (7),

(8) $$\lim_{M \to \infty} \overline{\lim_{n \to \infty}} I(n,M,x) = 0.$$

Using (8) we can show that if, as before
$$b_n = \mathbf{P}_0[T > n] = R_n, \quad c_n = \mathbf{P}_x[\mathbf{T}_B = n; \mathbf{x}_{\mathbf{T}_B} = z],$$
then

(9) $$\lim_{n \to \infty} \frac{1}{b_n} \sum_{k=1}^{n} c_k b_{n-k} = \sum_{k=1}^{\infty} c_k.$$

This is clear because, given any $\epsilon > 0$, we can choose M such that

(10) $$\overline{\lim_{n \to \infty}} \frac{1}{b_n} \sum_{k=M}^{n-M} c_k b_{n-k} \le \overline{\lim_{n \to \infty}} \frac{1}{R_n} \sum_{k=M}^{n-M} \mathbf{P}_x[\mathbf{T}_z = k] R_{n-k}$$
$$= \overline{\lim_{n \to \infty}} I(n,M,x-z) < \epsilon.$$

And as for the tails, P1 implies that
$$\lim_{n \to \infty} \frac{1}{b_n} \sum_{k=1}^{M-1} c_k b_{n-k} = \sum_{k=1}^{M-1} c_k,$$
while

(11) $$\frac{1}{b_n} \sum_{k=n-M+1}^{n} c_k b_{n-k} \le \frac{1}{b_n} \sum_{k=n-M+1}^{n} c_k \le \frac{1}{b_n} \sum_{k=n-M+1}^{n} \mathbf{P}_x[\mathbf{T}_z = k]$$
$$= \frac{1}{b_n} \{\mathbf{P}_{x-z}[\mathbf{T} > n-M] - \mathbf{P}_{x-z}[\mathbf{T} > n]\}$$

which tends to zero by T1.

But according to (4)

(12) $$\lim_{n \to \infty} \frac{d_n}{b_n} = g_{B'}(z, \infty) = g$$

where $d_n = \mathbf{P}_z[\mathbf{T}_{B'} > n]$, as before. Decomposing

(13) $$\frac{1}{b_n} \sum_{k=1}^{n} c_k d_{n-k} = \frac{g}{b_n} \sum_{k=1}^{n} c_k b_{n-k}$$
$$+ \frac{1}{b_n} \sum_{k=1}^{n} c_k [d_{n-k} - g b_{n-k}],$$

it follows from (10), (11), and (12) that the last term in (13) tends to zero as n tends to infinity. And from (9) and (12) we conclude that

$$\lim_{n \to \infty} \frac{1}{b_n} \sum_{k=1}^{n} c_k d_{n-k} = g \sum_{k=1}^{\infty} c_k$$

$$= g_{B'}(z,\infty) \sum_{k=1}^{\infty} \mathbf{P}_x[\mathbf{T}_B = k; \mathbf{x}_{T_B} = z]$$

$$= g_{B'}(z,\infty) H_B(x,z).$$

This is equation (5) so that *we have established the sufficiency of condition (7) for (5) and hence for (1).*

E1 *For simple random walk in the plane*

$$R_n = \mathbf{P}_0[T > n] \sim \frac{\pi}{\ln n} \quad \text{as } n \to \infty.$$

Proof:[9] We have

$$U_{2n+1} = 0, \qquad U_{2n} = 4^{-2n} \binom{2n}{n}^2,$$

and Stirling's formula provides the estimate

$$U_{2n} \sim \frac{1}{n\pi}, \qquad n \to \infty.$$

Hence

$$U_0 + U_2 + \cdots + U_{2n} \sim \frac{1}{\pi} \ln n, \qquad n \to \infty.$$

Using the identity

$$\sum_{k=0}^{2n} U_k R_{2n-k} = 1, \qquad n \geq 0,$$

as in the proof of P1, one obtains

$$R_{2n-2k}[U_0 + \cdots + U_{2k}] + U_{2k+2} + \cdots + U_{2n} \geq 1.$$

Here we let k depend on n, choosing

$$k = k(n) = n - \left[\frac{n}{\ln n}\right].$$

Then

$$U_0 + \cdots + U_{2k} \sim \frac{\ln k}{\pi} \sim \frac{\ln (n-k)}{\pi}$$

as $n \to \infty$, and

$$U_{2k+2} + \cdots + U_{2n} \sim 0$$

[9] This proof is from [27].

as $n \to \infty$. Hence, for each $\epsilon > 0$,

$$\frac{\ln(n-k)}{\pi} R_{2n-2k} \geq 1 - \epsilon$$

for all large enough values of n, so that

$$\lim_{n \to \infty} \frac{\ln n}{\pi} R_n \geq 1.$$

On the other hand,

$$1 = \sum_{k=0}^{n} R_k U_{n-k} \geq R_n (U_0 + \cdots + U_n),$$

giving

$$\varlimsup_{n \to \infty} \frac{\ln n}{\pi} R_n \leq 1.$$

That completes the proof of E1. The important thing for our purpose is that

$$\lim_{n \to \infty} \frac{R_n}{R_{2n}} = \lim_{n \to \infty} \frac{\ln 2n}{\ln n} = 1.$$

Hence equation (7) holds and, in view of the discussion following the proof of T1, various extensions of T1 are valid for simple random walk. The next two examples are devoted to two of these. E2 concerns equation (1) which in view of E1 may be written as

$$\lim_{n \to \infty} \frac{\ln n}{\pi} \mathbf{P}_x[\mathbf{T}_B > n] = g_B(x, \infty), \qquad x \in R - B.$$

E2 *For simple random walk* we shall use the above result to obtain a novel version of the famous double point problem. Let A_n denote the event that $\mathbf{x}_0, \mathbf{x}_1, \ldots, \mathbf{x}_n$, the positions of the random walk at times $0, 1, \ldots,$ up to n, are *all distinct*. It is known that

$$2 < \gamma = \lim_{n \to \infty} \{\mathbf{P}[A_n]\}^{1/n} < 3$$

exists. As pointed out in section 10, it requires no great art to calculate γ quite accurately, but the difficulty of the problem is indicated by the fact that it is not even known whether

$$\lim_{n \to \infty} \frac{\mathbf{P}[A_{n+1}]}{\mathbf{P}[A_n]} = \lim_{n \to \infty} \mathbf{P}[A_{n+1} \mid A_n]$$

exists.[10] Of course this limit has the same value as the preceding limit, if it exists at all.

Imitating this problem, while *avoiding* some of its essential difficulties, we define B_n as the event that $\mathbf{x}_n \notin \{\mathbf{x}_0, \ldots, \mathbf{x}_{n-1}\}$. Thus B_n is the event

---

[10] See Kesten [S16] for the best results, including the existence of the limit of $\mathbf{P}[A_{n+2} \mid A_n]$, as $n \to \infty$.

that the random walk at time n finds itself *at a new point*, i.e., one which was not visited before. We shall show that

$$\lim_{n \to \infty} \mathbf{P}[B_{n+1} \mid B_n] = 1/2.$$

Proof:

$$\mathbf{P}[B_{n+1} \mid B_n] = \frac{\mathbf{P}_0[B_n B_{n+1}]}{\mathbf{P}_0[B_n]}.$$

By reversing the random walk it is seen that

$$\mathbf{P}_0[B_n] = \mathbf{P}_0[T > n] = R_n,$$

and similarly, that

$$\mathbf{P}_0[B_n B_{n+1}] = \sum_{x \in R} P(0,x) \mathbf{P}_x[T > n, T_x > n],$$

where

$$T = \min [k \mid 1 \le k, \mathbf{x}_k = 0], \qquad T_x = \min [k \mid 1 \le k, \mathbf{x}_k = x].$$

Now it is quite clear, from the symmetry of the simple random walk, that $\mathbf{P}_x[T > n, T_x > n]$ has the same value for each of the four points x where $P(0,x) > 0$. But we prefer to spurn this opportunity to make an apparent simplification, in order to prove the result for an arbitrary random walk in the plane which is aperiodic, symmetric ($P(x,y) = P(y,x)$), has the property that $P(0,0) = 0$ and finally is such that T1 holds for sets of cardinality two. (This is always true but we have proved it only for simple random walk.)

For each fixed $x \ne 0$ we let $B = \{0, x\}$. Then

$$\mathbf{P}_x[T > n; T_x > n] = \mathbf{P}_0[T > n; T_x > n]$$
$$= \sum_{y \in R-B} P(0,y) \mathbf{P}_y[T_B > n-1].$$

Hence

$$\lim_{n \to \infty} \frac{\mathbf{P}_x[T > n; T_x > n]}{R_n} = \sum_{y \in R-B} P(0,y) g_B(y, \infty).$$

Since $|B| = 2$ and the random walk is symmetric,

$$\mu(t) = \mu^*(t) = H_B(\infty, t) = 1/2, \qquad t \in B$$

(see E14.1), and using T14.3,

$$g_B(y, \infty) = \tfrac{1}{2} A(y, 0) + \tfrac{1}{2} A(y, x) - \frac{1}{K_B(\cdot\cdot)} = \tfrac{1}{2} [a(y) + a(y-x) - a(x)].$$

Hence we conclude from P13.3 that

$$\sum_{y \in R-B} P(0,y) g_B(y, \infty) = \sum_{y \in R} P(0,y) g_B(y, \infty)$$
$$= \tfrac{1}{2}[a(0) + 1 + a(x) - a(x)] = 1/2.$$

That gives

$$\lim_{n\to\infty} \frac{\mathbf{P}_0[B_n B_{n+1}]}{R_n} = \frac{1}{2}$$

which is the desired result, since $R_n = \mathbf{P}_0[B_n]$.

E3 Let

$$\mathbf{N}_n = \sum_{k=1}^{n} \delta(0, \mathbf{x}_k)$$

denote the *number of visits of the random walk to the origin in time n*. We propose to show, for every recurrent aperiodic random walk in the plane which satisfies equation (7) in the discussion following T1, that

$$\lim_{n\to\infty} \frac{\mathbf{P}_x[\mathbf{N}_n = m]}{R_n} = 1$$

for all x in R and for every integer $m \geq 1$. When $m = 1$, observe that

$$\frac{\mathbf{P}_x[\mathbf{N}_n = 1]}{R_n} = \frac{1}{R_n} \sum_{k=1}^{n} \mathbf{P}_x[\mathbf{T} = k] R_{n-k}.$$

Hence the desired result was proved in the course of deriving (8) and (9) from condition (7) in the discussion following the proof of T1. Nor is it difficult to extend the argument to the case $m > 1$. One has

$$\mathbf{P}_x[\mathbf{N}_n = m] = \sum_{k=1}^{n} \mathbf{P}_x[\mathbf{T} = k] \mathbf{P}_0[\mathbf{N}_{n-k} = m - 1],$$

which leads to a simple proof by induction; for if

$$f_n = \frac{\mathbf{P}_0[\mathbf{N}_n = m - 1]}{R_n} \to 1$$

as $n \to \infty$, then we truncate

$$\frac{\mathbf{P}_x[\mathbf{N}_n = m]}{R_n} = \frac{1}{R_n} \sum_{k=1}^{n-M} \mathbf{P}_x[\mathbf{T} = k] R_{n-k} f_{n-k}$$
$$+ \frac{1}{R_n} \sum_{k=n-M+1}^{n} \mathbf{P}_x[\mathbf{T} = k] R_{n-k} f_{n-k}.$$

In view of (11) the last term tends to 0 for each fixed $M > 0$. If M is chosen so that $|f_k - 1| < \epsilon$ when $k \geq M$, then we get

$$\overline{\lim_{n\to\infty}} \left| \frac{\mathbf{P}_x[\mathbf{N}_n = m]}{R_n} - 1 \right| < \epsilon,$$

which completes the proof since ϵ is arbitrary.

Problems

1. For an entirely arbitrary random walk (recurrent or transient) prove that
$$\Pi_A(x,x) = \Pi_A(y,y),$$
if the set A consists of the *two distinct points* x and y.

2. For transient random walk show that
$$g_{(0)}(x,y) = G(x,y) - \frac{G(x,0)G(0,y)}{G(0,0)}.$$

3. In this and the next two problems we develop the theory of *recurrent aperiodic symmetric random walk in the plane* by following a different route from that in sections 11, 12, and 13. The proof of P12.1 is taken as the basis of further developments. It yields
$$a(x) = (2\pi)^{-2} \int \frac{1 - \cos x \cdot \theta}{1 - \phi(\theta)} d\theta.$$

Now use the monotone convergence theorem (this is not possible if the random walk is unsymmetric!) to conclude that for every $x \in R$
$$\sum_{t \in R} P(x,t)a(t) = (2\pi)^{-2} \int \frac{1 - \phi(\theta) \cos x \cdot \theta}{1 - \phi(\theta)} d\theta = a(x) + \delta(x,0).$$

4. Prove that the operator $A = A(x,y)$, restricted to a finite set B, $|B| > 1$, has one simple positive eigenvalue, that its other eigenvalues are negative, and that its inverse K has the property
$$K(\cdot\cdot) = \sum_{x \in B} \sum_{y \in B} K(x,y) \neq 0.$$

Hint: Following the author [93], who imitated Kac [50], suppose that λ_1 and λ_2 are two distinct positive eigenvalues of A, with eigenfunctions $u_1(x)$ and $u_2(x)$, $x \in B$. Since A is symmetric, u_1 and u_2 may be taken real and such that $(u_1,u_1) = (u_2,u_2) = 1$, $(u_1,u_2) = 0$. Here $(f,g) = \sum_{x \in B} f(x)g(x)$. One may choose real constants α_1, α_2, not both zero, such that $v(x) = \alpha_1 u_1(x) + \alpha_2 u_2(x)$ satisfies $\sum_{x \in B} v(x) = 0$. Now show by direct computation that
$$(v,Av) = \alpha_1^2 \lambda_1 + \alpha_2^2 \lambda_2 > 0.$$

On the other hand, using the definition of $a(x) = A(x,0)$ in problem 3, show that
$$(v,Av) = -(2\pi)^{-2} \int \left| \sum_{x \in B} v(x) e^{ix \cdot \theta} \right|^2 [1 - \phi(\theta)]^{-1} d\theta \leq 0$$
with equality if and only if $v(x) \equiv 0$ on B.

5. According to P13.2 the hitting probability function $H_B(x,y)$ is uniquely determined by its properties

(a) $\quad \sum_{t \in R} P(x,t) H_B(t,y) - H_B(x,y) = 0, \quad x \in R - B$

(b) $\quad H_B(x,y) = \delta(x,y), \quad x \in B$

(c) $\quad |H_B(x,y)| \leq M$ for some $M < \infty$.

Show that the function
$$\frac{K_B(\cdot y)}{K_B(\cdot \cdot)} + \sum_{t \in B} A(x,t) \left[K_B(t,y) - \frac{K_B(t \cdot) K_B(\cdot y)}{K_B(\cdot \cdot)} \right]$$
has these three properties. This may be done by using only the results of problems 3, 4 and of P12.2 and P13.2. Thus T11.1 may be proved for symmetric random walk without the manipulations in P11.1 through P11.8.

6. How must the conclusion of P12.3 be modified if one drops the "isotropy" assumption (c) to the effect that $Q(\theta) = \sigma^2 |\theta|^2$?

7. Verify the properties of capacity, stated without proof at the end of section 14.

8. Let $x_1, x_2, \ldots, x_n$ denote n fixed distinct points of the plane R, and let B denote the set $B = \{x_1, x_2, \ldots, x_n, y\}$. For two-dimensional aperiodic recurrent symmetric random walk prove that
$$\lim_{|y| \to \infty} H_B(\infty, y) = 1/2.$$

9. *Simple random walk in the plane.* The random walk, starting at $\mathbf{x}_0 = x \neq 0$ must visit either one, two, three, or all four of the four neighbors $i, -i, 1, -1$ of the origin before its first visit to 0. Calling $\mathbf{N}$ the exact number of neighbors visited, calculate
$$p_n = \lim_{|x| \to \infty} \mathbf{P}_x[\mathbf{N} = n], \quad n = 1, 2, 3, 4.$$
(It seems amusing that the ratios $p_1:p_2:p_3:p_4$ are very close to, but not quite, $4:3:2:1$.)

10. Does there exist a triangle, whose vertices are lattice points x_1, x_2, x_3 in the plane such that for simple random walk
$$H_{(x_1,x_2,x_3)}(\infty, x_i) = 1/3, \text{ for } i = 1, 2, 3?$$

11. Explain why it would have been disconcerting, at the end of section 15, to obtain any answer other than $2\alpha/\pi$.

Hint: The probability $\rho(z)$ of absorption on the real axis satisfies
$$(P - I)\rho(z) = 0 \text{ when } \text{Re}(z) > 0, \text{ Im}(z) > 0,$$
with boundary values $\rho(k) = 1$ for $k \geq 1$, $\rho(ik) = 0$ for $k \geq 1$. Hence the continuous analogue of ρ is a function which is bounded and harmonic

in the first quadrant, with boundary values 1 and 0 on the positive real and imaginary axes, respectively. (The solution is the actual hitting probability of the positive real axis for two-dimensional *Brownian motion*—see Itô and McKean [47], Ch. VII, or Lévy [72], Ch. VII.)

12. For *simple random walk in the plane* let A_n denote the event that $\mathbf{x}_n \neq \mathbf{x}_k$ for $k = 0, 1, \ldots, n-1$, i.e., that a *new* point is visited at time n. The point $\mathbf{x}_n + (\mathbf{x}_n - \mathbf{x}_{n-1})$ is the point which the random walk would visit at time $n+1$ if it continued in the same direction as that from $\mathbf{x}_{n-1}$ to $\mathbf{x}_n$. If B_n is the event that $\mathbf{x}_n + (\mathbf{x}_n - \mathbf{x}_{n-1}) \neq \mathbf{x}_k$ for $k = 0, 1, \ldots, n$, show that

$$\lim_{n \to \infty} \mathbf{P}_0[B_n \mid A_n] = 2 - \frac{4}{\pi}.$$

13. For *simple random walk in the plane*, let $\mathbf{T}$ denote the time of the first double point (the time of the first visit of $\mathbf{x}_n$ to a point which was occupied for some $n \geq 0$). Let

$$f(x) = \mathbf{P}_0[\mathbf{x}_\mathbf{T} = x], \qquad h(x) = \mathbf{P}_0[\mathbf{T}_x < \mathbf{T}],$$

where $\mathbf{T} = \min [k \mid k \geq 0, \mathbf{x}_k = x]$, and prove that

$$h(x) - \sum_{y \in R} h(y) P(y, x) = \delta(x, 0) - f(x), \qquad x \in R,$$

$$h(x) = \sum_{y \in R} a(x - y) f(y) - a(x), \qquad x \in R,$$

and finally observe that

$$\mathbf{E}_0[a(\mathbf{x}_\mathbf{T})] = 1, \qquad \mathbf{E}_0[|\mathbf{x}_\mathbf{T}|^2] = \mathbf{E}_0[\mathbf{T}] \mathbf{E}_0[|\mathbf{x}_1|^2].$$

14. Generalize the results of problem 13 to arbitrary aperiodic recurrent random walk in the plane with $\mathbf{E}_0[|\mathbf{x}_1^2|] < \infty$, and simultaneously to the following class of stopping times: $\mathbf{T}$ must have the property that, with probability one, the random walk visits each point at most once before time $\mathbf{T}$. Note that the generality of this result is precisely the reason why one cannot expect to use problem 13 to obtain the numerical value of $\mathbf{E}_0[\mathbf{T}]$ for simple random walk.

15. *Simple random walk in the plane.* Let A be a finite subset of R, and ∂A the set of those points which lie in $R - A$ but have one or more of their four neighbors in A. If a function φ is given on ∂A, then the *interior Dirichlet problem* consists in finding a function f on $A \cup \partial A$ such that $f = \varphi$ on ∂A and $Pf(x) = f(x)$ for all $x \in A$. Show that this problem has a unique solution, namely

$$f(x) = \mathbf{E}_x[\varphi(\mathbf{x}_\mathbf{T})], \qquad x \in A \cup \partial A,$$

where $\mathbf{T} = \min [k \mid k \geq 0, \mathbf{x}_k \in \partial A]$. Does this result remain valid when $|A| = \infty$? Can it be generalized to arbitrary aperiodic random walk (by defining ∂A as $R - A$)?

Chapter IV

RANDOM WALK ON A HALF-LINE[1]

17. THE HITTING PROBABILITY OF THE RIGHT HALF-LINE

For one-dimensional random walk there is an extensive theory concerning a very special class of infinite sets. These sets are half-lines, i.e., semi-infinite intervals of the form $a \leq x < \infty$ or $-\infty < x \leq a$, where a is a point (integer) in R. When $B \subset R$ is such a set it goes without saying that one can define the functions

$$Q_n(x,y), \quad H_B(x,y), \quad g_B(x,y), \quad x,y \text{ in } R,$$

just as in section 10, Chapter III. Of course the identities discovered there remain valid—their proof having required no assumptions whatever concerning the dimension of R, the periodicity or recurrence of the random walk, or the cardinality of the set B.

In this chapter, then, the random walk will be assumed to be *genuinely one dimensional,* i.e., according to D7.1 the zero-dimensional random walk with $P(0,0) = 1$ is excluded but no other assumptions

---

[1] A large literature is devoted to the results of this chapter, usually dealing with the more general situation of arbitrary, i.e., not integer valued, sums of random variables. In 1930 Pollaczek [82] solved a difficult problem in the theory of queues (mass service systems) which was only much later recognized to be a special case of the one-sided absorption problem (in section 17) and of the problem of finding the maximum of successive partial sums (in section 19). The basic theorems in this chapter were first proved by elementary combinatorial methods devised in [91] and [92]. For the sake of a brief unified treatment we shall instead employ the same Fourier analytical approach as Baxter in [2] and Kemperman in [57]. Kemperman's book contains a bibliography of important theoretical papers up to 1960. The vast applied literature is less accessible since identical probability problems often arise in the context of queueing, inventory or storage problems, particle counters, traffic congestion, and even in actuarial science.

are made concerning periodicity, which plays a very secondary role; only occasionally in results like P18.8, P19.4 which involve application of the renewal theorem will it be essential that the random walk be aperiodic. Similarly, the recurrence of the random walk will not be so important here as in Chapter III. When $B \subset R$ is a finite set it is clear that

$$\text{(1)} \qquad \sum_{y \in B} H_B(x,y) = 1, \qquad x \in R,$$

at least in the aperiodic case, if and only if the random walk is recurrent. But when B is a half-line it will turn out that the criterion for whether (1) holds or not is quite a different one. Let B be a right half-line, i.e., $B = [x \mid a \leq x < \infty]$, and consider Bernoulli random walk with $P(0,1) = p$, $P(0,-1) = q = 1 - p$. When $p = q$ the random walk is recurrent so that (1) holds since every point is then visited with probability one, and therefore also every subset B of R. When $p > q$ the random walk is transient, but (1) is still true since every point to the right of the starting point is visited with probability one. Hence (1) can hold for transient as well as recurrent random walk. For a case when (1) fails one has of course to resort to a transient random walk, and indeed Bernoulli random walk with $p < q$ provides an example. These examples indicate that it is not quite trivial to determine when (1) holds. We shall see, and this is not surprising, that (1) is equivalent to the statement that the random walk visits the half-line B infinitely often with probability one, regardless of the starting point, and an extremely simple necessary and sufficient condition for (1) to hold will be given in Theorem T1 of this section.

The object of our study is to gain as much information as possible about the hitting probabilities $H_B(x,y)$ and the Green function $g_B(x,y)$ of a half-line. To start with the hitting probabilities, one might approach the problem by trying to solve the exterior Dirichlet problem

$$\text{(2)} \qquad \sum_{t \in R} P(x,t) H_B(t,y) - H_B(x,y) = 0 \text{ if } x \in R - B, \quad y \in B$$

together with the boundary condition

$$H_B(x,y) = \delta(x,y) \text{ when } x \in B \text{ and } y \in B.$$

These equations are obviously correct in view of P10.1, part (a).

Another possibility would be to utilize part (c) of the same proposition, namely

(3) $$H_B(x,y) = \sum_{t \in R-B} g_B(x,t) P(t,y), \qquad x \in R - B, \quad y \in B.$$

The methods we will actually use will be based on an identity very much like (3). In the process we shall require simple but quite powerful methods of Fourier analysis, or alternatively of complex variable theory. To avoid convergence difficulties, which one might encounter if one tried to form Fourier series like

$$\sum_{y \in R} g_B(x,y) e^{iy\theta}$$

(a series which need not converge, as later developments will show) one is tempted to introduce generating functions. In other words,

$$g_B(x,y) = \sum_{n=0}^{\infty} Q_n(x,y)$$

is replaced by the series

$$\sum_{n=0}^{\infty} t^n Q_n(x,y),$$

with $0 \leq t < 1$. It will be unnecessary to give this series a name, however, since far more drastic changes in notation are both convenient, and in accord with tradition in this particular subject. We shall, in short, switch back to the notation and terminology in D3.1, describing the random walk $\mathbf{x}_n$ with $\mathbf{x}_0 = 0$, as

$$\mathbf{x}_n = \mathbf{X}_1 + \cdots + \mathbf{X}_n = \mathbf{S}_n,$$

the $\mathbf{X}_i$ being independent identically distributed integer valued random variables with $\mathbf{P}[\mathbf{X}_i = x] = P(0,x)$. Thus we shall arrange matters so that $\mathbf{x}_0 = 0$ as a general rule, and exceptions from this general rule will be carefully noted. The principal new definition we need for now is

D1 $\qquad \mathbf{T} = \min [n \mid 1 \leq n \leq \infty; \mathbf{S}_n > 0],$
$\qquad\qquad \mathbf{T}' = \min [n \mid 1 \leq n \leq \infty; \mathbf{S}_n \geq 0].$

Thus $\mathbf{T}$ is the first time that the random walk $\mathbf{x}_n$ is in the set $B = [1,\infty)$. It is infinite if there is no first time. In the notation of Chapter III,

$$\mathbf{P}[\mathbf{T} < \infty] = \sum_{y=1}^{\infty} H_B(0,y),$$
$$\mathbf{P}[\mathbf{S}_\mathbf{T} = y; \mathbf{T} < \infty] = H_B(0,y) \text{ for } y > 0 \text{ if } B = [1,\infty).$$

Similarly

$$\mathbf{P}[\mathbf{T}' < \infty] = \sum_{y=0}^{\infty} \Pi_{B'}(0,y),$$

$\mathbf{P}[\mathbf{S}_{\mathbf{T}'} = y, \mathbf{T}' < \infty] = \Pi_{B'}(0,y)$ for $y \geq 0$ if $B' = [0,\infty)$.

The real work of this chapter now begins with a lemma which imitates and extends equation (3) above.

P1
$$\mathbf{E}\left[\sum_{k=0}^{\mathbf{T}-1} t^k e^{i\theta S_k}\right] = \sum_{k=0}^{\infty} t^k \mathbf{E}[e^{i\theta S_k}; \mathbf{T} > k]$$
$$= [1 - t\phi(\theta)]^{-1}\{1 - \mathbf{E}[t^{\mathbf{T}} e^{i\theta S_{\mathbf{T}}}]\},$$

for $0 \leq t < 1$, and $-\infty < \theta < \infty$. The same result holds with $\mathbf{T}$ replaced throughout by $\mathbf{T}'$.

Remark (a): When $\mathbf{T} = \infty$, $t^{\mathbf{T}} = 0$, so that it is not necessary to define $e^{i\theta S_{\mathbf{T}}}$ when $\mathbf{T} = \infty$.

Remark (b): We notice that P1 extends (3) as follows.

$$\sum_{k=0}^{\infty} t^k \mathbf{E}[e^{i\theta S_k}; \mathbf{T} > k] = \sum_{k=0}^{\infty} t^k \sum_{y=-\infty}^{0} Q_k(0,y) e^{i\theta y}$$

is a Fourier series, its Fourier coefficients being

$$\sum_{k=0}^{\infty} t^k Q_k(0,y) \text{ when } y \leq 0, \; 0 \text{ when } y > 0.$$

Now we write P1 in the form

$$[1 - t\phi(\theta)] \sum_{k=0}^{\infty} t^k \mathbf{E}[e^{i\theta S_k}; \mathbf{T} > k] = 1 - \mathbf{E}[t^{\mathbf{T}} e^{i\theta S_{\mathbf{T}}}].$$

Each side is a Fourier series, and we may equate Fourier coefficients. On the left-hand side one uses the convolution theorem (P6.3), keeping in mind that $\phi(\theta)$ is the Fourier series of $P(0,x)$. For $y > 0$ (and this is the most interesting case) one gets

$$-\sum_{k=0}^{\infty} t^{k+1} \sum_{x=-\infty}^{0} Q_k(0,x) P(x,y) = -\sum_{k=1}^{\infty} t^k \mathbf{P}[\mathbf{T} = k; \mathbf{S}_{\mathbf{T}} = y], \quad y > 0.$$

Changing signs, and letting $t \nearrow 1$ (i.e., t approaches one from below),

$$\sum_{k=0}^{\infty} \sum_{x=-\infty}^{0} Q_k(0,x) P(x,y) = \mathbf{P}[\mathbf{S}_{\mathbf{T}} = y], \quad y > 0.$$

But if $B = [1,\infty)$ this is nothing but
$$\sum_{x=-\infty}^{0} g_B(0,x)P(x,y) = H_B(0,y),$$
which we recognize as equation (3).

Proof of P1: Although the foregoing considerations can obviously be made to yield a proof of P1, we proceed more quickly as follows.[2] The first identity in P1 is just a matter of definition:
$$E[e^{i\theta S_k}; T > k] = E[e^{i\theta S_k} A_k]$$
where A_k is a random variable which is one or zero according as $T > k$ or $T \leq k$. So we go on to the important part of P1, writing
$$E \sum_{0}^{T-1} t^k e^{i\theta S_k} = E \sum_{0}^{\infty} t^k e^{i\theta S_k} - E \sum_{T}^{\infty} t^k e^{i\theta S_k}.$$
There are no convergence difficulties, since $|t| < 1$. Now
$$E \sum_{0}^{\infty} t^k e^{i\theta S_k} = [1 - t\phi(\theta)]^{-1},$$
while
$$E \sum_{k=T}^{\infty} t^k e^{i\theta S_k} = \sum_{k=0}^{\infty} E[t^{T+k} e^{i\theta S_{T+k}}] = \sum_{k=0}^{\infty} t^k E[t^T e^{i\theta S_T} e^{i\theta(S_{T+k} - S_T)}].$$
But $S_{T+k} - S_T$ is independent of T and S_T—here is a typical application of P3.1(b) in section 3. One obtains
$$E \sum_{0}^{T-1} t^k e^{i\theta S_k} = [1 - t\phi(\theta)]^{-1} - \sum_{k=0}^{\infty} t^k \phi^k(\theta) E[t^T e^{i\theta S_T}]$$
$$= [1 - t\phi(\theta)]^{-1} \{1 - E[t^T e^{i\theta S_T}]\}.$$

That concludes the proof of the first part, and the argument for T' is omitted, being exactly the same. In fact, it is interesting to observe that *no use at all* was made of the definition of T, other than its property of being a stopping time.

Our observation that P1 is valid for any stopping time T should have a sobering influence indeed. As we have so far used none of the properties of our special stopping time T it would appear that we are still far from our goal—which is to calculate $E[t^T e^{i\theta S_T}]$, and similar characteristic functions, as explicitly as possible. Indeed P1 is of no help at all for arbitrary stopping times and whatever success we shall have will depend on the special nature of T and T' as hitting times of a half-line.

[2] For an even shorter proof of P1 and of P4 below in the spirit of Fourier analysis, see Dym and McKean [S7; pp. 184–187].

17. HITTING PROBABILITY OF THE RIGHT HALF-LINE

The results from classical analysis that we now introduce will be helpful in working with the very special type of Fourier series, which we have already encountered—namely Fourier series of the form

$$\sum_{k=0}^{\infty} a_k e^{ik\theta} \quad \text{or} \quad \sum_{k=-\infty}^{0} a_k e^{ik\theta}.$$

D2 *If a Fourier series* $\psi(\theta) = \sum_{k=-\infty}^{\infty} c_k e^{ik\theta}$ *is absolutely convergent, i.e., if* $\sum_{k=-\infty}^{\infty} |c_k| < \infty$, *it will be called exterior if in addition* $c_k = 0$ *for* $k > 0$, *and interior if* $c_k = 0$ *for* $k < 0$. *If a complex valued function* $f(z)$ *is analytic in the open disc* $|z| < 1$ *and continuous on the closure* $|z| \le 1$, *it is called an inner function. If* $f(z)$ *is analytic in* $|z| > 1$ *and continuous and bounded on* $|z| \ge 1$, *then* $f(z)$ *is called an outer function. The symbols* ψ_i, f_i *will be used to denote interior Fourier series and inner functions, while* ψ_e, f_e *will denote exterior Fourier series and outer functions.*

P2 *Given an interior Fourier series* $\psi_i(\theta)$, *there is a unique inner function* $f_i(z)$ *such that* $f_i(e^{i\theta}) = \psi_i(\theta)$ *for real* θ. *Similarly* $\psi_e(\theta)$ *can be extended uniquely to* $f_e(z)$.

Proof: The extension of

$$\psi_i(\theta) = \sum_{k=0}^{\infty} c_k e^{ik\theta}$$

is

$$f_i(z) = \sum_{k=0}^{\infty} c_k z^k, \quad |z| \le 1,$$

which is an inner function. Uniqueness follows from the fact that an analytic function in $|z| < 1$ is determined by its boundary values.[3] The proof for ψ_e and f_e is the same, with $|z| \ge 1$ replacing $|z| \le 1$.

P3 *Given a pair* f_i *and* f_e *such that*

$$f_i(z) = f_e(z) \quad \text{when} \quad |z| = 1,$$

one can conclude that there is a constant c such that

$$f_i(z) = c \text{ for } |z| \le 1, \quad f_e(z) = c \text{ for } |z| \ge 1.$$

Proof: Define $g(z)$ as $f_i(z)$ when $|z| < 1$, as $f_e(z)$ when $|z| > 1$, and as the common value of f_i and f_e when $|z| = 1$. Then g is analytic everywhere except possibly on the unit circle. However

---

[3] If there were two functions, their difference would have a maximum somewhere in $|z| < 1$. This contradicts the maximum modulus principle. If necessary, consult [1] on this point and concerning the theorems of Morera and Liouville which are used below.

g is continuous everywhere and also bounded. The continuity of g suffices to apply Morera's theorem: one integrates g around a point z_0 with $|z_0| = 1$ and verifies that the integral is zero. Morera's theorem then guarantees that g is analytic at z_0. Since z_0 is arbitrary, g is analytic everywhere. But such a function, if in addition bounded, reduces to a constant by Liouville's theorem. Thus $g(z) \equiv c$ which implies P3.

A pair of useful inner and outer functions are f_i and f_e defined by

D3
$$f_i(t;z) = e^{-\sum_1^\infty \frac{t^k}{k} \mathbf{E}[z^{\mathbf{S}_k}; \mathbf{S}_k > 0]}$$
$$f_e(t;z) = e^{-\sum_1^\infty \frac{t^k}{k} \mathbf{E}[z^{\mathbf{S}_k}; \mathbf{S}_k < 0]}$$
$$c(t) = e^{-\sum_1^\infty \frac{t^k}{k} \mathbf{P}[\mathbf{S}_k = 0]}.$$

Their crucial importance in our work is due to

P4 When $z = e^{i\theta}$, θ real, $0 \le t < 1$
$$1 - t\phi(\theta) = c(t) f_i(t;z) f_e(t;z).$$

Proof: First observe that $f_i(t;z)$ as given in D3 is really an inner function of z for $0 \le t < 1$. One has the estimate
$$\left| \sum_1^\infty \frac{t^k}{k} \mathbf{E}[z^{\mathbf{S}_k}; \mathbf{S}_k > 0] \right| \le \sum_1^\infty \frac{t^k}{k} \mathbf{P}[\mathbf{S}_k > 0] \le \ln \frac{1}{1-t}$$
when $|z| \le 1$ so that $|f_i(t;z)| \le 1 - t$ for $|z| \le 1$, and similarly $|f_e(t;z)| \le 1 - t$ for $|z| \ge 1$. Analyticity is also obvious since
$$\sum_1^\infty \frac{t^k}{k} \mathbf{E}[z^{\mathbf{S}_k}; \mathbf{S}_k > 0] = \sum_{n=1}^\infty z^n \sum_{k=1}^\infty \frac{t^k}{k} \mathbf{P}[\mathbf{S}_k = n],$$
and so is the fact that f_i and f_e satisfy the appropriate continuity and boundedness conditions.

To check P4, note that in view of $|t\phi(\theta)| < 1$,
$$1 - t\phi(\theta) = e^{-\sum_1^\infty \frac{t^k}{k} \phi^k(\theta)},$$
$$\sum_{k=1}^\infty \frac{t^k}{k} \phi^k(\theta) = \sum_{k=1}^\infty \frac{t^k}{k} \mathbf{E}[e^{i\theta \mathbf{S}_k}]$$
$$= \sum_{k=1}^\infty \frac{t^k}{k} \mathbf{E}[e^{i\theta \mathbf{S}_k}; \mathbf{S}_k > 0]$$
$$+ \sum_{k=1}^\infty \frac{t^k}{k} \mathbf{E}[e^{i\theta \mathbf{S}_k}; \mathbf{S}_k < 0] + \sum_{k=1}^\infty \frac{t^k}{k} \mathbf{P}[\mathbf{S}_k = 0].$$

That gives the desired *factorization* of $1 - t\phi$.

Now we are ready for the first results of probabilistic interest.

P5 (a) $\qquad 1 - \mathbf{E}[t^\mathbf{T} z^{\mathbf{S_T}}] = f_i(t;z),$
(b) $\qquad 1 - \mathbf{E}[t^{\mathbf{T'}} z^{\mathbf{S_{T'}}}] = c(t)f_i(t;z),$
(c) $\qquad \sum_{k=0}^{\infty} t^k \mathbf{E}[z^{\mathbf{S}_k}; \mathbf{T} > k] = [c(t)f_e(t;z)]^{-1},$
(d) $\qquad \sum_{k=0}^{\infty} t^k \mathbf{E}[z^{\mathbf{S}_k}; \mathbf{T'} > k] = [f_e(t;z)]^{-1}.$

Here $0 \le t < 1$, $|z| \le 1$ *in* (a) *and* (b), *while* $|z| \ge 1$ *in* (c) *and* (d).

Proof: To obtain parts (a) and (c) (the proofs of (b) and (d) are quite similar and will be omitted) we let

$$g_e(z) = \sum_{k=0}^{\infty} t^k \mathbf{E}[z^{\mathbf{S}_k}; \mathbf{T} > k],$$
$$h_i(z) = 1 - \mathbf{E}[t^\mathbf{T} z^{\mathbf{S_T}}].$$

One easily verifies that $g_e(z)$ is an outer function, according to D2, and $h_i(z)$ an inner function. An easy way to do this, say in the case of $g_e(z)$, is to look at

$$g_e(e^{i\theta}) = \sum_{k=0}^{\infty} t^k \mathbf{E}[e^{i\theta \mathbf{S}_k}; \mathbf{T} > k] = \sum_{y=-\infty}^{0} e^{i\theta y} \sum_{k=0}^{\infty} t^k \mathbf{P}[\mathbf{S}_k = y; \mathbf{T} > k]$$

which clearly is an exterior Fourier series. In view of P2, $g_e(z)$ is an exterior function.

Now we can write P1 in the form

$$g_e(e^{i\theta}) = [1 - t\phi(\theta)]^{-1} h_i(e^{i\theta}),$$

and in view of P4

$$c(t) f_i(t;z) f_e(t;z) g_e(z) = h_i(z)$$

when $|z| = 1$. Since f_i, f_e, and $c(t)$ are never zero (they are exponentials!),

$$g_e(z) f_e(t;z) = \frac{h_i(z)}{c(t) f_i(t;z)}, \qquad |z| = 1.$$

By P3 there is a constant k (it may depend on t but not on z) such that

$$g_e(z) f_e(t;z) = k \text{ for } |z| \ge 1$$
$$h_i(z) = k c(t) f_i(t;z) \text{ for } |z| \le 1.$$

To determine this constant let $z \to 0$ in the last identity (this is one of several easy ways of doing it). Clearly

$$\lim_{z \to 0} h_i(z) = \lim_{z \to 0} f_i(t;z) = 1,$$

so that $k = [c(t)]^{-1}$. Hence

$$h_i(z) = f_i(t;z),$$

which is part (a) of P5 and

$$g_e(z) = [c(t)f_e(t;z)]^{-1}$$

which is part (c).

E1 For *one-dimensional symmetric random walk*, certain simplifications of the formulas in D3, P4, and P5 are worth examining in detail. From $P(x,y) = P(y,x)$ it follows that $\mathbf{P}[\mathbf{S}_n = k] = \mathbf{P}[\mathbf{S}_n = -k]$, or

$$\mathbf{E}[e^{i\theta \mathbf{S}_n}; \mathbf{S}_n > 0] = \mathbf{E}[e^{-i\theta \mathbf{S}_n}; \mathbf{S}_n < 0].$$

In other words,

(1) $\qquad f_i(t;z) = \overline{f_e(t;z)}$ for $0 \le t < 1$ and $|z| = 1$.

P4 then gives

(2) $\qquad\qquad 1 - t\phi(\theta) = c(t) \, | f_i(t;e^{i\theta})|^2,$

and by P5 (a) this becomes

(3) $\qquad\qquad 1 - t\phi(\theta) = c(t) \, | 1 - \mathbf{E}[t^{\mathbf{T}} e^{i\theta \mathbf{S}_{\mathbf{T}}}]|^2.$

Equations (2) and (3) constitute a very interesting form of a classical theorem in Fourier analysis. In its first and simplest form it is due to Fejér [30], 1915, and concerns the representation of non-negative trigonometric polynomials. It states that if

$$\psi(\theta) = a_0 + 2 \sum_{k=1}^{n} [a_k \cos k\theta + b_k \sin k\theta]$$

is non-negative for all real values of θ, then there exists a polynomial of degree n

$$f(z) = \sum_{k=0}^{n} c_k z^k,$$

such that

(4) $\qquad\qquad \psi(\theta) = | f(e^{i\theta})|^2$ for all real θ.

Equation (3) of course constitutes a generalization of (4) in the sense that $1 - t\phi(\theta)$ may be an infinite trigonometric series rather than a trigonometric polynomial. But we get an actual example of (4) by considering a random walk such that $P(0,x) = 0$ when $|x| > n$ and $P(0,n) > 0$. Then

$$1 - t\phi(\theta) = 1 - tP(0,0) - 2t \sum_{k=1}^{n} P(0,k) \cos k\theta,$$

and since $\phi(\theta)$ is real and $|\phi(\theta)| \leq 1$

$$1 - t\phi(\theta) = \psi(\theta)$$

is a non-negative trigonometric polynomial of degree n. Setting

$$f(z) = \sqrt{c(t)}\,\{1 - \mathbf{E}[t^T z^{S_T}]\},$$

equation (3) shows that

$$1 - t\phi(\theta) = \psi(\theta) = |f(z)|^2 \quad \text{when } z = e^{i\theta},$$

and we have an example of the Fejér representation in equation (4) if we can demonstrate that $f(z)$ is a polynomial of degree n in z. But this is obvious since

$$f(z) = \sum_{k=0}^{\infty} c_k z^k$$

with

$$c_k = \sqrt{c(t)}\,\delta(k,0) - \sqrt{c(t)} \sum_{j=1}^{\infty} \mathbf{P}[\mathbf{T} = j;\, \mathbf{S_T} = k]t^j,$$

and $c_k = 0$ for $k > n$ since $P(0,x) = 0$ for $k > n$, which implies that

$$\mathbf{P}[\mathbf{T} = j;\, \mathbf{S_T} = k] \leq \mathbf{P}[\mathbf{S_T} = k] = 0 \quad \text{for } k > n.$$

Let us apply this result to the determination of $\mathbf{E}[t^T]$ for *simple random walk!* It is probabilistically obvious that $\mathbf{S_T} = 1$. Thus

$$\psi(\theta) = 1 - t\phi(\theta) = 1 - t\cos\theta,$$

while

$$f(z) = \sqrt{c(t)}\,\{1 - z\mathbf{E}[t^T]\}.$$

Now we shall solve the equation

$$1 - t\cos\theta = c(t)\,|\,1 - e^{i\theta}\mathbf{E}[t^T]\,|^2, \quad -\infty < \theta < \infty$$

for $c(t)$ and $\mathbf{E}[t^T]$. Expanding the right-hand side,

$$[c(t)]^{-1}(1 - t\cos\theta) = 1 - 2\cos\theta\,\mathbf{E}[t^T] + \{\mathbf{E}[t^T]\}^2,$$

and since θ is arbitrary,

$$t[c(t)]^{-1} = 2\mathbf{E}[t^T],$$
$$[c(t)]^{-1} = 1 + \{\mathbf{E}[t^T]\}^2.$$

There is only one solution if one requires that $0 \le \mathbf{E}[t^T] < 1$ when $0 \le t < 1$, and it is

$$\mathbf{E}[t^T] = \frac{1 - \sqrt{1-t^2}}{t},$$

$$c(t) = \frac{1 + \sqrt{1-t^2}}{2}.$$

For *arbitrary symmetric random walk* one must of course be satisfied with somewhat less. Taking $0 \le t < 1$ and $0 \le z = r < 1$, P5 gives

$$\sqrt{c(t)}\,[1 - \mathbf{E}[t^T r^{S_T}]] = e^{-\sum_{k=1}^{\infty} \frac{t^k}{k}\{\frac{1}{2}\mathbf{P}[S_k=0] + \mathbf{E}[r^{S_k};S_k>0]\}}$$
$$= e^{-\frac{1}{2}\sum_{k=1}^{\infty} \frac{t^k}{k}\mathbf{E}[r^{|S_k|}]},$$

where we have used the symmetry property in concluding that

$$\tfrac{1}{2}\mathbf{P}[S_k = 0] + \mathbf{E}[r^{S_k};S_k > 0] = \tfrac{1}{2}\mathbf{E}[r^{|S_k|}].$$

One can write

$$\sum_{k=1}^{\infty} \frac{t^k}{k} \mathbf{E}[r^{|S_k|}] = \sum_{n=-\infty}^{\infty} r^{|n|} \sum_{k=1}^{\infty} \frac{t^k}{k} \mathbf{P}[S_k = n].$$

The sequence $r^{|n|}$ has the Fourier series

$$\sum_{n=-\infty}^{\infty} r^{|n|} e^{in\theta} = \frac{1 - r^2}{1 + r^2 - 2r\cos\theta}, \qquad 0 \le r < 1,$$

and the sequence

$$c_n = \sum_{k=1}^{\infty} \frac{t^k}{k} \mathbf{P}[S_k = n]$$

has the Fourier series

$$\sum_{n=-\infty}^{\infty} c_n e^{in\theta} = -\ln[1 - t\phi(\theta)], \qquad 0 \le t < 1.$$

As both sequences are absolutely convergent one can apply Parseval's formula (P6.2) and conclude

$$\sum_{n=-\infty}^{\infty} c_n r^{|n|} = -\frac{1}{2\pi} \int_{-\pi}^{\pi} \frac{1-r^2}{1+r^2-2r\cos\theta} \ln[1 - t\phi(\theta)]\,d\theta.$$

Putting the pieces together, it is clear that we have

(5) $\quad \sqrt{c(t)}\,\{1 - \mathbf{E}[t^T r^{S_T}]\}$
$$= e^{\frac{1}{4\pi}\int_{-\pi}^{\pi} \frac{1-r^2}{1+r^2-2r\cos\theta} \ln[1-t\phi(\theta)]\,d\theta}, \qquad 0 \le t,r < 1.$$

Finally, setting $r = 0$ in (5) gives

(6) $$c(t) = e^{\frac{1}{2\pi}\int_{-\pi}^{\pi} \ln[1-t\phi(\theta)]\,d\theta}, \qquad 0 \leq t < 1.$$

One can of course go further on the basis of (5) and (6). To indicate the possibilities we shall let r tend to one in equation (5). The *Poisson kernel*

$$K(r,\theta) = \frac{1}{2\pi}\frac{1-r^2}{1+r^2-2r\cos\theta}$$

has the important property (cf. [1] or [105]) that

$$\lim_{r \nearrow 1} \int_{-\pi}^{\pi} K(r,\theta)\psi(\theta)\,d\theta = \psi(0)$$

if the function $\psi(\theta)$ is continuous at $\theta = 0$. And of course

$$\lim_{r \nearrow 1} \mathbf{E}[t^{\mathbf{T}} r^{S_\mathbf{T}}] = \mathbf{E}[t^{\mathbf{T}}],$$

(regardless of whether $\mathbf{T} < \infty$ with probability one or not). Thus it follows from equation (5) that

(7) $$\sqrt{c(t)}\{1 - \mathbf{E}[t^{\mathbf{T}}]\} = \sqrt{1-t}, \qquad 0 \leq t < 1.$$

But now (7) will yield the answer to the question whether or not $\mathbf{T} < \infty$ with probability one. $\mathbf{P}[\mathbf{T} < \infty] = 1$ if and only if

$$\sum_{k=1}^{\infty} \mathbf{P}[\mathbf{T} = k] = \lim_{t \nearrow 1} \mathbf{E}[t^{\mathbf{T}}] = 1.$$

According to D3

$$\lim_{t \nearrow 1} c(t) = e^{-\sum_{1}^{\infty} \frac{1}{k}\mathbf{P}[S_k = 0]},$$

and this limit is strictly positive by P7.6 which tells us that for some $A > 0$

$$\sum_{1}^{\infty}\frac{1}{k}\mathbf{P}[S_k = 0] \leq A\sum_{1}^{\infty} k^{-3/2} < \infty.$$

Therefore equation (7) gives

$$\lim_{t \nearrow 1}\{1 - \mathbf{E}[t^{\mathbf{T}}]\} = \mathbf{P}[\mathbf{T} = \infty] = 0,$$

so that $\mathbf{T} < \infty$ *with probability one for every symmetric random walk on the line* (provided, of course, that $P(0,0) < 1$).

E2 There is another type of random walk for which very explicit results may be obtained,[4] defined in D2.3 as *left-continuous random walk*.

---

[4] A far more complete discussion of special classes of random walk which permit explicit calculation is given by Kemperman [57], section 20.

It is characterized by the property of the transition function $P(x,y)$ that

$$P(0,x) = 0 \text{ when } x < -1, \quad P(0,-1) > 0.$$

It will be useful to introduce two analytic functions, to wit

$$P(z) = \sum_{n=0}^{\infty} P(0, n-1)z^n, \quad |z| \leq 1,$$

$$f_t(z) = z - tP(z), \quad |z| \leq 1, \quad 0 \leq t < 1.$$

In terms of $P(z)$ and $f_t(z)$ one obtains

(1) $$1 - t\phi(\theta) = e^{-i\theta}[e^{i\theta} - tP(e^{i\theta})] = e^{-i\theta} f_t(e^{i\theta}),$$

where $\phi(\theta)$ is the characteristic function of the random walk.

Everything that follows depends on the property of $f_t(z)$ that *for each t, $0 \leq t < 1$, it has exactly one simple zero in the disc $|z| \leq 1$. This zero, denoted by $r = r(t)$, is on the positive real axis.* This statement is relatively easy to verify. First we look for a zero r satisfying $0 \leq r < 1$. This means that $r = 0$ when $t = 0$, and

$$\frac{r}{t} = P(r) = P(0,-1) + \sum_{k=1}^{\infty} P(0,k-1)r^k,$$

when $0 < t < 1$. Thus we are looking for a point of intersection between the straight line r/t and the graph of $P(r)$, plotted against r. That there is such a point with $0 < r = r(t) < 1$ is clear from the intermediate value theorem: at $r = 0$ the straight line is below $P(r)$ since $P(0,-1) > 0$, and at $r = 1$ the straight line is above, since $t^{-1} > P(1) = 1$. To prove that this is the only zero in the unit disk, and that it is simple, we use Rouché's theorem (cf. [1]). This useful lemma from the theory of analytic functions will yield the desired conclusion if we can exhibit a function $g(z)$, analytic in $|z| < 1$ and with exactly one simple zero there, such that

$$|f_t(z) + g(z)| < |g(z)|$$

on every sufficiently large circle with center at the origin and radius less than one. (The content of Rouché's theorem is that the validity of the above inequality on a simple closed curve implies that $f_t(z)$ and $g(z)$ have the same number of zeros inside this curve.) A very simple choice for $g(z)$ works, namely $g(z) = -z$, for then the inequality reduces to

$$t |P(z)| < |z|.$$

This is true for $0 \leq t < 1$ and $t < |z| \leq 1$, since

$$t|P(z)| \leq t\sum_{n=0}^{\infty} P(0,n-1)|z|^n \leq tP(0,-1) + |z|[1 - P(0,-1)]$$
$$< |z| \text{ when } t < |z| \leq 1.$$

In terms of the function $r(t)$ our main results concerning left continuous random walk are

$$f_e(t;z) = 1 - \frac{r(t)}{z}, \qquad |z| \geq 1, \quad 0 \leq t < 1, \tag{2}$$

$$c(t)f_i(t;z) = \frac{z - tP(z)}{z - r(t)}, \qquad |z| \leq 1, \quad 0 \leq t < 1, \tag{3}$$

$$c(t) = \frac{tP(0,-1)}{r(t)}. \tag{4}$$

The proof parallels very closely that of P5. Equation (1) may be expressed as

$$1 - t\phi(\theta) = u_i(t;e^{i\theta})u_e(t;e^{i\theta}), \tag{5}$$

where

$$u_i(t;z) = \frac{z - tP(z)}{z - r(t)}, \qquad u_e(t;z) = 1 - \frac{r(t)}{z}.$$

It is quite easy to verify that u_i and u_e are inner and outer functions according to the definition D2. The singularity of $u_i(t;z)$ at the point $z = r(t)$ is only an apparent (removable) one since this point is a zero of the numerator $z - tP(z)$. Now one uses P4 to go from equation (5) to

$$c(t)f_i(t;z)f_e(t;z) = u_i(t;z)u_e(t;z), \qquad |z| = 1.$$

Just as in the proof of P5 one has to divide by u_i and by f_e. This is possible since $u_i \neq 0$ in $|z| \leq 1$, $r(t)$ being a simple zero. One gets a pair of functions

$$h_i(z) = \frac{c(t)f_i(t;z)}{u_i(t;z)}, \qquad |z| \leq 1,$$

$$h_e(z) = \frac{u_e(t;z)}{f_e(t;z)}, \qquad |z| \geq 1,$$

with the desired properties: h_i is an inner function, h_e an outer function, and $h_i(z) = h_e(z)$ on $|z| = 1$. By P3 both functions are constant, and have the same value k. To determine k, let $|z| \to \infty$. One finds

$$\lim_{|z| \to \infty} u_e(t;z) = \lim_{|z| \to \infty} f_e(t;z) = 1,$$

so that $k = 1$. Therefore

$$u_e(t;z) = f_e(t;z), \qquad |z| \geq 1,$$

which gives equation (2), and similarly (3) comes from the fact that h_i is one in $|z| \leq 1$. Finally equation (4) is obtained by letting $z \to 0$ in equation (3).

Now it is time to investigate the stopping time **T**, in the hope of establishing the highly reasonable result that $P[T < \infty] = 1$ if and only if $\mu \geq 0$, where μ is the mean of the random walk (D6.4). Using P5 and equations (3) and (4),

$$(6) \qquad 1 - E[t^T z^{S_T}] = \frac{r(t)}{tP(0,-1)} \cdot \frac{z - tP(z)}{z - r(t)}.$$

Letting $z \nearrow 1$ (along the real axis)

$$(7) \qquad 1 - E[t^T] = \frac{r(t)}{tP(0,-1)} \cdot \frac{1-t}{1-r(t)}.$$

Thus the question of whether $T < \infty$ with probability one hinges upon the behavior of $r(t)$ as $t \nearrow 1$. But $r(t)$ is defined as the root of the equation $r/t = P(r)$. It is clear from this definition of $r(t)$ that it is a monotone increasing function of t. To get more information we consider three separate cases, (i) and (iii) being illustrated in the graph below.

Case (i): $\qquad \mu = \sum_{k=-1}^{\infty} kP(0,k) = P'(1) - 1 > 0.$

Graph of $P(r)$

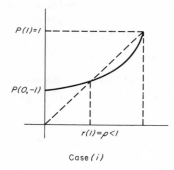
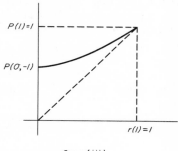

Case (i)　　　　　　　　　　Case (iii)

In this case

$$\lim_{t \nearrow 1} r(t) = \rho < 1,$$

so that by equation (7)

$$\lim_{t \nearrow 1} \{1 - E[t^T]\} = P[T = \infty] = 0.$$

Case (ii): $\qquad \mu = 0.$

Now

$$\lim_{t \nearrow 1} r(t) = 1,$$

but, using the definition of $r(t)$

$$\frac{1-t}{1-r(t)} = 1 - t\frac{1 - P[r(t)]}{1 - r(t)} = 1 - \frac{r(t)}{P[r(t)]} \cdot \frac{1 - P[r(t)]}{1 - r(t)}.$$

Hence (7) gives

$$\lim_{t \nearrow 1} \{1 - \mathbf{E}[t^\mathbf{T}]\} = \frac{1}{P(0,-1)} \left[1 - \lim_{r \nearrow 1} \frac{r}{P(r)} \frac{1-P(r)}{1-r} \right],$$

and

$$\lim_{r \nearrow 1} \frac{r}{P(r)} \frac{1-P(r)}{1-r} = \lim_{r \nearrow 1} \frac{1-P(r)}{1-r} = P'(1) = \mu + 1 = 1,$$

so that again

$$\mathbf{P}[\mathbf{T} = \infty] = 0.$$

Case (iii): $\quad\quad\quad\quad\quad \mu < 0.$

Again $r(t)$ tends to one as $t \nearrow 1$ and one can follow the argument in case (ii) to obtain

$$\lim_{t \nearrow 1} \{1 - \mathbf{E}[t^\mathbf{T}]\} = \mathbf{P}[\mathbf{T} = \infty] = \frac{1}{P(0,-1)} \left[1 - \lim_{r \nearrow 1} \frac{1-P(r)}{1-r} \right] = \frac{-\mu}{P(0,-1)}.$$

In Case (iii) it is also of interest to compute $\mathbf{P}[\mathbf{T}' = \infty]$. Because the random walk is left continuous,

$$\mathbf{P}[\mathbf{T}' = \infty] = 1 - F,$$

F being the probability of a return to 0 (D1.4). One gets, using P5 and equation (4)

$$1 - F = \lim_{t \nearrow 1} c(t) \mathbf{P}[\mathbf{T} = \infty] = -\mu.$$

Observe that this is equivalent to saying that $G = (-\mu)^{-1}$, a result which was obtained by entirely different methods in E3.1 of Chapter I.

The general question of when the hitting time $\mathbf{T}$ of the right half-line is finite is answered by

T1 *The following four statements are equivalent for genuinely one-dimensional random walk.*

(a) $\mathbf{P}[\mathbf{T} < \infty] = 1,$ \quad\quad (b) $\mathbf{P}[\mathbf{T}' < \infty] = 1,$

(c) $\sum_1^\infty \frac{1}{k} \mathbf{P}[S_k > 0] = \infty,$ \quad (d) $\sum_1^\infty \frac{1}{k} \mathbf{P}[S_k \geq 0] = \infty,$

so that either (c) *or* (d) *serves as a necessary and sufficient condition for* $\mathbf{T}$ *and* $\mathbf{T}'$ *to be finite. In the particular case when the first absolute moment* $m = \sum |x| P(0,x)$ *exists,* (a) *through* (d) *hold if and only if*

$$\mu = \sum x P(0,x) \geq 0.$$

Proof: Setting $z = 1$ in the first two statements of P5, one has

$$1 - \mathbf{E}[t^\mathbf{T}] = e^{-\sum_1^\infty \frac{t^k}{k} \mathbf{P}[S_k > 0]},$$

$$1 - \mathbf{E}[t^{\mathbf{T}'}] = e^{-\sum_1^\infty \frac{t^k}{k} \mathbf{P}[S_k \geq 0]}.$$

This shows that (a) is equivalent to (c) and (b) to (d). But (c) and (d) will be equivalent if we can show that the series $\sum k^{-1}\mathbf{P}[S_k = 0]$ converges. This follows from the estimate in P7.6 in Chapter II, but we also offer the following direct proof. From (b) in P5, setting $z = 0$,

$$1 - \mathbf{E}[t^{\mathbf{T}'}; \mathbf{S}_{\mathbf{T}'} = 0] = e^{-\sum_1^\infty \frac{t^k}{k} \mathbf{P}[S_k = 0]}.$$

If the series $\sum k^{-1}\mathbf{P}[S_k = 0]$ were divergent, we could conclude that

$$\mathbf{P}[\mathbf{T}' < \infty; \mathbf{S}_{\mathbf{T}'} = 0] = 1,$$

but this implies that $\mathbf{P}[\mathbf{X}_k > 0] = \sum_{x=1}^\infty P(0,x) = 0$. By applying the same argument to the reversed random walk it follows that also $\mathbf{P}[\mathbf{X}_k < 0] = 0$ so that $P(0,0) = 1$. But this is a degenerate case which we have excluded so that $\sum_{k=1}^\infty k^{-1}\mathbf{P}[S_k = 0]$ converges and (a) through (d) are equivalent.

For the last part of the theorem we use results from Chapter I. If $\mu > 0$, then $\lim_{n\to\infty} n^{-1}S_n = \mu > 0$ with probability one (this is P3.4), so that $\mathbf{T} < \infty$. When $\mu = 0$, the random walk is recurrent by P2.8 so that every point is visited, and a fortiori every half-line. Finally let $\mu < 0$. Then it is a consequence of the strong law of large numbers that $\lim_{n\to\infty} S_n = -\infty$. Now suppose that we had $\mathbf{T} < \infty$ with probability one. Then, for reasons which must be evident, $\mathbf{P}[\mathbf{T}_B < \infty] = 1$ for every half-line $B = [x \mid b \leq x < \infty]$. It follows that $\overline{\lim}_{n\to\infty} S_n = +\infty$, and this contradiction completes the proof of T1.

18. RANDOM WALK WITH FINITE MEAN

For the purpose of this section only, let it be assumed that the random walk has finite first absolute moment

$$m = \sum_{x=-\infty}^\infty |x| P(0,x) < \infty.$$

18. RANDOM WALK WITH FINITE MEAN

In the event that $\mu \geq 0$, it follows from T17.1 of the last section that the hitting time $\mathbf{T}$ of the right half-line $[1,\infty)$ is a random variable, in other words, that $\mathbf{T} < \infty$ with probability one. In this case the expectation of $\mathbf{T}$ and also of $\mathbf{S_T}$ is of interest—in fact of far greater interest for the theory in the next three chapters than one might guess at this point. We begin with

P1 *If $\mu = 0$ then $\mathbf{E}[\mathbf{T}] = \infty$, and if $\mu > 0$, then*

$$\mathbf{E}[\mathbf{T}] = e^{\sum_1^\infty \frac{1}{k} \mathbf{P}[\mathbf{S}_k \leq 0]} < \infty.$$

Proof:

$$\mathbf{E}[\mathbf{T}] = \sum_{k=1}^\infty k\mathbf{P}[\mathbf{T} = k] = \lim_{t \nearrow 1} \sum_{k=0}^\infty t^k \mathbf{P}[\mathbf{T} > k] = \lim_{t \nearrow 1} \frac{1 - \mathbf{E}[t^\mathbf{T}]}{1 - t},$$

as is verified by straightforward calculation. This limit may be finite or infinite. From part (a) of P17.5

$$1 - \mathbf{E}[t^\mathbf{T}] = e^{-\sum_1^\infty \frac{t^k}{k} \mathbf{P}[\mathbf{S}_k > 0]},$$

so that

$$\frac{1 - \mathbf{E}[t^\mathbf{T}]}{1 - t} = e^{\sum_1^\infty \frac{t^k}{k} \mathbf{P}[\mathbf{S}_k \leq 0]}$$

and

$$\mathbf{E}[\mathbf{T}] = \lim_{t \nearrow 1} e^{\sum_1^\infty \frac{t^k}{k} \mathbf{P}[\mathbf{S}_k \leq 0]} = e^{\sum_1^\infty \frac{1}{k} \mathbf{P}[\mathbf{S}_k \leq 0]} \leq \infty.$$

By Theorem T17.1, the series in the last exponent is $+\infty$ (diverges) when $\mu = 0$, whereas it is finite when $\mu > 0$. (Strictly speaking the theorem in question gives this result for the series whose general term is $k^{-1}\mathbf{P}[\mathbf{S}_k \geq 0]$ when $\mu < 0$, and we arrive at the desired conclusion by considering the reversed random walk $P^*(x,y) = P(y,x)$ whose partial sums are $-\mathbf{S}_k$.) That completes the proof.

P2 *When $\mu > 0$, $\mathbf{E}[\mathbf{S_T}] = \mu \mathbf{E}[\mathbf{T}]$.*

Proof: This result in fact holds for a large class of stopping times, as is hardly surprising since it is of the form

$$\mathbf{E}[\mathbf{X}_1 + \cdots + \mathbf{X_T}] = \mathbf{E}[\mathbf{X}_1]\mathbf{E}[\mathbf{T}]$$

which says that $\mathbf{T}$ may be treated as if it were independent of the sequence of identically distributed random variables $\mathbf{X}_i$. The usual proof is based on Wald's lemma in sequential analysis, which in turn may be derived from a simple form of a Martingale system theorem

(Doob [23], p. 350). But for the sake of completeness, here is a direct proof.

From P17.1

$$\sum_{k=0}^{\infty} t^k \mathbf{E}[e^{i\theta S_k}; \mathbf{T} > k] = \frac{-i\theta}{1 - t\phi(\theta)} \cdot \frac{1 - \mathbf{E}[t^{\mathbf{T}} e^{i\theta S_{\mathbf{T}}}]}{-i\theta}.$$

Since $\mu > 0$ we know from P1 that

$$\sum_{k=0}^{\infty} \mathbf{P}[\mathbf{T} > k] = \mathbf{E}[\mathbf{T}] < \infty.$$

Thus it appears reasonable to let $t \nearrow 1$, and to avoid difficulty on the right-hand side we do this only for such θ that $\phi(\theta) \neq 1$. Then

$$\sum_{k=0}^{\infty} \mathbf{E}[e^{i\theta S_k}; \mathbf{T} > k] = \frac{-i\theta}{1 - \phi(\theta)} \cdot \frac{1 - \mathbf{E}[e^{i\theta S_{\mathbf{T}}}]}{-i\theta},$$

when $\phi(\theta) \neq 1$. Since the random walk is nondegenerate there is a deleted neighborhood of $\theta = 0$ where $\phi(\theta) \neq 1$. Letting $\theta \to 0$, assuming only values in this neighborhood,

$$\lim_{\theta \to 0} \frac{-i\theta}{1 - \phi(\theta)} = \frac{1}{\mu},$$

by P6.4. Thus we conclude

$$\mathbf{E}[\mathbf{T}] = \sum_{k=0}^{\infty} \mathbf{P}[\mathbf{T} > k] = \frac{1}{\mu} \lim_{\theta \to 0} \frac{1 - \mathbf{E}[e^{i\theta S_{\mathbf{T}}}]}{-i\theta},$$

But $S_{\mathbf{T}}$ is a positive random variable and by P6.5 the existence of

$$\lim_{\theta \to 0} \frac{1 - \mathbf{E}[e^{i\theta S_{\mathbf{T}}}]}{-i\theta}$$

implies that $\mathbf{E}[S_{\mathbf{T}}]$ is finite and equal to this limit. That gives $\mathbf{E}[\mathbf{T}] = \mu^{-1}\mathbf{E}[S_{\mathbf{T}}]$ which was to be proved.

One problem concerning first moments remains to be treated, and that turns out to be the most interesting and difficult one. It concerns $\mathbf{E}[S_{\mathbf{T}}]$ for random walk with mean $\mu = 0$. To illustrate several aspects of this problem we return to the examples of section 17.

E1 The random walk is *symmetric*, so the mean is automatically zero, the first absolute moment being assumed to exist. However the variance

$$\sigma^2 = \sum_{x=-\infty}^{\infty} x^2 P(0,x)$$

may or may not be finite, and that will turn out to be the crux of the problem. We start with equation (3) in Example E17.1, and let $t \nearrow 1$ so that

(1) $$1 - \phi(\theta) = c \, | \, 1 - \mathbf{E}[e^{i\theta S_T}] \, |^2,$$

where

$$0 < c = \lim_{t \nearrow 1} c(t) = \lim_{t \nearrow 1} e^{\frac{1}{2\pi}\int_{-\pi}^{\pi} \ln[1 - t\phi(\theta)] \, d\theta} < \infty.$$

This much is clear about c from (3) in E17.1. Although one can also show without much difficulty that $\ln[1 - \phi(\theta)]$ is integrable on the interval $[-\pi, \pi]$ so that

$$c = e^{\frac{1}{2\pi}\int_{-\pi}^{\pi} \ln[1 - \phi(\theta)] \, d\theta},$$

this fact will not be needed here (see problem 1).

Now we divide equation (1) by θ^2 and write

(2) $$\frac{1 - \phi(\theta)}{\theta^2} = c \left| \frac{1 - \mathbf{E}[e^{i\theta S_T}]}{-i\theta} \right|^2, \qquad \theta \neq 0.$$

The point of this exercise is that when θ approaches zero,

$$\lim_{\theta \to 0} \frac{1 - \phi(\theta)}{\theta^2} = \frac{\sigma^2}{2}.$$

We knew this from P6.4 when $\sigma^2 < \infty$, and a simple calculation confirms that the identity remains correct when $\sigma^2 = \infty$, in the sense that the limit is then also infinite.

Considering the two cases separately, we have

(i) when $\sigma^2 < \infty$

$$\frac{\sigma^2}{2c} = \lim_{\theta \to 0} \left| \frac{1 - \mathbf{E}[e^{i\theta S_T}]}{-i\theta} \right|^2.$$

If we only knew a little more, namely that also the limit

$$\lim_{\theta \to 0} \frac{1 - \mathbf{E}[e^{i\theta S_T}]}{-i\theta}$$

exists and is positive, then by P6.5, $\mathbf{E}[S_T]$ would exist and equal this limit. This would give the result

(3) $$\frac{\sigma}{\sqrt{2c}} = \mathbf{E}[S_T].$$

This conclusion could be justified by careful Fourier analysis including some weakening of the hypotheses in P6.5, but it seems hardly worth the trouble. The additional difficulties we shall encounter in the nonsymmetric case will definitely call for a more sophisticated approach. The result, in T1 below, will of course contain the truth of (3) as a special case.

In case

(ii) when $\sigma^2 = \infty$, heuristic reasoning along the above lines very strongly suggests that

$$\mathbf{E}[\mathbf{S_T}] = \infty.$$

That is easy to prove, for if we had $\mathbf{E}[\mathbf{S_T}] < \infty$, then from P6.4 one would get the contradictory conclusion that

$$\sigma^2/2c = \{\mathbf{E}[\mathbf{S_T}]\}^2 < \infty.$$

A slightly different formulation of the problem is of considerable interest. By proper use of P17.5 one can rewrite equation (1) as

$$1 - \phi(\theta) = \{1 - \mathbf{E}[e^{i\theta \mathbf{S_T}}]\}\{1 - \mathbf{E}[e^{-i\theta \mathbf{S_{T'}}}]\}.$$

Again dividing by θ^2 and letting $\theta \to 0$, one is now led to the conjecture that

(4) $$\frac{\sigma^2}{2} = \mathbf{E}[\mathbf{S_T}]\mathbf{E}[\mathbf{S_{T'}}],$$

and of course it looks as though both expectations on the right would be finite if and only if σ^2 is finite.

E2 For the *left-continuous* random walk of example E17.2 we found (in equation (6)) that

$$1 - \mathbf{E}[t^T z^{\mathbf{S_T}}] = \frac{r(t)}{tP(0,-1)} \cdot \frac{z - tP(z)}{z - r(t)}.$$

When $\mu > 0$, we had $r(t) \to r(1) = \rho < 1$ as $t \nearrow 1$, so that a simple computation gives

$$\mathbf{E}[T] = \frac{\rho}{(1-\rho)P(0,-1)}.$$

Also

$$1 - \mathbf{E}[z^{\mathbf{S_T}}] = \frac{\rho}{P(0,-1)} \cdot \frac{z - P(z)}{z - \rho},$$

and differentiation with respect to z followed by letting $z \to 1$ gives $\mathbf{E}[\mathbf{S_T}] = \mu \mathbf{E}[T]$ which illustrates P2.

Now suppose $\mu = 0$. Then we know that $r(t) \to 1$ as $t \nearrow 1$, so that

$$1 - \mathbf{E}[z^{\mathbf{S_T}}] = \frac{1}{P(0,-1)} \cdot \frac{z - P(z)}{z - 1}.$$

In this case it is quite easy to show that $E[S_T] < \infty$ if and only if

$$\sigma^2 = \sum_{k=-1}^{\infty} k^2 P(0,k) = P''(1) < \infty,$$

but we omit the details.

In one important respect the two preceding examples were misleading. It is possible to have a random walk with $m < \infty$, $\mu = 0$, $\sigma^2 = \infty$ so that nevertheless $E[S_T] < \infty$. To find the simplest possible example, consider right-continuous random walk, i.e., $P(0,1) > 0$, $P(0,x) = 0$ for $x > 2$. It is clearly possible to choose $P(0,k)$ for $k = 1, 0, -1, -2, \ldots$ in such a way that $\mu = 0$ and at the same time $\sigma^2 = \infty$. But if $\mu = 0$ we know that $T < \infty$. Now S_T is a random variable which can assume only the value one. Therefore $E[S_T] = 1 < \infty$, even though $\sigma^2 = \infty$.

To describe the actual state of affairs in an intuitively pleasing fashion it is necessary to consider the behavior of the random walk in regard to the left as well as the right half-line. Some new notation is unfortunately required for this purpose.

D1 $T^* = \min [n \mid 1 \leq n \leq \infty; S_n < 0]$,

$T'^* = \min [n \mid 1 \leq n \leq \infty; S_n \leq 0]$,

$Z = S_T$, and

$\bar{Z} = S_{T'^*}$, Z and $\bar{Z}$ being defined only when T and T'^* are finite with probability one.

What motivates the notation in D1 is of course that T^* plays the role of T for the reversed random walk, and T'^* the role of T'. Z and $\bar{Z}$ were introduced mainly to avoid cumbersome notation. They were first used (1952) by Blackwell[5] and later Feller aptly christened them the *ladder random variables* of the random walk. The "ladder" referred to is a sequence $Z_1, Z_1 + Z_2, \ldots, Z_1 + \cdots + Z_n$, the Z_i being independent and distributed as Z. The terms in the ladder represent the first positive value of the random walk, the first value exceeding the first positive value, and so on. In a similar way $\bar{Z}$ determines a "ladder" to the left of the origin. In the context in which such ladders are useful mathematical constructions it is usually essential to know whether the "rungs" of the ladder, i.e., the random variables Z and $\bar{Z}$ have finite expectation. And the answer to this question is given by

[5] See Blackwell [4] and Feller [31], Vol. I, p. 280.

T1 *For arbitrary nondegenerate one-dimensional random walk (it is not assumed that $m < \infty$), the following two statements are equivalent:*

(A) $\mathbf{T} < \infty$, $\mathbf{T'^*} < \infty$, $\mathbf{E}[\mathbf{Z}] < \infty$, and $\mathbf{E}[-\overline{\mathbf{Z}}] < \infty$;

and

(B) $\mu = 0$, and $\sigma^2 < \infty$.

Proof: It is fairly easy to show that (A) implies (B). After doing so we shall pause and prove several lemmas, namely P3 through P5, and the proof of P5 will complete the proof of the theorem.[6]

Using P17.4, P17.5 of section 17 and the notation in D1

$$1 - t\phi(\theta) = \{1 - \mathbf{E}[t^{\mathbf{T}} e^{i\theta \mathbf{Z}}]\}\{1 - \mathbf{E}[t^{\mathbf{T'^*}} e^{i\theta \overline{\mathbf{Z}}}]\}$$

for $0 \leq t < 1$, θ real. Now we let $t \nearrow 1$ (this causes no concern) and just as in E1 we divide by θ^2 when $\theta \neq 0$. As we are assuming (A) it follows from P6.4 that

$$\lim_{\theta \to 0} \frac{1 - \mathbf{E}[e^{i\theta \mathbf{Z}}]}{i\theta} = \mathbf{E}[\mathbf{Z}]$$

and that a similar result holds for $\overline{\mathbf{Z}}$. This gives us

$$\lim_{\theta \to 0} \frac{1 - \phi(\theta)}{\theta^2} = \mathbf{E}[\mathbf{Z}]\mathbf{E}[-\overline{\mathbf{Z}}] < \infty.$$

Hence

$$\operatorname{Re} \frac{1 - \phi(\theta)}{\theta^2} = \sum P(0,x) \frac{1 - \cos x\theta}{\theta^2}$$

is a bounded function of θ. Therefore

$$\sum_{x=-M}^{M} x^2 P(0,x) = 2 \lim_{\theta \to 0} \sum_{x=-M}^{M} P(0,x) \frac{1 - \cos x\theta}{\theta^2} \leq 2\mathbf{E}[\mathbf{Z}]\mathbf{E}[-\overline{\mathbf{Z}}]$$

for every $M > 0$. It follows that the random walk has finite variance, and we conclude from P6.4 that the mean $\mu = 0$, and the variance

$$\sigma^2 = 2\mathbf{E}[\mathbf{Z}]\mathbf{E}[-\overline{\mathbf{Z}}] < \infty.$$

That completes the proof that (A) implies (B).

For the converse we require (for the first time!) a form of the Central Limit Theorem (P6.8).

---

[6] See problem 6 at the end of this chapter for a simple probabilistic proof, due to H. Kesten, of the fact that $\mu = 0$, $\sigma^2 < \infty$ implies $\mathbf{E}(\mathbf{Z}) < \infty$; in fact $\mu = 0$ and $m_{k+1} < \infty$, $k \geq 1$, implies $\mathbf{E}(\mathbf{Z}^k) < \infty$.

P3 *Suppose that $\mu = 0$ and $\sigma^2 < \infty$. Then*

$$\lim_{n \to \infty} \frac{1}{\sqrt{n}} \mathbf{E}[\mathbf{S}_n; \mathbf{S}_n \geq 0] = \frac{\sigma}{\sqrt{2\pi}},$$

$$\lim_{t \nearrow 1} \sqrt{1-t} \sum_{n=1}^{\infty} \frac{t^n}{n} \mathbf{E}[\mathbf{S}_n; \mathbf{S}_n \geq 0] = \frac{\sigma}{\sqrt{2}}.$$

Proof: An examination of the proof will show that one may assume, without loss of generality, that $\sigma^2 = 1$. Defining

$$F_n(x) = \sum_{t=-\infty}^{[xn^{1/2}]} P_n(0,t) = \mathbf{P}[\mathbf{S}_n \leq \sqrt{n}x],$$

$$F(x) = \frac{1}{\sqrt{2\pi}} \int_{-\infty}^{x} e^{-\frac{t^2}{2}} dt,$$

we find that the content of the Central Limit Theorem is the convergence of $F_n(x)$ to $F(x)$ at every real x. For every $A > 0$, $k > 0$

$$\left| \int_0^{\infty} x \, dF_k(x) - \int_0^{\infty} x \, dF(x) \right|$$
$$\leq \left| \int_0^A x \, dF_k(x) - \int_0^A x \, dF(x) \right| + \int_A^{\infty} x \, dF(x) + \int_A^{\infty} x \, dF_k(x).$$

The last integral is less than

$$\frac{1}{A} \int_A^{\infty} x^2 \, dF_k(x) = \frac{1}{A} \mathbf{E}\left[\frac{\mathbf{S}_k^2}{k}; \mathbf{S}_k^2 > Ak\right] \leq \frac{1}{A} \mathbf{E}\left[\frac{\mathbf{S}_k^2}{k}\right] = \frac{1}{A}.$$

Given any $\epsilon > 0$ we now choose A so large that simultaneously

(1) $$\frac{1}{A} < \frac{\epsilon}{3}, \quad \int_A^{\infty} x \, dF(x) < \frac{\epsilon}{3}.$$

Using this A, one can find N such that for $k > N$

(2) $$\left| \int_0^A x \, dF_k(x) - \int_0^A x \, dF(x) \right| < \frac{\epsilon}{3}.$$

(As an easy way of verifying the last statement, use integration by parts. Assuming that A is not a point of discontinuity of $F_k(x)$ for any k, one gets

$$\int_0^A x \, dF_k(x) - \int_0^A x \, dF(x) = A[F_k(A) - F(A)]$$
$$- \int_0^A [F_k(x) - F(x)] \, dx,$$

which gives (2) by use of the dominated convergence theorem). Combining (1) and (2) we have

$$\text{(3)} \qquad \left| \int_0^\infty x \, dF_k(x) - \int_0^\infty x \, dF(x) \right| < \epsilon$$

for large enough k. But

$$\int_0^\infty x \, dF_k(x) = \frac{1}{\sqrt{k}} \, \mathbf{E}[\mathbf{S}_k; \mathbf{S}_k \geq 0],$$

$$\int_0^\infty x \, dF(x) = \frac{1}{\sqrt{2\pi}},$$

and that proves the first part of P3. The second part reduces to showing that if a sequence c_n has a limit as $n \to \infty$, then

$$\lim_{t \nearrow 1} \sqrt{1-t} \sum_{n=1}^\infty \frac{c_n t^n}{\sqrt{n}} = \sqrt{\pi} \lim_{n \to \infty} c_n.$$

A typical Abelian argument accomplishes the proof, using the fact (derivable from Stirling's formula) that

$$\binom{-1/2}{n} \sim \frac{1}{\sqrt{n\pi}}, \qquad n \to \infty,$$

so that as t tends to one

$$\sqrt{1-t} \sum_{n=1}^\infty \frac{c_n t^n}{\sqrt{n}} \sim \sqrt{\pi} \left(\lim_{n \to \infty} c_n \right) \sqrt{1-t} \sum_{n=0}^\infty \binom{-1/2}{n} t^n$$
$$= \sqrt{\pi} \lim_{n \to \infty} c_n.$$

In the course of proving that (B) implies (A) in T1 we shall be able to conclude that

$$\lim_{t \nearrow 1} \sum_1^\infty \frac{t^k}{k} \{\tfrac{1}{2} - \mathbf{P}[\mathbf{S}_k > 0]\}$$

exists and is finite. In view of the Central Limit Theorem this is a power series of the form

$$\sum_1^\infty a_k t^k, \text{ with } a_k = o\!\left(\frac{1}{k}\right),$$

the "little o" meaning that $ka_k \to 0$ as $k \to \infty$. To such power series one can apply Tauber's theorem [39], the simplest and historically the first (1897) of many forms of the converse of Abel's theorem, to conclude that

P4 $$\lim_{t \nearrow 1} \sum_1^\infty \frac{t^k}{k}\left\{\frac{1}{2} - \mathbf{P}[S_k > 0]\right\} = \sum_1^\infty \frac{1}{k}\left\{\frac{1}{2} - \mathbf{P}[S_k > 0]\right\}.$$

Now we are ready to resume the proof of T1, and by following a rather elaborate route we shall obtain

P5 *If $\mu = 0$, $\sigma^2 < \infty$, then*

(a) $$0 < \mathbf{E}[Z] = \frac{\sigma}{\sqrt{2}}e^\alpha < \infty$$

(b) $$0 < \mathbf{E}[-\bar{Z}] = \frac{\sigma}{\sqrt{2}}e^{-\alpha} < \infty,$$

where α is the sum of the convergent[7] series

(c) $$\alpha = \sum_1^\infty \frac{1}{k}\left\{\frac{1}{2} - \mathbf{P}[S_k > 0]\right\}.$$

Proof: First of all it is clear that P5 gives more information than necessary to complete the proof of T1. (If (B) holds, then $m < \infty$ and $\mu = 0$, so that $\mathbf{T} < \infty$ and $\mathbf{T}'^* < \infty$ by T17.1; further $\sigma^2 < \infty$, so that P5 not only shows that Z and $\bar{Z}$ have finite moments, but also gives their explicit values. Hence a proof of P5 will complete the proof of T1.)

Next we note that it will suffice to prove (a) and (c) in P5. Once we have (a) and (c) the same proof which gave (a) and (c) can obviously be applied to the reversed random walk. It has for Z the random variable $-S_{T^*}$. Hence $\mathbf{E}[-S_{T^*}] < \infty$ and since $0 \leq -\bar{Z} \leq -S_{T^*}$ we have $\mathbf{E}[-\bar{Z}] < \infty$. But then we can apply the first part of theorem T1, which was already proved, and which yielded the information that

$$\sigma^2 = 2\mathbf{E}[Z]\mathbf{E}[-\bar{Z}].$$

Combined with (a) and (c) this shows that (b) holds.

---

[7] We shall prove only that the series in (c) converges. However, using delicate Fourier analytical methods, Rosén [88] has shown that this series converges absolutely.

Now we can decompose $\mathbf{E}[Z]$, which may be finite or infinite for all we know, into

$$(1) \qquad \mathbf{E}[Z] = \lim_{n\to\infty} \sum_{k=1}^{n} \mathbf{E}[S_k; \mathbf{T} = k]$$

$$= \lim_{n\to\infty} \sum_{k=1}^{n} \{\mathbf{E}[S_k; \mathbf{T} > k-1] - \mathbf{E}[S_k; \mathbf{T} > k]\}$$

$$= -\lim_{n\to\infty} \mathbf{E}[S_n; \mathbf{T} > n].$$

Using P17.5, part (c), we find that

$$\sum_{n=0}^{\infty} t^n \mathbf{E}[r^{S_n}; \mathbf{T} > n] = e^{\sum_{1}^{\infty} \frac{t^k}{k} \mathbf{E}[r^{S_k}; S_k \leq 0]},$$

for $0 \leq t < 1$, $r > 1$. To obtain first moments we differentiate with respect to r and subsequently let $r \searrow 1$. That gives

$$(2) \qquad \sum_{n=0}^{\infty} t^n \mathbf{E}[S_n; \mathbf{T} > n] = \sum_{1}^{\infty} \frac{t^k}{k} \mathbf{E}[S_k; S_k \leq 0] \, e^{\sum_{1}^{\infty} \frac{t^k}{k} \mathbf{P}[S_k \leq 0]}.$$

By Abel's theorem [39], applied to (1)

$$\mathbf{E}[Z] = -\lim_{t \nearrow 1} (1-t) \sum_{n=0}^{\infty} t^n \mathbf{E}[S_n; \mathbf{T} > n] \leq \infty.$$

Therefore we obtain from (2), after multiplying (2) by $(1-t)$ and letting $t \nearrow 1$,

$$(3) \qquad \mathbf{E}[Z] = \lim_{t \nearrow 1} F(t) G(t) \leq \infty,$$

where $F(t)$ and $G(t)$ denote

$$F(t) = \sqrt{1-t} \, e^{\sum_{1}^{\infty} \frac{t^k}{k} \mathbf{P}[S_k \leq 0]},$$

$$G(t) = -\sqrt{1-t} \sum_{1}^{\infty} \frac{t^k}{k} \mathbf{E}[S_k; S_k \leq 0].$$

From P3 we know that

$$(4) \qquad \lim_{t \nearrow 1} G(t) = \frac{\sigma}{\sqrt{2}}.$$

(The inequality sign in P3 points the wrong way, but this makes no difference at all; just reverse the random walk!) The next step is to look at $F(t)$, and for $0 \leq t < 1$

$$F(t) = e^{\sum_{1}^{\infty} \frac{t^k}{k} \{\frac{1}{2} - \mathbf{P}[S_k > 0]\}}.$$

In view of (3) and (4)

$$(5) \qquad \mathbf{E}[Z] = \frac{\sigma}{\sqrt{2}} \lim_{t \nearrow 1} F(t) \leq \infty.$$

Now there are two possibilities. Either $\lim_{t \nearrow 1} F(t)$ in (5) is finite, in which case we can apply P4 to conclude that

$$\mathbf{E}[Z] = \frac{\sigma}{\sqrt{2}} e^{\sum_1^\infty \frac{1}{k}\{\frac{1}{2} - \mathbf{P}[S_k > 0]\}} = \frac{\sigma}{\sqrt{2}} e^\alpha < \infty,$$

so that both (a) and (c) in P5 are proved. Or else (this is the only other possibility)

(6) $$\lim_{t \nearrow 1} F(t) = +\infty.$$

Observe that

$$\tfrac{1}{2} - \mathbf{P}[S_k > 0] = \mathbf{P}[S_k = 0] - \{\tfrac{1}{2} - \mathbf{P}[S_k < 0]\}.$$

Since

$$\sum_1^\infty \frac{1}{k} \mathbf{P}[S_k = 0] < \infty,$$

as observed in the proof of T17.1, (6) implies that

(7) $$\lim_{t \nearrow 1} \sum_1^\infty \frac{t^k}{k} \left\{\frac{1}{2} - \mathbf{P}[S_k < 0]\right\} = -\infty.$$

Now we retrace all the steps in the present proof, up to equation (5) applied to the reversed random walk. Its first positive partial sum is $-S_{T^*}$ of the present random walk (in the sense that it has the same probability distribution). But then (7) leads to the inevitable conclusion that $\mathbf{E}[-S_{T^*}] = 0$ which is impossible since $-S_{T^*}$ is by definition a random variable whose only possible values are the positive integers. Thus (6) is impossible, (a) and (c) hold, P5 is proved, and therefore also T1 is true.

As we hinted before, T1 is a powerful tool when used in conjunction with the renewal theorem. To prepare for the applications in the next section we make some convenient definitions.

D2 $$c(1) = c = e^{-\sum_1^\infty \frac{1}{k} \mathbf{P}[S_k = 0]}, \text{ and for } |z| < 1$$

$$U(z) = \frac{1}{\sqrt{c}} e^{\sum_1^\infty \frac{1}{k} \mathbf{E}[z^{-S_k}; S_k < 0]},$$

$$V(z) = \frac{1}{\sqrt{c}} e^{\sum_1^\infty \frac{1}{k} \mathbf{E}[z^{S_k}; S_k > 0]},$$

$$U(z) = \sum_{n=0}^\infty u(n) z^n, \qquad V(z) = \sum_{n=0}^\infty v(n) z^n.$$

To justify the definition one needs to remark that $U(z)$ and $V(z)$ are analytic functions in $|z| < 1$ without any assumptions concerning the random walk. It suffices to show that the exponent in $U(z)$ is analytic, and the proof for $V(z)$ is then the same. Clearly

$$\sum_{1}^{\infty} \frac{1}{k} \mathbf{E}[z^{-S_k}; S_k < 0] = \sum_{n=1}^{\infty} z^n \sum_{k=1}^{\infty} \frac{1}{k} \mathbf{P}[S_k = -n],$$

and by P7.6 the coefficients are bounded uniformly in n, since for some $A > 0$

$$\sum_{k=1}^{\infty} \frac{1}{k} \mathbf{P}[S_k = -n] \leq A \sum_{k=1}^{n} k^{-3/2}.$$

The power series $U(z)$ and $V(z)$ are identified with analytic functions discussed in section 17 by

P6 $\qquad U(z) = \lim_{t \nearrow 1} \{\sqrt{c(t)} f_e(t; z^{-1})\}^{-1}, \qquad |z| < 1,$

$\qquad\qquad V(z) = \lim_{t \nearrow 1} \{\sqrt{c(t)} f_i(t; z)\}^{-1}, \qquad |z| < 1.$

The proof of P6 is immediate from D2 above and D17.3.

Now we shall connect $U(z)$ and $V(z)$ with the ladder random variables, but we will of course discuss $\mathbf{Z}$ only when $\mathbf{T} < \infty$ and $\overline{\mathbf{Z}}$ only when $\mathbf{T}'^* < \infty$.

P7 *When* $\mathbf{T} < \infty$

(a) $\qquad\qquad V(\xi) = \frac{1}{\sqrt{c}} \frac{1}{1 - \mathbf{E}[\xi^Z]}, \qquad |\xi| < 1,$

and when $\mathbf{T}'^* < \infty$

(b) $\qquad\qquad U(\xi) = \sqrt{c} \frac{1}{1 - \mathbf{E}[\xi^{-\overline{Z}}]}, \qquad |\xi| < 1.$

In particular, when $\mu > 0$, *(a) holds; when* $\mu < 0$, *(b) holds, and when* $\mu = 0$, *both (a) and (b) hold.*

Proof: One simply applies part (a) of P17.5 to P6 in order to obtain (a) and then part (b) of P17.5 to the reversed random walk to get (b). The criteria in terms of the mean μ are immediate from T17.1 of the preceding section.

Now we are ready to apply the renewal theorem. It will yield

P8 *Let the random walk be aperiodic. If* $T < \infty$ *and* $E[Z] < \infty$, *then*

(a) $$\lim_{n \to \infty} v(n) = \frac{1}{\sqrt{c}\, E[Z]} = \sqrt{\frac{2}{c}}\, \frac{e^{-\alpha}}{\sigma},$$

and if $T'^* < \infty$ *and* $E[-\overline{Z}] < \infty$, *then*

(b) $$\lim_{n \to \infty} u(n) = \frac{\sqrt{c}}{E[-\overline{Z}]} = \sqrt{2c}\, \frac{e^{\alpha}}{\sigma}.$$

In particular, if $\mu = 0$ *and* $\sigma^2 < \infty$, *then both* (a) *and* (b) *hold and*

$$\lim_{n \to \infty} u(n) \cdot \lim_{n \to \infty} v(n) = \frac{2}{\sigma^2}.$$

Proof: To apply the renewal theorem, P9.3, it is helpful to think of a random walk

$$\mathbf{y}_n = \mathbf{Z}_1 + \mathbf{Z}_2 + \cdots + \mathbf{Z}_n, \qquad \mathbf{y}_0 = 0,$$

where $\mathbf{Z}_1, \mathbf{Z}_2, \ldots$ are independent with the same distribution as the ladder random variable $\mathbf{Z}$. If this random walk is aperiodic, and if $E[Z] < \infty$, then the renewal theorem gives

$$\lim_{x \to \infty} \sum_{n=0}^{\infty} \mathbf{P}_0[\mathbf{y}_n = x] = \frac{1}{E[Z]}.$$

But that is precisely the result desired in part (a), since it follows from P7 that

$$v(x) = \frac{1}{\sqrt{c}} \sum_{n=0}^{\infty} \mathbf{P}_0[\mathbf{y}_n = x].$$

Finally, the evaluation of the limit in part (a) of P7 comes from P5. Therefore it remains only to check that the random walk $\mathbf{y}_n$ is aperiodic if the given random walk has this property. The converse is obvious since the possible values of $\mathbf{y}_n$ are a subset of the group $\overline{R}$ of possible values of the given random walk. However the statement we need is not hard to obtain. In fact, it turns out that plus one is a possible value of $\mathbf{y}_n$, i.e., that $\mathbf{P}[Z = 1] > 0$. This fact depends on the observation that there exists an $x \leq 0$ such that $P(x, 1) > 0$. For this point x there also exists a path, i.e., a sequence,

$$x_1 \leq 0,\ x_2 \leq 0, \ldots, x_{m-1} \leq 0,\ x_m = x$$

such that $P(0,x_1)P(x_1,x_2)\ldots P(x_{m-1},x) = p > 0$. But then $\mathbf{P}[Z = 1] \geq pP(x,1) > 0$. This construction fails only in one case, namely if $P(0,x) = 0$ for all $x \leq 0$, but in this case the random walk $\mathbf{y}_n$ is exactly the original random walk.

That was the proof of part (a) and in part (b) there is nothing new. The criterion in terms of μ and σ^2 comes verbatim from the main theorem (T1) of this section.

E3 The case of *symmetric random walk* deserves special mention. It is clear from D2 that in this case

$$U(z) = V(z), \qquad u(n) = v(n).$$

Hence also the limits in P8 should be the same, if they exist. That does not look obvious, but is true because the definition of α in P5 gives

$$\sqrt{c}\, e^\alpha = e^{\sum_1^\infty \frac{1}{k}\{-\frac{1}{2}\mathbf{P}[S_k = 0] + \frac{1}{2} - \mathbf{P}[S_k > 0]\}} = 1.$$

Specializing still further (as far as possible) we consider *symmetric simple random walk*. In this case the result of P8 is *exact* in the sense that (see E17.1)

$$U(z) = V(z) = \frac{\sqrt{2}}{1-z}, \qquad u(n) = v(n) = \frac{\sqrt{2}}{\sigma} = \sqrt{2}.$$

In order to explain the usefulness of P8, let us anticipate a little. In P19.3 of the next section it will be proved that the Green function $g_B(x,y)$ of the left half-line $B = [x \mid x \leq -1]$ can be represented by

$$(1) \qquad g_B(x,y) = \sum_{n=0}^{\min(x,y)} u(x-n)v(y-n),$$

for x and y in $R - B$. This representation will be shown to hold for every one-dimensional random walk, but in those cases when P8 applies one can expect to get useful asymptotic estimates. The results of Chapter III where we studied $g_B(x,y)$ when B was a single point or a finite set can serve as a model; for there we had the remarkable formula

$$(2) \qquad g_{\{0\}}(x,y) = a(x) + a(-y) - a(x-y),$$

the statement

$$(3) \qquad \lim_{|x| \to \infty} [a(x+y) - a(x)] = 0$$

was the analogue of P8, and many interesting results were obtained by applying (3) to (2).

19. THE GREEN FUNCTION AND THE GAMBLER'S RUIN PROBLEM

The maximum of $0 = \mathbf{x}_0, \mathbf{x}_1, \mathbf{x}_2, \ldots, \mathbf{x}_n$, or in other words max $[\mathbf{S}_0, \mathbf{S}_1, \ldots, \mathbf{S}_n]$ where $\mathbf{S}_0 = 0$, can be studied in terms of the theory developed so far. It has independent probabilistic interest, it may, for example, be related to the gambler's ruin problem mentioned in the section heading. In fact many other seemingly different problems in applied probability theory may be reformulated as questions concerning the random variable $\mathbf{M}_n = \max [0, \mathbf{S}_1, \ldots, \mathbf{S}_n]$. Furthermore our study of $\mathbf{M}_n$ will lead quickly and painlessly to the expression for the Green function of a half-line mentioned in equation (1) at the end of the last section.

D1 $\qquad \mathbf{M}_n = \max\limits_{0 \le k \le n} \mathbf{S}_k, \qquad n \ge 0.$

$\qquad \mathbf{T}_n = \min [k \mid 0 \le k \le n, \mathbf{S}_k = \mathbf{M}_n], \qquad n \ge 0.$

The random variables $\mathbf{T}_n$ denote the first time at which the random walk attains its maximum $\mathbf{M}_n$ during the first n steps. It will play an auxiliary role in deriving the characteristic function of $\mathbf{M}_n$. For a result of independent interest concerning $\mathbf{T}_n$, see problem 7 at the end of the chapter.

One can decompose, for each $n \ge 0$, $|z| \le 1$, $|w| \le 1$,

$$E[z^{\mathbf{M}_n} w^{\mathbf{M}_n - \mathbf{S}_n}] = \sum_{k=0}^{n} E[z^{\mathbf{M}_n} w^{\mathbf{M}_n - \mathbf{S}_n}; \mathbf{T}_n = k]$$

$$= \sum_{k=0}^{n} E[z^{\mathbf{S}_k} w^{\mathbf{S}_k - \mathbf{S}_n}; \mathbf{T}_n = k]$$

$$= \sum_{k=0}^{n} E[z^{\mathbf{S}_k}; \mathbf{S}_0 < \mathbf{S}_k, \mathbf{S}_1 < \mathbf{S}_k, \ldots, \mathbf{S}_{k-1} < \mathbf{S}_k;$$

$$w^{\mathbf{S}_k - \mathbf{S}_n}; \mathbf{S}_{k+1} \le \mathbf{S}_k, \mathbf{S}_{k+2} \le \mathbf{S}_k, \ldots, \mathbf{S}_n \le \mathbf{S}_k].$$

The last expectation can be simplified as it is clear that the terms

$$z^{\mathbf{S}_k}; \mathbf{S}_0 < \mathbf{S}_k, \ldots, \mathbf{S}_{k-1} < \mathbf{S}_k;$$

up to the second semicolon, are independent of the terms that follow. Therefore we get a product of expectations, and if in the second one of these one relabels $S_{k+1} - S_k = S_1$, $S_{k+2} - S_k = S_2$, and so forth, the result will appear as

$$E[z^{M_n} w^{M_n - S_n}] = \sum_{k=0}^{n} E[z^{S_k}; S_0 < S_k, \ldots, S_{k-1} < S_k] E[w^{-S_{n-k}};$$
$$S_0 \leq 0, \ldots, S_{n-k} \leq 0]$$
$$= \sum_{k=0}^{n} E[z^{S_k}; T'^* > k] E[w^{-S_{n-k}}; T > n - k].$$

In the last step we recalled, and used, the definition of the hitting times T and T'^* in D17.1 and D18.1, respectively.

At this point it is convenient to introduce generating functions. The representation of our expectation as a convolution quite strongly suggests this, and so for $0 \leq t < 1$ one obtains

$$\sum_{0}^{\infty} t^n E[z^{M_n} w^{M_n - S_n}] = \sum_{n=0}^{\infty} t^n E[z^{S_n}; T'^* > n] \cdot \sum_{n=0}^{\infty} t^n E[w^{-S_n}; T > n].$$

It remains only to look up P17.5 to verify that we have proved

P1 *For* $0 \leq t < 1$, $|z| \leq 1$, $|w| \leq 1$,

$$\sum_{n=0}^{\infty} t^n E[z^{M_n} w^{M_n - S_n}] = [c(t) f_i(t;z) f_e(t;w^{-1})]^{-1}.$$

To illustrate the use of P1 we shall set $w = 1$ in P1 to obtain some simple results concerning M_n, and concerning a limiting random variable M, which may or may not exist.

D2 *We say that* $M = \max_{k \geq 0} S_k$ *exists if* $\lim_{n \to \infty} M_n = M < \infty$ *with probability one.*

Since $M_n \leq M_{n+1}$ there is a very simple criterion for whether M exists or not. It is quite obvious that a monotone sequence of measurable functions converges almost everywhere if and only if it converges in distribution so that the condition in D2 will be satisfied if and only if the probabilities $P[M_n = k]$ have a limit as $n \to \infty$, and the sum of these limits is one. Alternatively, the convergence of the characteristic functions to a characteristic function is of course also necessary and sufficient. With this in mind it is easy to prove

P2 **M** *exists if and only if*

(a) $$\sum_{1}^{\infty} \frac{1}{k} \mathbf{P}[S_k > 0] < \infty,$$

and if (a) *holds, then*

(b) $$\mathbf{E}[z^{\mathbf{M}}] = e^{-\sum_{1}^{\infty} \frac{1}{k} \mathbf{E}[1 - z^{S_k}; S_k > 0]}, \qquad |z| \leq 1.$$

Proof: According to theorem T17.1, $\mathbf{T} = \infty$ with positive probability if and only if (a) holds. But now it should be intuitively clear that $\mathbf{T} = \infty$ with positive probability if and only if **M** exists: one verifies that $\lim_{n \to \infty} S_n = -\infty$, so that $\mathbf{M} < \infty$ with probability one if $\mathbf{T} = \infty$ with positive probability, while $\overline{\lim}_{n \to \infty} S_n = +\infty$ with probability one otherwise. We omit the details because it seems more amusing to give an alternative purely analytic proof of P2, based on the remarks preceding its statement.

First one verifies, using P1 and the relevant definitions, that

(1) $$(1 - t) \sum_{n=0}^{\infty} t^n \mathbf{E}[z^{\mathbf{M}_n}] = e^{-\sum_{1}^{\infty} \frac{t^k}{k} \mathbf{E}[1 - z^{S_k}; S_k > 0]}$$

for $0 \leq t < 1$, $|z| \leq 1$. Now let us assume that **M** exists. Then

$$\lim_{n \to \infty} \mathbf{E}[z^{\mathbf{M}_n}] = \mathbf{E}[z^{\mathbf{M}}]$$

exists and by Abel's theorem

(2) $$\mathbf{E}[z^{\mathbf{M}}] = \lim_{t \nearrow 1} (1 - t) \sum_{n=0}^{\infty} t^n \mathbf{E}[z^{\mathbf{M}_n}].$$

Now set $z = 0$ in (1) and then let $t \nearrow 1$. Then one gets from (1) and (2)

(3) $$\mathbf{P}[\mathbf{M} = 0] = \lim_{t \nearrow 1} e^{-\sum_{1}^{\infty} \frac{t^k}{k} \mathbf{P}[S_k > 0]} > 0,$$

which shows first of all that (a) holds, and secondly that one can also let $t \nearrow 1$ in equation (1) with arbitrary $|z| \leq 1$ to conclude that (b) holds. (We have been a little careless in taking the inequality in (3) for granted. But it is really easy to see that the existence of **M** implies that $\mathbf{P}[\mathbf{M} = 0] > 0$. It is equivalent to showing that **M** does not exist when **T** is finite with probability one.)

Finally we suppose that (a) holds. Then, letting t increase to one in (1)

$$\lim_{t \nearrow 1} (1 - t) \sum_{n=0}^{\infty} t^n \mathbf{E}[z^{\mathbf{M}_n}] = e^{-\sum_{1}^{\infty} \frac{1}{k} \mathbf{E}[1 - z^{S_k}; S_k > 0]}.$$

In view of the monotonicity of the sequence $\mathbf{M}_n$ this implies that $\mathbf{E}[z^{\mathbf{M}_n}]$ has a limit, which is given by (b); and, as remarked immediately prior to the statement of P2, we can conclude that $\mathbf{M}$ exists (is finite) with probability one.

E1 Consider the *left-continuous random walk* of E17.2. It is necessary and sufficient for the existence of $\mathbf{M}$ that the mean μ be negative. In equations (2), (3), and (4) of E17.2 we calculated $f_i(t;z)$, $f_e(t;z)$ and $c(t)$. Applying P1 one gets

$$\sum_{n=0}^{\infty} t^n \mathbf{E}[z^{\mathbf{M}_n}] = [c(t)f_i(t;z)f_e(t,1)]^{-1} = \frac{1}{1-r(t)} \cdot \frac{z-r(t)}{z-tP(z)},$$

where

$$P(z) = \sum_{n=0}^{\infty} P(0, n-1)z^n.$$

Using P2 we have, assuming $\mu < 0$,

$$\mathbf{E}[z^{\mathbf{M}}] = \lim_{t \nearrow 1} \frac{1-t}{1-r(t)} \cdot \frac{z-r(t)}{z-tP(z)}.$$

When $\mu < 0$ we saw in section 17 that $r(1) = 1$, and that

$$\lim_{t \nearrow 1} \frac{1-t}{1-r(t)} = 1 - \lim_{r \nearrow 1} \frac{1-P(r)}{1-r} = -\mu,$$

so that one obtains

$$\mathbf{E}[z^{\mathbf{M}}] = -\mu \frac{z-1}{z-P(z)}, \qquad \mu < 0.$$

There is one more question of obvious interest. To find the first moment of $\mathbf{M}$, we differentiate the generating function of $\mathbf{M}$ to get

$$\mathbf{E}[\mathbf{M}] = -\mu \lim_{z \to 1} \frac{d}{dz}\left(\frac{z-1}{z-P(z)}\right).$$

Because $P'(1) = \mu + 1$, one finds

$$\mathbf{E}[\mathbf{M}] = -\frac{1}{2\mu} P''(1) \leq \infty,$$

finite or infinite, depending on whether

$$P''(1) = \sum_{n=0}^{\infty} n(n-1)P(0, n-1)$$

is finite or infinite. Thus $\mathbf{E}[\mathbf{M}]$ exists in the case under consideration if and only if the random walk has a finite second moment. Indeed it has been shown[8] that this phenomenon is quite general:

For every one-dimensional random walk with $m < \infty$ and $\mu < 0$,

$$E[M^k] < \infty \text{ if and only if } m^+{}_{k+1} = \sum_{n=0}^{\infty} n^{k+1}P(0,n) < \infty.$$

Finally we consider the Green function of the half-line and some of the very simplest applications.

D3 If $B = [x \mid -\infty < x \leq -1]$, $g_B(x,y) = g(x,y)$ is the Green function of B as defined in D10.1.

Remark: $g(x,y) = 0$ when x or y or both are in B, according to D10.1, and $g(x,y) < \infty$ for all x and y, provided the random walk is aperiodic, according to P10.1, part (c). But in view of the special nature of our set B one can make a stronger statement. The Green function $g(x,y)$ is finite for all x and y regardless of whether the random walk is periodic or not. If the random walk is transient, then $g(x,y) \leq G(x,y) < \infty$. If it is recurrent, then

$$g_B(x,y) \leq g_{\{b\}}(x,y)$$

for every point b in B. But even in the periodic case, provided $P(0,0) < 1$, it is always possible to select some b in B such that the random walk, starting at $\mathbf{x}_0 = x$, has a positive probability of visiting the point b before visiting y. For every b which has been so chosen it is easy to see that $g_{\{b\}}(x,y) < \infty$, and that concludes the remark.

P3 For complex a, b, $|a| < 1$, $|b| < 1$,

$$(1 - ab) \sum_{x=0}^{\infty} \sum_{y=0}^{\infty} a^x b^y g(x,y) = U(a)V(b),$$

or

$$g(x,y) = \sum_{n=0}^{\min(x,y)} u(x-n)v(y-n),$$

where $U(a)$, $V(b)$ and their coefficients $u(n)$ and $v(n)$ are defined in D18.2.

Proof: In the necessary formal manipulations we shall use the notation of partial sums $\mathbf{S}_n$, and their maxima $\mathbf{M}_n$. It will even be necessary to consider the reversed random walk, so that $\mathbf{S}_n{}^*$ and $\mathbf{M}_n{}^*$ will denote the partial sums and maxima of the random walk with transition function $P^*(x,y) = P(y,x)$.

[8] Cf. Kiefer and Wolfowitz [63].

$$(1 - ab) \sum_{x=0}^{\infty} \sum_{y=0}^{\infty} a^x b^y g(x,y)$$

$$= (1 - ab) \sum_{n=0}^{\infty} \sum_{x=0}^{\infty} \sum_{y=0}^{\infty} a^x b^y Q_n(x,y)$$

$$= (1 - ab) \sum_{n=0}^{\infty} \sum_{x=0}^{\infty} \sum_{y=0}^{\infty} a^x b^y \mathbf{P}[x + \mathbf{S}_n = y;$$
$$x + \mathbf{S}_j \geq 0 \text{ for } j = 0, 1, \ldots, n]$$

$$= (1 - ab) \sum_{n=0}^{\infty} \sum_{x=0}^{\infty} \sum_{y=0}^{\infty} a^x b^y \mathbf{P}[\mathbf{S}_n^* = x - y;$$
$$\mathbf{S}_j^* \leq x \text{ for } j = 0, 1, \ldots, n]$$

$$= (1 - ab) \sum_{n=0}^{\infty} \sum_{x=0}^{\infty} \sum_{y=0}^{\infty} a^x b^y \mathbf{P}[\mathbf{M}_n^* \leq x, \mathbf{M}_n^* - \mathbf{S}_n^* \leq y;$$
$$\mathbf{S}_n^* = x - y]$$

$$= \sum_{n=0}^{\infty} \sum_{x=0}^{\infty} \sum_{y=0}^{\infty} a^x b^y \mathbf{P}[\mathbf{M}_n^* = x; \mathbf{M}_n^* - \mathbf{S}_n^* = y]$$

$$= \sum_{n=0}^{\infty} \mathbf{E}[a^{\mathbf{M}_n^*} b^{\mathbf{M}_n^* - \mathbf{S}_n^*}].$$

If the "stars" were not present we would now have from P1

$$(1 - ab) \sum_{x=0}^{\infty} \sum_{y=0}^{\infty} a^x b^y g(x,y) = [c(1) f_i(1;a) f_e(1;b^{-1})]^{-1}.$$

But we have stars. According to the definitions of f_i and f_e this merely amounts to an interchange of a and b. Another difficulty, that of letting $t \nearrow 1$ in f_i and f_e was discussed following D18.2 in section 18. Using this definition, and P18.6 which followed it, we have

$$(1 - ab) \sum_{x=0}^{\infty} \sum_{y=0}^{\infty} a^x b^y g(x,y)$$
$$= [c(1) f_i(1;b) f_e(1;a^{-1})]^{-1}$$
$$= U(a) V(b)$$
$$= (1 - ab) \sum_{x=0}^{\infty} \sum_{y=0}^{\infty} a^x b^y \sum_{n=0}^{\min(x,y)} u(x-n) v(y-n).$$

That completes the proof.

19. THE GREEN FUNCTION AND GAMBLER'S RUIN PROBLEM

By part (c) of P10.1 the hitting probabilities $H_B(x,y)$ of the half-line B are given by

$$H_B(x,y) = \sum_{t=0}^{\infty} g(x,t)P(t,y), \qquad x \in R - B, \quad y \in B.$$

P3 together with P18.8 in the last section gives a way of finding the limit of $H_B(x,y)$ as $x \to +\infty$. But at this point it is necessary to assume that the random walk is aperiodic.

P4 *For aperiodic random walk with mean 0 and finite variance σ^2, the hitting probability of a half-line has a limit at infinity. If $B = [x \mid -\infty < x \leq -1]$, then*

$$\lim_{x \to +\infty} H_B(x,y) = \sqrt{2c}\, \frac{e^\alpha}{\sigma} \sum_{t=0}^{\infty} [v(0) + v(1) + \cdots + v(t)]P(t,y), \qquad y \in B.$$

Proof: As $u(x)$ and $v(x)$ both tend to a limit when $x \to +\infty$ (by P18.8) one has $g(x,y) \leq A\,[\min(x,y) + 1]$ for some $A > 0$. The random walk has finite second moment, so that one can apply the dominated convergence theorem to conclude

$$\lim_{x \to +\infty} H_B(x,y) = \lim_{x \to +\infty} \sum_{t=0}^{\infty} g(x,t)P(t,y) = \sum_{t=0}^{\infty} \lim_{x \to +\infty} g(x,t)P(t,y),$$

and P4 then follows from P18.8 applied to P3.

Curiously P4 does not give the simplest possible representation for the limit of the hitting probabilities. We shall encounter this problem again in Chapter VI, where a far more direct approach (also based on the renewal theorem) will yield the formula

$$\lim_{x \to -\infty} H_A(x,y) = \frac{1}{E[Z]} P[Z \geq y],$$

where A is the half-line $A = [x \mid x \geq 1]$ and Z the ladder random variable of D18.1. This result will turn out to hold also when $E[Z] = +\infty$, the right-hand side being then interpreted as zero.

Thus P4 is an incomplete analogue of T14.1 in Chapter III which will be completed in P24.7 of Chapter VI. Although we shall not attempt to develop the potential theory for half-lines in any detail, the possibility of doing so is evident and now we will present just one more crucial step in this direction. For B a single point (the origin), Poisson's equation was found in P13.3 to be

$$\sum_{y \in R - B} P(x,y)a(y) = a(x), \qquad x \in R - B.$$

There was also an adjoint equation, too evident to comment upon,

$$\sum_{y \in R-B} a(-y)P(y,x) = a(-x), \qquad x \in R - B.$$

Here B was the origin. In the case of the half-line (say $B = [x \mid x \leq -1]$), these equations are

D4 (a) $\qquad \sum_{y=0}^{\infty} P(x,y)f(y) = f(x), \qquad x \geq 0,$

(b) $\qquad \sum_{y=0}^{\infty} g(y)P(y,x) = g(x), \qquad x \geq 0.$

In a slightly different context (a) and (b) are known as the *Wiener-Hopf equation*,[9] the singular integral equation

$$\int_0^{\infty} P(x,y)f(y) \, dy = f(x), \qquad x > 0$$

where

$$P(x,y) = p(y - x), \qquad \int_{-\infty}^{\infty} |p(x)| \, dx < \infty.$$

We shall now, in P5, exhibit solutions of equations (a) and (b). Later, in E27.3 of Chapter VI, potential theoretical methods will enable us also to investigate the uniqueness of these solutions. For aperiodic recurrent random walk it will be seen that the solutions in P5 are the only *non-negative* solutions.

P5 *For any one-dimensional recurrent random walk the Wiener-Hopf equation* (a) *has the solution*

$$f(x) = u(0) + u(1) + \cdots + u(x), \qquad x \geq 0,$$

and (b) *has the solution*

$$g(x) = v(0) + v(1) + \cdots + v(x), \qquad x \geq 0.$$

Proof: The theorem is actually valid if and only if both $T < \infty$ and $T^* < \infty$ with probability one (see problem 11 in Chapter VI). Therefore we shall use only the recurrence in the following way. As $T^* < \infty$, the left half-line $B = [x \mid x \leq -1]$ is sure to be visited, and

$$\sum_{y \in B} H_B(0,y) = \sum_{y=-\infty}^{-1} \sum_{t=0}^{\infty} g(0,t)P(t,y) = 1.$$

[9] See [81] for the first rigorous study; for further reading in this important branch of harmonic analysis the articles of Kreĭn [69] and Widom [105] are recommended.

Observing that $g(0,t) = u(0)v(t)$, and that $u(0) = v(0)$, one obtains

(1) $$\frac{1}{v(0)} = \sum_{y=-\infty}^{-1} \sum_{t=0}^{\infty} v(t)P(t,y).$$

It will be convenient to take an arbitrary $x \geq 0$, and to rewrite (1) in the form

(2) $$\frac{1}{v(0)} = \sum_{t=0}^{\infty} v(t) \sum_{y=-\infty}^{-1} P(t-y+x,x) = \sum_{t=0}^{\infty} v(t) \sum_{k=t+x+1}^{\infty} P(k,x).$$

We are now able to prove part (b). It follows from the definition of $g(x,y)$ that

(3) $$g(x,y) = \sum_{t=0}^{\infty} g(x,t)P(t,y) + \delta(x,y), \qquad x \geq 0, y \geq 0.$$

Setting $x = 0$ in (3)

$$g(0,y) = u(0)v(y) = \sum_{t=0}^{\infty} u(0)v(t)P(t,y) + \delta(0,y),$$

and summing on y from 0 to x one obtains

(4) $$v(0) + \cdots + v(x) = \sum_{t=0}^{\infty} v(t) \sum_{y=0}^{x} P(t,y) + \frac{1}{v(0)}$$
$$= \sum_{t=0}^{\infty} v(t) \sum_{k=t}^{t+x} P(k,x) + \frac{1}{v(0)}.$$

Now we substitute into equation (4) the expression for $[v(0)]^{-1}$ obtained in (2). Thus

$$v(0) + \cdots + v(x) = \sum_{t=0}^{\infty} v(t) \left[\sum_{k=t}^{t+x} P(k,x) + \sum_{k=t+x+1}^{\infty} P(k,x) \right].$$

That proves part (b) of P5, as

$$g(x) = v(0) + \cdots + v(x) = \sum_{t=0}^{\infty} v(t) \sum_{k=t}^{\infty} P(k,x)$$
$$= \sum_{k=0}^{\infty} \sum_{t=0}^{k} v(t)P(k,x) = \sum_{k=0}^{\infty} g(k)P(k,x), \qquad x \geq 0.$$

The proof of part (a) may be dismissed with the remark that equation (a) is equation (b) for the reversed random walk. The above proof therefore applies since the recurrence of a random walk is maintained when it is reversed.

E2 Consider *Bernoulli random walk* with $P(0,1) = p < P(0,-1) = q$. To calculate $g(x,y)$ as painlessly as possible, observe that $g(x,0) = u(x)v(0)$ is the expected number of visits from x to 0 before leaving the right half-line $x \geq 0$. Since the random walk must pass through 0, this expectation is independent of x. Thus $u(x)$ is constant. (The same argument shows, incidentally, that $u(x)$ is constant for every left-continuous random walk with mean $\mu \leq 0$.) By a slightly more complicated continuity argument one can also determine $v(x)$, but for the sake of variety, we shall determine $v(x)$ from P5. Equation (b) in P5 is

$$qg(1) = g(0), \qquad pg(n-1) + qg(n+1) = g(n), \qquad n \geq 1.$$

There is evidently a unique solution, except for the constant $g(0)$, namely

$$g(n) = \left[\frac{q}{q-p} - \frac{p}{q-p}\left(\frac{p}{q}\right)^n\right]g(0), \qquad n \geq 0,$$

so that, in view of P5, $v(0) = g(0)$ and

$$v(n) = g(n) - g(n-1) = v(0)\left(\frac{p}{q}\right)^n \text{ for } n \geq 1.$$

E3 As an amusing application of E2 consider the following random variables:

(a) the maximum **W** of the Bernoulli random walk with $p < q$ before the first visit to $(-\infty, -1]$, i.e.,

$$\mathbf{W} = \max_{0 \leq n \leq \mathbf{T}^*} \mathbf{S}_n,$$

(b) the maximum **W'** of the same random walk before the first return to 0 or to the set $(-\infty, 0]$, that is to say

$$\mathbf{W'} = \max_{0 \leq n \leq \mathbf{T'}^*} \mathbf{S}_n.$$

An easy calculation gives

$$\mathbf{E}[\mathbf{W'}] = p + p\mathbf{E}[\mathbf{W}] = p\sum_{n=0}^{\infty}\frac{g(0,n)}{g(n,n)} = p\sum_{n=0}^{\infty}\frac{v(n)}{v(0) + v(1) + \cdots + v(n)}$$

$$= (q-p)\sum_{n=1}^{\infty}\frac{(p/q)^n}{1-(p/q)^n} = (q-p)\sum_{n=1}^{\infty} d(n)\left(\frac{p}{q}\right)^n.$$

Here $d(1) = 1$, $d(2) = d(3) = 2$, $d(4) = 3$, in short $d(n)$ is the number of divisors of the integer n.

Now we go on to study a class of random walks which exhibit approximately the same behavior as the Bernoulli random walk with $p < q$ in E2 and E3. We shall state conditions, in P6, under which $u(x)$ is *approximately constant*, and $v(x)$ *approximately a geometric sequence.* (Of course there will be an analogous class of random walks with $\mu > 0$, for which one gets similar results by reversing the process.)

We impose two conditions. Aperiodicity is one of these, being essential for the renewal theorem to apply. The other is a growth restriction on $P(0,x)$ for large $|x|$. It will be satisfied in particular for *bounded* random walk, i.e., when $P(0,x) = 0$ for sufficiently large $|x|$. Such a growth restriction ((b) and (c) below) is essential. Without it, one can show that $v(x)$ will still tend to zero as $x \to +\infty$, but *not exponentially.*

Thus we consider one-dimensional random walk satisfying

(a) $P(0,x)$ aperiodic, $m < \infty$, $\mu < 0$,
(b) *for some positive real number r*

$$\sum_{x \in R} r^x P(0,x) = 1,$$

(c) *for the number r in* (b)

$$0 < \sum_{x \in R} x r^x P(0,x) = \mu^{(r)} < \infty.$$

P6 *For random walk satisfying* (a), (b), (c), *r is uniquely determined and greater than one. There are two positive constants k_1, k_2 such that*

$$\lim_{x \to +\infty} u(x) = k_1, \quad \lim_{x \to +\infty} v(x) r^x = k_2.$$

Proof: First we obtain the limit behavior of $u(x)$ which is quite independent of conditions (b) and (c). One simply goes back to P18.8 to obtain

$$\lim_{x \to +\infty} u(x) = \frac{\sqrt{c}}{E[-\overline{Z}]} = k_1 > 0,$$

provided that $E[-\overline{Z}] < \infty$. But by P18.2 $E[Z] < \infty$ for random walk with finite positive mean, and therefore $E[-\overline{Z}]$ is finite for random walk satisfying (a).

The proof of the second part depends strongly on (b) and (c). First consider

$$f(\rho) = \sum_{n=-\infty}^{\infty} \rho^n P(0,n).$$

Condition (b) implies that $f(\rho)$ is finite for all ρ in the closed interval

I_r with endpoints 1 and r (we still have to prove that $r > 1$, so that $I_r = [1,r]$). Now

$$f'(\rho) = \sum_{n=-\infty}^{\infty} n\rho^{n-1} P(0,n) \quad \text{for } \rho \in I_r,$$

and

$$f''(\rho) = \sum_{n=-\infty}^{\infty} n(n-1)\rho^{n-2} P(0,n) > 0$$

for all ρ in the interior of I_r. The last sum is positive since $P(0,n) > 0$ for some $n < 0$ (for otherwise (a) could not hold). Hence $f(\rho)$ is convex in the interior of I_r so that condition (c), which states that $f'(r) > 0$, implies $1 < r$. Finally r is unique, for if there are two values r_1 and r_2 satisfying (b) and (c), with $1 < r_1 \leq r_2$, then $f(\rho)$ is strictly convex on $(1,r_2)$. But by (a) we have $f(1) = f(r_1) = f(r_2)$ which implies $r_1 = r_2$.

The rest of the proof depends on the elegant device of defining a new, auxiliary random walk. Let

$$P^{(r)}(x,y) = P(x,y)r^{y-x}, \qquad x, y \in R.$$

Condition (b) implies that $P^{(r)}$ is a transition function of an aperiodic random walk, as it gives

$$\sum_{y \in R} P^{(r)}(x,y) = 1, \qquad x \in R,$$

and obviously $P^{(r)}$ is a difference kernel, it is non-negative, and the aperiodicity of P is also preserved. The random walk defined by $P^{(r)}$ has positive mean $\mu^{(r)}$ in view of (c). Now let $g^{(r)}(x,y)$ be the Green function of the half-line $B = [x \mid x \leq -1]$ for $P^{(r)}$. Naturally we also write

$$g^{(r)}(x,y) = \sum_{n=0}^{\min(x,y)} u^{(r)}(x-n) v^{(r)}(y-n),$$

where $u^{(r)}(x)$ and $v^{(r)}(y)$ are defined in terms of $P^{(r)}$ as usual.

What relation is there, if any, between $g^{(r)}$ and g, $u^{(r)}$ and u, $v^{(r)}$ and v? It turns out to be a very simple one. First of all

(1) $$g^{(r)}(x,y) = g(x,y)r^{y-x}, \qquad x,y \in R.$$

To prove (1) one has to go back to the formal definition of g in D10.1, the identity (1) being formal in nature and seemingly completely devoid of any obvious probability interpretation. Using the obvious notation one writes

$$g(x,y) = \sum_{n=0}^{\infty} Q_n(x,y),$$

$$g^{(r)}(x,y) = \sum_{n=0}^{\infty} Q_n^{(r)}(x,y).$$

For x,y in $R - B$ (otherwise there is nothing to prove),

$$Q_1^{(r)}(x,y) = P(x,y)r^{y-x},$$

and

$$Q_{n+1}^{(r)}(x,y) = \sum_{t \in R-B} Q_1^{(r)}(x,t) Q_n^{(r)}(t,y)$$

gives

$$Q_n^{(r)}(x,y) = Q_n(x,y) r^{y-x},$$

which implies (1).

Proceeding from (1),

(2) $\quad u^{(r)}(x) v^{(r)}(0) = g^{(r)}(x,0) = r^{-x} g(x,0) = r^{-x} u(x) v(0),$

(3) $\quad u^{(r)}(0) v^{(r)}(y) = g^{(r)}(0,y) = g(0,y) r^y = u(0) v(y) r^y.$

We shall use only (3). The random walk $P^{(r)}$ is aperiodic with positive mean. To its ladder random variable $\mathbf{Z}$ we give the name $\mathbf{Z}^{(r)}$. Since we know by P18.2 that $\mathbf{E}[\mathbf{Z}^{(r)}] < \infty$, P18.8 gives

$$\lim_{y \to +\infty} v^{(r)}(y) = \frac{1}{\sqrt{c^{(r)} \mathbf{E}[\mathbf{Z}^{(r)}]}} = k > 0,$$

where $c^{(r)}$ is $c(1) = c$ for the $P^{(r)}$ random walk. Finally equation (3) implies

$$\lim_{y \to +\infty} v(y) r^y = \frac{u^{(r)}(0)}{u(0)} k = k_2 > 0.$$

completing the proof of P6.

E4 Let us paraphrase P6 in a form first considered by Täcklind [98], 1942, in the context of the *gambler's ruin problem*. Suppose that a gambler starting with initial capital x, plays a "favorable game," or, in random walk terminology, that we observe a random walk $\mathbf{x}_n$, with mean $\mu > 0$, starting at $\mathbf{x}_0 = x > 0$. What is the *probability of ruin*

$$f(x) = \mathbf{P}_x[\mathbf{x}_n \leq 0 \text{ for some } n \geq 0]?$$

This problem can be reduced to P6 above by reversing the random walk. If $\mathbf{y}_n$ is the reversed random walk, and if we take $\mathbf{y}_0 = 0$, then $\mathbf{y}_n$ has negative mean $-\mu$, and

$$f(x) = \mathbf{P}_0[\mathbf{y}_n \geq x \text{ for some } n \geq 0].$$

This is a maximum problem of the type treated in P2, and comparing the generating function in P2 with the formula for $V(z)$ in D18.2 of the last section, one finds that

$$f(x) - f(x+1) = \mathbf{P}_0\left[\max_{n \geq 0} \mathbf{y}_n = x\right]$$

is a constant multiple of the function $v(x)$ for the random walk $\mathbf{y}_n$. So if one computes r for this random walk in accord with P6, the result is

$$\lim_{x \to +\infty} [f(x) - f(x+1)]r^x = k_3 > 0,$$

which of course implies the weaker result

$$\lim_{x \to +\infty} f(x)r^x = k_4 = k_3\left(1 - \frac{1}{r}\right)^{-1} > 0.$$

This shows that, *in a favorable game, the gambler's probability of ultimate ruin decreases exponentially, as his starting capital increases.*

20. FLUCTUATIONS AND THE ARC-SINE LAW

This section is devoted to a part of the theory of one-dimensional random walk which chronologically preceded and led to the development of the results in the first three sections. In 1950 Sparre Andersen discovered a very surprising result concerning the number $\mathbf{N}_n$ of positive terms in the sequence $\mathbf{S}_0, \mathbf{S}_1, \ldots, \mathbf{S}_n$, $n \geq 0$. He proved [91] that regardless of the distribution of $\mathbf{X}_1 = \mathbf{S}_1$ (in fact when $\mathbf{X}_1$ is an arbitrary real valued random variable)

$$\mathbf{P}[\mathbf{N}_n = k] = \mathbf{P}[\mathbf{N}_k = k]\mathbf{P}[\mathbf{N}_{n-k} = 0], \quad 0 \leq k \leq n.$$

He also found the generating functions and was able to show that the limiting distribution of $\mathbf{N}_n$ is

$$\lim_{n \to \infty} \mathbf{P}[\mathbf{N}_n \leq nx] = \frac{2}{\pi} \arcsin \sqrt{x}, \quad 0 \leq x \leq 1$$

under very general conditions. These conditions (see T2 below and also problems 8, 9, and 13) are satisfied for example for symmetric random walk. To understand why this should be surprising, one must appreciate the importance of P. Lévy's earlier work ([72], Ch. VI) on Brownian motion. It was in this context that the arc-sine law first saw the light of day—and thus the myth was firmly established that its domain of validity was random walk with finite variance —random walk in other words which by the Central Limit Theorem has the Brownian movement process as a limiting case.

D1 *For real* x, $\theta(x) = 1$ *if* $x > 0$ *and* 0 *otherwise*. $\mathbf{N}_n = \sum_{k=1}^{n} \theta(\mathbf{S}_k)$ *when* $n \geq 1$ *and* $\mathbf{N}_0 = 0$, *where* $\mathbf{S}_k$ *is the random walk* $\mathbf{x}_k$ *with* $\mathbf{x}_0 = 0$. *We shall assume that* $\mathbf{x}_n$ *is arbitrary one-dimensional random walk, excluding only the trivial case* $P(0,0) = 1$.

In this notation, Sparre Andersen's theorem is

T1 (a) $\quad \mathbf{P}[\mathbf{N}_n = k] = \mathbf{P}[\mathbf{N}_k = k]\mathbf{P}[\mathbf{N}_{n-k} = 0], \quad 0 \leq k \leq n,$

and, for $0 \leq t < 1$

(b) $$\sum_{j=0}^{\infty} \mathbf{P}[\mathbf{N}_j = 0]t^j = e^{\sum_{1}^{\infty} \frac{t^k}{k} \mathbf{P}[\mathbf{S}_k \leq 0]},$$

(c) $$\sum_{j=0}^{\infty} \mathbf{P}[\mathbf{N}_j = j]t^j = e^{\sum_{1}^{\infty} \frac{t^k}{k} \mathbf{P}[\mathbf{S}_k > 0]}.$$

Proof: Equations (b) and (c) follow immediately from earlier results. To get (b) observe that

$$\mathbf{P}[\mathbf{N}_j = 0] = \mathbf{P}[T > j]$$

so that

$$\sum_{j=0}^{\infty} \mathbf{P}[\mathbf{N}_j = 0]t^j = \sum_{j=0}^{\infty} t^j \mathbf{P}[T > j] = \frac{1 - \mathbf{E}[t^T]}{1 - t},$$

and from P17.5(a) with $z = 1$

$$\sum_{j=0}^{\infty} \mathbf{P}[\mathbf{N}_j = 0]t^j = (1 - t)^{-1} f_i(t;1)$$

$$= (1 - t)^{-1} e^{-\sum_{1}^{\infty} \frac{t^k}{k} \mathbf{P}[\mathbf{S}_k > 0]} = e^{\sum_{1}^{\infty} \frac{t^k}{k} \mathbf{P}[\mathbf{S}_k \leq 0]}.$$

The proof of (c) is similar, being based on the identity

$$\mathbf{P}[\mathbf{N}_j = j] = \mathbf{P}[T'^* > j].$$

To obtain (a) we resort to an analysis of the same type as, but slightly deeper than, that which gave P17.5. Recall the notions of absolutely convergent Fourier series, and exterior and interior Fourier series introduced in D17.2. Let us call these three classes of functions $\mathscr{A}$, $\mathscr{A}_e$, $\mathscr{A}_i$. Note also that each of these three function spaces is closed under multiplication. For example if ϕ and ψ are in $\mathscr{A}_i$, then for real θ,

$$\phi(\theta) = \sum_0^\infty a_k e^{ik\theta}, \quad \sum_0^\infty |a_k| < \infty,$$

$$\psi(\theta) = \sum_0^\infty b_k e^{ik\theta}, \quad \sum_0^\infty |b_k| < \infty.$$

But then the product is

$$\phi(\theta)\psi(\theta) = \sum_{-\infty}^\infty c_k e^{ik\theta}$$

where $c_k = 0$ for $k \leq -1$,

$$c_n = \sum_{k=0}^n a_k b_{n-k} \text{ for } n \geq 0, \quad \sum_0^\infty |c_n| < \infty,$$

so that the product $\phi\psi$ is again in $\mathscr{A}_i$.

Following Baxter's treatment [2] of fluctuation theory, we introduce the "+ operator" and the "− operator." For arbitrary

$$\phi(\theta) = \sum_{-\infty}^\infty a_k e^{ik\theta} \text{ in } \mathscr{A}$$

we define

$$\phi^+(\theta) = \sum_{k=1}^\infty a_k e^{ik\theta},$$

$$\phi^-(\theta) = \sum_{k=-\infty}^0 a_k e^{ik\theta}.$$

Thus ϕ^+ is in $\mathscr{A}_i$ and ϕ^- in $\mathscr{A}_e$ whenever ϕ is in $\mathscr{A}$. In other words, the " + " and " − " operators are projections of $\mathscr{A}$ into the subspaces $\mathscr{A}_i$ and $\mathscr{A}_e$ of $\mathscr{A}$. We list the obvious algebraic properties of the " + " operator. It is linear, i.e., for arbitrary ϕ, ψ in $\mathscr{A}$,

$$(a\phi + b\psi)^+ = a\phi^+ + b\psi^+.$$

Since it is a projection, $(\phi^+)^+ = \phi^+$, $(\phi^+)^- = 0$, and since $\mathscr{A}^+$ (the set of all ψ such that $\psi = \phi^+$ for some $\phi \in \mathscr{A}$) is closed under multiplication, it is clear that

$$(\phi^+\psi^+)^+ = \phi^+\psi^+ \text{ when } \phi, \psi \text{ are in } \mathscr{A}.$$

It is also convenient to define $\mathscr{A}^-$ in the obvious manner. (Note that $\mathscr{A}^- = \mathscr{A}_e$, whereas $\mathscr{A}^+$ is not $\mathscr{A}_i$ but a subspace of $\mathscr{A}_i$.) $\mathscr{A}^+$

and $\mathscr{A}^-$ are disjoint, so that an arbitrary ϕ in $\mathscr{A}$ can be decomposed uniquely into $\phi = \phi_1 + \phi_2$ with ϕ_1 in $\mathscr{A}^+$ and ϕ_2 in $\mathscr{A}^-$. Of course $\phi_1 = \phi^+$, $\phi_2 = \phi^-$.

Now we are ready to define certain useful Fourier series, and leave to the reader the slightly tedious but easy task of checking that they are all in $\mathscr{A}$. They depend on the parameters s, t which will be real with $0 \le s < 1$, $0 \le t < 1$. Let

$$\phi_n(s;\theta) = \sum_{k=0}^{n} s^k \mathbf{E}[e^{i\theta \mathbf{S}_n}; \mathbf{N}_n = k], \qquad n \ge 0,$$

$$\psi(s,t;\theta) = \sum_{n=0}^{\infty} t^n \phi_n(s;\theta),$$

and

$$\phi(\theta) = \sum_{n=-\infty}^{\infty} P(0,n)e^{in\theta},$$

as usual.

We require the important identity

(1) $$s(\phi\phi_n)^+ + (\phi\phi_n)^- = \phi_{n+1}, \qquad n \ge 0.$$

The variables s, t, and θ have been suppressed, but (1) is understood to be an identity for each θ (real), $0 \le s < 1$, $0 \le t < 1$. To prove it we write

$$\phi\phi_n = \sum_{k=0}^{n} s^k \mathbf{E}[e^{i\theta \mathbf{S}_{n+1}}; \mathbf{N}_n = k] = \sum_{k=0}^{n} s^k \mathbf{E}[e^{i\theta \mathbf{S}_{n+1}}; \mathbf{N}_{n+1} = k+1; \mathbf{S}_{n+1} > 0] + \sum_{k=0}^{n} s^k \mathbf{E}[e^{i\theta \mathbf{S}_{n+1}}; \mathbf{N}_{n+1} = k; \mathbf{S}_{n+1} \le 0],$$

where we have decomposed the event that $\mathbf{N}_n = k$ in an obvious manner. It is easily recognized that this happens to be a decomposition of $\phi\phi_n$ into its projections on $\mathscr{A}^+$ and $\mathscr{A}^-$. Hence, using the uniqueness of such decompositions,

$$(\phi\phi_n)^+ = \sum_{k=0}^{n} s^k \mathbf{E}[e^{i\theta \mathbf{S}_{n+1}}; \mathbf{N}_{n+1} = k+1; \mathbf{S}_{n+1} > 0]$$

and a comparison of the sum on the right with the definition of ϕ_{n+1} shows that

$$s(\phi\phi_n)^+ = \phi_{n+1}^+.$$

Similarly one recognizes

$$(\phi\phi_n)^- = \phi_{n+1}^-,$$

and upon adding the last two identities one has (1). Multiplying (1) by t^{n+1} and summing over $n \geq 0$,

$$1 + st(\phi\psi)^+ + t(\phi\psi)^- = \psi,$$

or equivalently

$$[\psi(1 - st\phi)]^+ = [\psi(1 - t\phi) - 1]^- = 0.$$

This says that $\psi(1 - st\phi) \in \mathscr{A}_e$, $\psi(1 - t\phi) \in \mathscr{A}_i$, and, by P17.2, these functions can be extended uniquely, $\psi(1 - st\phi)$ to an outer function g_e, and $\psi(1 - t\phi)$ to an inner function g_i. Thus

$$\psi(1 - st\phi) = g_e(z), \quad \psi(1 - t\phi) = g_i(z), \quad \text{for } |z| = 1.$$

Using the factorization

$$1 - t\phi = c(t)f_i(t;z)f_e(t;z), \quad |z| = 1$$

of P17.4,

$$g_e(z) = g_i(z)\frac{1 - st\phi}{1 - t\phi} = g_i(z)\frac{c(st)}{c(t)}\frac{f_i(st;z)f_e(st;z)}{f_i(t;z)f_e(t;z)},$$

and

$$\frac{c(st)}{c(t)}\frac{f_i(st;z)}{f_i(t;z)}g_i(z) = g_e(z)\frac{f_e(t;z)}{f_e(st;z)} = \text{constant},$$

since both sides together determine a bounded analytic function. The constant (which may depend on s and t) is determined by checking that

$$g_i(0) = f_i(st;0) = f_i(t;0) = 1.$$

This yields

$$g_i(z) = \frac{f_i(t;z)}{f_i(st;z)}$$

and from $g_i = \psi(1 - t\phi)$ one gets

(2) $$\psi = \sum_{n=0}^{\infty} t^n \mathbf{E}[e^{i\theta S_n} s^{N_n}] = [c(t)f_i(st;e^{i\theta})f_e(t;e^{i\theta})]^{-1}.$$

Equation (2) is still far more general than the proof of T1 requires. Thus we specialize, setting $\theta = 0$, to find

(3) $$\sum_{n=0}^{\infty} t^n \mathbf{E}[s^{N_n}] = [c(t)f_i(st;1)f_e(t;1)]^{-1}.$$

To complete the proof of T1 one has to show that (a) holds. Now equations (b) and (c), which have already been proved, may be written

(4) $$\sum_{j=0}^{\infty} \mathbf{P}[\mathbf{N}_j = 0] t^j = [c(t) f_e(t;1)]^{-1},$$

(5) $$\sum_{j=0}^{\infty} \mathbf{P}[\mathbf{N}_j = j](st)^j = [f_i(st;1)]^{-1}.$$

Combining (3), (4), and (5)

$$\sum_{n=0}^{\infty} t^n \mathbf{E}[s^{\mathbf{N}_n}] = \sum_{j=0}^{\infty} \mathbf{P}[\mathbf{N}_j = 0] t^j \cdot \sum_{j=0}^{\infty} s^j \mathbf{P}[\mathbf{N}_j = j] t^j$$

$$= \sum_{n=0}^{\infty} t^n \sum_{k=0}^{n} \mathbf{P}[\mathbf{N}_{n-k} = 0] s^k \mathbf{P}[\mathbf{N}_k = k].$$

Identifying coefficients of $t^n s^k$ completes the proof of (a) and thus of T1.

E1 *Symmetric random walk.* As shown in Example E17.1, equation (7),

$$1 - \mathbf{E}[t^T] = \sqrt{\frac{1-t}{c(t)}},$$

so that

(1) $$\sum_{j=0}^{\infty} \mathbf{P}[\mathbf{N}_j = 0] t^j = \frac{1 - \mathbf{E}[t^T]}{1 - t} = \frac{(1-t)^{-1/2}}{\sqrt{c(t)}}.$$

It is equally easy to get

(2) $$\sum_{j=0}^{\infty} \mathbf{P}[\mathbf{N}_j = j] t^j = \sqrt{c(t)} (1-t)^{-1/2},$$

for example, by setting $z = 1$ in P17.4, which gives

$$(1 - t) = c(t) f_i(t;1) f_e(t;1) = \left\{ \sum_{j=0}^{\infty} \mathbf{P}[\mathbf{N}_j = 0] t^j \right\}^{-1} \left\{ \sum_{j=0}^{\infty} \mathbf{P}[\mathbf{N}_j = j] t^j \right\}^{-1}.$$

For *simple random walk* it was shown in E17.1 that

$$c(t) = \frac{1 + \sqrt{1 - t^2}}{2},$$

so that

$$\sqrt{c(t)} = \frac{t}{\sqrt{1+t} - \sqrt{1-t}} = \frac{\sqrt{1+t} + \sqrt{1-t}}{2}.$$

Therefore one can express $P[N_n = k]$ in terms of binomial coefficients. With c_n defined by

$$\sum_0^\infty c_n t^n = \sqrt{\frac{1+t}{1-t}},$$

one finds

$$P[N_n = 0] = c_{n+1}, \quad P[N_n = n] = \tfrac{1}{2}c_n + \tfrac{1}{2}\delta(n,0), \quad n \geq 0.$$

There is no doubt what causes the slight but ugly asymmetry in the distribution of N_n. It is the slight but unpleasant difference between positive and non-negative partial sums. As indicated in problems 13 and 14, the corresponding result for random variables with a continuous symmetric distribution function is therefore much more elegant. However even in the case of simple random walk one can obtain formally elegant results by a slight alteration in the definition of $N_n = \theta(S_0) + \cdots + \theta(S_n)$. One takes $\theta(S_k) = 1$ if $S_k > 0$, $\theta(S_k) = 0$ if $S_k < 0$, but when $S_k = 0$, $\theta(S_k)$ is given the same value as $\theta(S_{k-1})$. In other words, a zero partial sum S_k is counted as positive or negative according as S_{k-1} is positive or negative. With this definition it turns out[10] that

$$P[N_{2n} = 2k+1] = 0, \quad P[N_{2n} = 2k] = 2^{-2n}\binom{2k}{k}\binom{2n-2k}{n-k}.$$

Now we turn to the asymptotic study of N_n and show that even in the limit theorems the symmetric random walk continues to exhibit a certain amount of asymmetry due to our definition of N_n. We shall go to some trouble to exhibit the same kind of almost symmetric behavior also for random walk with mean zero and finite variance.

P1 *For symmetric genuinely one-dimensional random walk* (i.e., when $P(x,y) = P(y,x)$, $P(0,0) < 1$)

(a) $$\lim_{n \to \infty} \sqrt{n\pi}\, P[N_n = 0] = \frac{1}{\sqrt{c}}$$

(b) $$\lim_{n \to \infty} \sqrt{n\pi}\, P[N_n = n] = \sqrt{c}.$$

For genuinely one-dimensional random walk with mean zero and finite variance σ^2 the limits in (a) and (b) also exist, but their values are now e^α in equation (a) and $e^{-\alpha}$ in equation (b). The constant c is defined in D18.2 and α in P18.5.

Proof: According to D18.2 and P18.5,

$$c = e^{-\sum_1^\infty \frac{1}{k} P[S_k = 0]}, \quad \alpha = \sum_1^\infty \frac{1}{k}\left\{\frac{1}{2} - P[S_k > 0]\right\}.$$

[10] Cf. Feller [31], Vol. I, Chapter III, and Rényi [84], Chapter VIII, §11.

The series defining c always converges, and that defining α was shown to converge in the proof of P18.5, when $\mu = 0$ and $\sigma^2 < \infty$.

We shall worry only about the asymptotic behavior of $\mathbf{P}[\mathbf{N}_n = 0]$, since the evaluation of the other limit follows a similar path. Using part (b) of T1 one gets

$$\sqrt{1-t} \sum_{n=0}^{\infty} \mathbf{P}[\mathbf{N}_n = 0] t^n = e^{\sum_{1}^{\infty} \frac{t^k}{k} \{\mathbf{P}[\mathbf{S}_k \leq 0] - \frac{1}{2}\}} = e^{\sum_{1}^{\infty} \frac{t^k}{k} \{\frac{1}{2} - \mathbf{P}[\mathbf{S}_k > 0]\}}.$$

As $t \nearrow 1$ the right-hand side has a finite limit. This limit is clearly e^α in the case when $\mu = 0$ and $\sigma^2 < \infty$, and by using the fact that $\mathbf{P}[\mathbf{S}_n > 0] = \mathbf{P}[\mathbf{S}_n < 0]$ it is seen to be $\sqrt{c}$ when the random walk is symmetric. From this point on the proof for both cases can follow a common pattern. It will make use of only one simple property of the sequence $\mathbf{P}[\mathbf{N}_n = 0] = p_n$. That is its monotonicity and non-negativity: $p_0 \geq p_1 \geq \cdots \geq p_n \geq p_{n+1} \cdots \geq 0$. Thus P1 will follow from

P2 *If p_n is a monotone nonincreasing, non-negative sequence such that*

(1) $$\lim_{t \nearrow 1} \sqrt{1-t} \sum_{n=0}^{\infty} p_n t^n = 1,$$

then

(2) $$\lim_{n \to \infty} \sqrt{n\pi}\, p_n = 1.$$

Proof: We proceed in two stages. The first step uses only the non-negativity of p_n. For a non-negative sequence a special case of a well known theorem of Karamata[11] asserts that (1) implies

(3) $$\lim_{n \to \infty} \frac{1}{\sqrt{n}} (p_1 + \cdots + p_n) = \frac{2}{\sqrt{\pi}}.$$

This result, incidentally, is the converse of the Abelian theorem used in the proof of P18.3. It is not correct in general without the assumption that $p_n \geq 0$.

The next step goes from (3) to (2) and uses the monotonicity $p_n \geq p_{n+1}$. It is easier to appreciate the proof if we replace $\sqrt{n}$ by n^a, $0 < a < 1$, the limit by one, and try to show that

$$p_n \geq p_{n+1} \quad \text{and} \quad \frac{p_1 + \cdots + p_n}{n^a} \to 1$$

---

[11] See Hardy [39] for a complete discussion of this and other Tauberian theorems, and König [68] or Feller [31], Vol. 2, for an elegant proof of (3).

implies that $n^{1-a}p_n \to a$. The technique of the proof even suggests the further reduction of taking for $p(x)$ a monotone nonincreasing function such that $p(x) = p_n$ when $x = n$, and such that

$$(4) \qquad \lim_{x \to +\infty} x^{-a} \int_0^x p(t)\, dt = 1.$$

This is of course easy to do, and it remains to show that

$$(5) \qquad \lim_{x \to +\infty} x^{1-a} p(x) = a.$$

For each $c > 1$, the monotonicity of $p(x)$ gives

$$xp(cx) \le \frac{1}{c-1} \int_x^{cx} p(t)\, dt \le xp(x).$$

Thus

$$\frac{xp(cx)}{\int_0^x p(t)\, dt} \le \frac{\int_0^{cx} p(t)\, dt - \int_0^x p(t)\, dt}{(c-1) \int_0^x p(t)\, dt} \le \frac{xp(x)}{\int_0^x p(t)\, dt}.$$

Letting $x \to +\infty$, and using the hypothesis (4), the second inequality gives

$$\frac{c^a - 1}{c - 1} \le \varlimsup_{x \to \infty} \frac{xp(x)}{\int_0^x p(t)\, dt} = \varlimsup_{x \to \infty} x^{1-a} p(x).$$

Using also the left inequality one obtains

$$c^{a-1} \varlimsup_{x \to \infty} x^{1-a} p(x) \le \frac{c^a - 1}{c - 1} \le \varlimsup_{x \to \infty} x^{1-a} p(x).$$

This holds for arbitrary $c > 1$, so we let $c \searrow 1$ to obtain

$$\lim_{x \to \infty} x^{1-a} p(x) = \lim_{c \searrow 1} \frac{c^a - 1}{c - 1} = a,$$

proving (5) and hence P2 and P1.

We conclude the chapter with a form of the arc-sine law.

T2 *Under the conditions of* P1 (*either symmetry or mean zero and finite variance, but not necessarily both*)

(a) $\qquad \sqrt{k(n-k)}\, \mathbf{P}[N_n = k] = \dfrac{1}{\pi} + o(k,n)$

where $o(k,n)$ tends to zero uniformly in k and n as $\min(k, n-k) \to \infty$.

(b) For $0 \le x \le 1$

$$\lim_{n\to\infty} \mathbf{P}[\mathbf{N}_n \le nx] = \frac{2}{\pi} \arcsin \sqrt{x}.$$

Proof: In view of P1 and part (a) of T1,

$$\sqrt{k(n-k)}\,\mathbf{P}[\mathbf{N}_n = k] = \sqrt{k}\,\mathbf{P}[\mathbf{N}_k = k]\sqrt{n-k}\,\mathbf{P}[\mathbf{N}_{n-k} = 0]$$

$$= \left[\frac{1}{\sqrt{c\pi}} + o_1(k)\right]\left[\sqrt{\frac{c}{\pi}} + o_2(n-k)\right]$$

$$= \frac{1}{\pi} + o(k,n),$$

where $o(k,n)$ has the required property, given any $\epsilon > 0$ there is some N such that $o(k,n) < \epsilon$ when $k > N$ and $n - k > N$. Strictly speaking this was the proof for the symmetric case. In the other case it suffices to replace $\sqrt{c}$ by $e^{-\alpha}$.

Part (b) is the "classical" arc-sine law, which is valid under far weaker conditions than those for (a). (See problems 8 and 9 for a simple necessary and sufficient condition.) To derive (b) from (a) we may write

$$\mathbf{P}[\mathbf{N}_n \le nx] = \mathbf{P}[\mathbf{N}_n \le [nx]] = \frac{1}{n}\sum_{k=0}^{[nx]}\left[\frac{k}{n}\left(1 - \frac{k}{n}\right)\right]^{-1/2}\left(\frac{1}{\pi} + o(k,n)\right).$$

Interpreting the limit as the approximation to a Riemann integral,

$$\lim_{n\to\infty} \frac{1}{n\pi}\sum_{k=0}^{[nx]}\left[\frac{k}{n}\left(1 - \frac{k}{n}\right)\right]^{-1/2} = \frac{1}{\pi}\int_0^x \frac{dt}{\sqrt{t(1-t)}} = \frac{2}{\pi}\arcsin\sqrt{x}.$$

As the integral exists the error term must go to zero, and that proves T2.

E2 Consider *left-continuous random walk* with $m < \infty$, $\mu = 0$, $\sigma^2 = \infty$. This is an excellent example for the purpose of discovering pathologies in fluctuation theory. A particularly ill-behaved specimen of this sort was studied by Feller,[12] and it, or rather its reversed version, is known as an *unfavorable fair game*. We shall show that every random walk of this type is sufficiently pathological so that it fails by a wide margin to exhibit the limiting behavior in P1 and T2. In particular it will follow that the hypotheses in P1 and T2 cannot be weakened.

[12] See [31], Vol. I, p. 246.

From previous discussions of left-continuous random walk in E17.2, E18.2, and E19.1 we need only the following facts ((1), (2), and (3) below).

(1) $$1 - r(t) = e^{-\sum_{1}^{\infty} \frac{t^k}{k} \mathbf{P}[S_k < 0]}, \quad 0 \le t < 1,$$

where $r(t)$ is the unique positive solution (less than one) of the equation

(2) $$r(t) = tP[r(t)].$$

Here $P(z)$ is defined as

$$P(z) = \sum_{n=0}^{\infty} P(0, n-1) z^n,$$

where $P(x,y)$ is the transition function of the random walk. Among the obvious properties of $P(z)$ are

(3) $$P(1) = P'(1) = 1, \quad P''(1) = \infty,$$

where $P'(1)$ and $P''(1)$ denote the limits of $P'(z)$ and $P''(z)$ as z approaches one through the reals less than one.

Our analysis will make use of a theorem of Hardy and Littlewood, the forerunner of Karamata's theorem (a special case of which was P2). It asserts that when $a_n \ge 0$

(4) $$\lim_{t \nearrow 1} (1-t) \sum_{0}^{\infty} a_n t^n = 1 \text{ implies } \lim_{n \to \infty} \frac{1}{n} (a_1 + \cdots + a_n) = 1.$$

Using (1), (2), and (3) we shall study the behavior of

(5) $$A(t) = (1-t) \sum_{0}^{\infty} t^n \mathbf{P}[S_n < 0]$$

as $t \nearrow 1$. We shall show that for any constant a in the interval $\frac{1}{2} \le a \le 1$ it is possible to find a left-continuous random walk with mean zero and $\sigma^2 = \infty$ such that

(6) $$\lim_{t \nearrow 1} A(t) = a.$$

But by (4) this implies

(7) $$\lim_{n \to \infty} \frac{1}{n} \sum_{k=1}^{n} \mathbf{P}[S_k < 0] = a.$$

It will even be possible to find a left-continuous random walk such that the limit in (6), and hence, by an Abelian argument, the limit in (7) fails to exist altogether.

Suppose now that the limit in (7) exists with some $a > \frac{1}{2}$ or that it does not exist at all. In this case the arc-sine law in part (b) of P4 cannot be satisfied—for if it were, one would necessarily have

$$\lim_{n \to \infty} \mathbf{E}\left[\frac{N_n}{n}\right] = \lim_{n \to \infty} \frac{1}{n} \sum_{k=1}^{n} \mathbf{P}[S_k > 0] = \lim_{n \to \infty} \frac{1}{n} \sum_{k=1}^{n} \mathbf{P}[S_k < 0] = \frac{1}{2}.$$

20. FLUCTUATIONS AND THE ARC-SINE LAW

It is also clear, the random variables $\mathbf{N}_n/n$ being bounded, that the limit in (7) must exist if the limit of $\mathbf{P}[\mathbf{N}_n \leq nx]$ is to exist as $n \to \infty$ for every x. Thus if we carry out our program it will be clear that there are random walks for which $\mathbf{N}_n$ has no nondegenerate limiting distribution. The truth is actually still more spectacular: in problems 8 and 9 at the end of the chapter it will be shown that if there is some $F(x)$, defined on $0 < x < 1$, such that

$$\lim_{n \to \infty} \mathbf{P}[\mathbf{N}_n \leq nx] = F(x), \qquad 0 < x < 1,$$

then (7) must hold with some a between zero and one. If $a = 1$, then $F(x) \equiv 1$; if $a = 0$, then $F(x) \equiv 0$; and if $0 < a < 1$, then

(8) $$F(x) = \lim_{n \to \infty} \mathbf{P}[\mathbf{N}_n \leq nx] = \frac{\sin \pi a}{\pi} \int_0^x t^{-a}(1-t)^{a-1}\, dt.$$

The result of (8) with $a = \frac{1}{2}$ was obtained in T2. If we show, as planned, that one can get (8) with $a > \frac{1}{2}$ for suitable left-continuous random walks, then of course one can also get (8) with $a < \frac{1}{2}$. One simply considers the reversed, right-continuous random walk.

It remains, then, to investigate the limit in (6). Differentiating (1) for $0 \leq t < 1$

(9) $$\frac{r'(t)}{1 - r(t)} = \sum_1^\infty t^{k-1} \mathbf{P}[S_k < 0].$$

By equation (2)

(10) $$r'(t) = P[r(t)] + tr'(t)P'[r(t)].$$

This fact, together with (2) will be used to express $A(t)$ in equation (5) as a function of $r(t) = r$. Clearly $r(t)$ is a monotone function of t so that this defines t as a function of r. Moreover $r(t) \to 1$ as $t \to 1$, because the random walk has mean zero, as was pointed out in E17.2.

Using equations (2), (5), and (10), one obtains

(11) $$A(t) = t(1-t)\frac{r'(t)}{1 - r(t)} = \frac{P(r) - r}{1 - r} \cdot \frac{r}{P(r) - rP'(r)}$$

$$= \frac{r}{(1-r)\left[1 + r\dfrac{1 - P'(r)}{P(r) - r}\right]}.$$

It is therefore clear that (6) will hold with a certain value of a if and only if

(12) $$\lim_{r \nearrow 1} \frac{1}{1-r} \frac{P(r) - r}{1 - P'(r)} = a.$$

It is not hard to construct a left-continuous random walk for which the limit in (12) will have any desired value $\frac{1}{2} \leq a < 1$. The limit depends only on the behavior of $P(r)$ and $P'(r)$ near $r = 1$ and that is determined by the behavior of the coefficients $P(0,n)$ for large n. Suppose we consider a random walk such that for some integer $N > 0$, and some $c > 0$

(13) $$P(0,n) = \frac{c}{n^{\alpha+2}} \text{ for } n > N, \quad 0 < \alpha \leq 1.$$

The constant c is assumed to be adjusted so that (3) holds, and keeping this in mind, an easy calculation shows that (12) holds, and that the limit is

(14) $$\frac{1}{2} \leq a = \frac{1}{\alpha+1} < 1.$$

To get (12) with the limit $a = 1$ one can of course take $\alpha = 0$, and get the right result, since the random walk with $\alpha = 0$ will be transient, having infinite first moment, so that $\mathbf{N}_n/n \to +1$ with probability one. If one wants an example where $m < \infty$, so that the random walk remains recurrent, one can take

$$P(0,n) = \frac{c}{(n \ln n)^2}$$

for large enough n, but this is now a little tedious to check.

Finally, to produce an example where the limit in (12) fails to exist, there is a procedure familiar to all those whose work requires delicate asymptotic estimates. One simply decomposes the positive integers into blocks or intervals,

$$I_1 = [1,n_1), \quad I_2 = [n_1,n_2), \ldots, \quad I_k = [n_{k-1},n_k), \ldots.$$

Then one takes two numbers $0 < \alpha_1 < \alpha_2 < 1$, and chooses

$$P(0,n) = \frac{c}{n^{\alpha_1+2}} \text{ when } n \in I_1 \bigcup I_3 \bigcup I_5 \bigcup \ldots,$$

$$= \frac{c}{n^{\alpha_2+2}} \text{ when } n \in I_2 \bigcup I_4 \bigcup I_6 \bigcup \ldots.$$

It can now be shown that

$$\frac{1}{\alpha_2+1} = \varliminf_{t \nearrow 1} A(t) < \varlimsup_{t \nearrow 1} A(t) = \frac{1}{\alpha_1+1},$$

provided that the obvious condition is met: the length of the kth interval, which is $n_k - n_{k-1}$, must tend to infinity sufficiently fast as $k \to \infty$. This can be arranged at no extra cost, but we omit the details of the calculation.

Remark: The last result of E2 amounted to showing that *there need not exist a limiting distribution for the occupation time $\mathbf{N}_n$ of the right half-line*. This phenomenon is typical of a larger class of results, called *occupation time theorems*. Consider arbitrary non-degenerate random walk, and let

(1) $\quad B \subset \bar{R}, \ |B| < \infty, \qquad \varphi(x) = 1$ for $x \in B$, 0 otherwise,

$$\mathbf{N}_n = \sum_{k=1}^{n} \varphi(\mathbf{S}_k).$$

$\mathbf{N}_n$ is then the occupation time of B. A typical question concerns the possibility of finding a positive sequence a_n such that

(2) $\qquad \lim_{n \to \infty} \mathbf{P}[\mathbf{N}_n \leq a_n x] = F(x)$

exists at all continuity points of $F(x)$, and such that $F(x)$ is a non-degenerate distribution (has more than one point of increase). Typical answers are [13]: If $d = 1$, $\mu = 0$, $0 < \sigma^2 < \infty$, then there is a limit distribution $F(x)$ if one takes $a_n = \sqrt{n}$ (a truncated normal distribution). But if $d = 1$, $m < \infty$, $\mu = 0$, $\sigma^2 = \infty$, the situation is comparable to that concerning the arc-sine law. There is a one-parameter family of distributions (the so-called Mittag-Leffler distributions) which do arise as limits in (2). But there are also random walks which cannot be normalized by any sequence a_n to give a nondegenerate limit distribution.

In two dimensions the situation is similar. When the mean is zero and the second moments exist, one can show that

(3) $\qquad \lim_{n \to \infty} \mathbf{P}[\mathbf{N}_n > x \ln n] = e^{-cx}, \qquad 0 \leq x,$

where the constant c is a function of the second moments and of the cardinality $|B|$. It is of course inversely proportional to $|B|$.

Problems

1. For aperiodic symmetric one-dimensional random walk use P7.6 to prove that $\ln [1 - \phi(\theta)]$ is Lebesgue integrable on the interval $-\pi \leq \theta \leq \pi$.

2. Continuation. Conclude from problem (1) that for $0 \leq r \leq 1$

$$1 - \mathbf{E}[r^{\mathbf{Z}}] = \frac{\exp\left\{\dfrac{1}{4\pi} \displaystyle\int_{-\pi}^{\pi} \dfrac{1 - r^2}{1 + r^2 - 2r \cos \theta} \ln [1 - \phi(\theta)] \, d\theta\right\}}{\exp\left\{\dfrac{1}{4\pi} \displaystyle\int_{-\pi}^{\pi} \ln [1 - \phi(\theta)] \, d\theta\right\}}.$$

[13] For an elementary discussion see [31], Vol. I, Ch. III, and [51], Ch. II. Complete results, including necessary and sufficient conditions for the existence of a nondegenerate limit distribution $F(x)$, may be found in a paper of Darling and Kac [20].

3. Calculate
$$E[t^T e^{i\theta Z}]$$
for the *two-sided geometric random walk*, given by
$$P(0,0) = p, \qquad P(0,x) = \tfrac{1}{2}p(1-p)^{|x|}, \qquad x \neq 0.$$

4. When $\mu = 0$, $\sigma^2 < \infty$, show that
$$E[M_n] = \sum_{k=1}^{n} \frac{1}{k} E[S_k; S_k > 0] \sim \sum_{k=1}^{n} \frac{1}{2k} E|S_k| \sim \sqrt{\frac{2n}{\pi}}\, \sigma \text{ as } n \to \infty.$$

It is also known that
$$\lim_{n \to \infty} P[M_n \leq x\, E[M_n]] = \frac{2}{\sqrt{\pi}} \int_0^{\frac{x}{\sqrt{\pi}}} e^{-t^2}\, dt, \qquad x \geq 0.$$

This limit theorem (cf. Erdös and Kac [29] and Darling [19]) may be obtained from P19.1 with the aid of the Tauberian theorem in P20.2.

5. Show that one-dimensional random walk has $m < \infty$ and $\mu = 0$ if and only if
$$\sum_{k=1}^{\infty} \frac{1}{k} P\!\left[\left|\frac{S_k}{k}\right| > \epsilon\right] < \infty$$
for every $\epsilon > 0$.[14]

6. For each $n \geq 1$, prove that $E[X_k] = 0$ and $E|X_k|^{n+1} = m_{n+1} < \infty$ imply $E[Z^n] < \infty$. Hint:
$$E[Z^n] = \sum_{k=1}^{\infty} k^n \sum_{x=1}^{0} g_{(0,\infty)}(0,x) P(x,k)$$
$$\leq \sum_{k=1}^{\infty} k^n \sum_{x=-\infty}^{0} g_{(1)}(0,x) P(x,k).$$

Here $g_A(0,x)$ is the expected number of visits of the random walk, starting at 0, to x, before the first visit to the set A. Now show that $g_{(1)}(0,x)$ is bounded, so that for some $M < \infty$

---

[14] The following elegant extension is due to Katz [53]. One-dimensional random walk has mean 0 and absolute moment
$$m_\alpha = \sum |x|^\alpha P(0,x) < \infty, \qquad \alpha \geq 1,$$
if and only if the series
$$\sum_{k=1}^{\infty} \frac{1}{k^{2-\alpha}} P\!\left[\left|\frac{S_k}{k}\right| > \epsilon\right]$$
converges for every $\epsilon > 0$.

$$\mathbf{E}[Z^n] \le M \sum_{k=1}^{\infty} k^n \sum_{x=-\infty}^{0} P(x,k) < \infty$$

when $m_{n+1} < \infty$. (Due to H. Kesten.)

7. Let $\mathbf{T}_n = \min [k \mid 0 \le k \le n; \mathbf{S}_k = \mathbf{M}_n]$ and prove that

$$\mathbf{P}[\mathbf{N}_n = k] = \mathbf{P}[\mathbf{T}_n = k] \text{ for } 0 \le k \le n.$$

8. Let $g_n^{(r)} = \mathbf{E}[(n - \mathbf{N}_n)^r]$, $r \ge 0$, $n \ge 0$ where $\mathbf{N}_n$ is the number of positive partial sums in time n. Prove that

$$g_n^{(r+1)} = ng_n^{(r)} - \sum_{m=0}^{n-1} a_{n-m} g_m^{(r)} \qquad n \ge 1, \quad r \ge 0,$$

where $a_n = \mathbf{P}[\mathbf{S}_n > 0]$. Using this recurrence relation prove (following Kemperman [57], p. 93) that

$$\lim_{n \to \infty} n^{-r} g_n^{(r)} = (1 - \alpha)\left(1 - \frac{\alpha}{2}\right) \cdots \left(1 - \frac{\alpha}{r}\right)$$

for all $r \ge 1$ if and only if

$$\lim_{n \to \infty} \frac{a_1 + \cdots + a_n}{n} = \alpha.$$

9. Using the "method of moments" (see P23.3 in Chapter V) conclude that

$$\lim_{n \to \infty} \mathbf{P}[\mathbf{N}_n \le nx] = F(x), \qquad -\infty < x < \infty$$

exists if and only if

$$\lim_{n \to \infty} \frac{a_1 + \cdots + a_n}{n} = \alpha, \qquad 0 \le \alpha \le 1$$

exists, and that $F(x)$ is then related to α by

$$F(x) = F_\alpha(x) = \frac{\sin \pi\alpha}{\pi} \int_0^x t^{\alpha-1}(1-t)^{-\alpha} \, dt, \text{ if } 0 < \alpha < 1,$$

$$F_0(x) = 0 \text{ if } x < 0, \ 1 \text{ if } x \ge 0,$$
$$F_1(x) = 0 \text{ if } x < 1, \ 1 \text{ if } x \ge 1.$$

10. For aperiodic symmetric random walk with finite third moment m_3, one can use the result of problem 6 to sharpen the statement in P8, to the effect that

$$\lim_{n \to \infty} u(n) = \lim_{n \to \infty} v(n) = \frac{\sqrt{2}}{\sigma}.$$

Show that one obtains

$$0 < \sum_{n=0}^{\infty} \left[\frac{\sigma}{\sqrt{2}} u_n - 1\right] = \sum_{n=0}^{\infty} \left[\frac{\sigma}{\sqrt{2}} v_n - 1\right]$$
$$= \frac{1}{4\pi} \int_{-\pi}^{\pi} \ln\left[\frac{1 - \cos\theta}{1 - \phi(\theta)} \sigma^2\right] \frac{d\theta}{1 - \cos\theta} < \infty.$$

11. Let $\mathbf{X}_1, \mathbf{X}_2, \ldots$ be independent identically distributed random variables, $\mathbf{S}_0 = 0$, $\mathbf{S}_n = \mathbf{X}_1 + \cdots + \mathbf{X}_n$, and $\mathbf{M}_n = \max[\mathbf{S}_0, \mathbf{S}_1, \ldots, \mathbf{S}_n]$, as in D19.1. Now we define two other sequences $\mathbf{Z}_n$ and $\mathbf{W}_n$ of random variables, with $n \geq 0$. The first of these is

$$\mathbf{Z}_0 = 0, \quad \mathbf{Z}_1 = (\mathbf{X}_1)^+, \ldots, \mathbf{Z}_{n+1} = (\mathbf{Z}_n + \mathbf{X}_{n+1})^+,$$

where $x^+ = x$ if $x > 0$, and 0 otherwise. The second sequence is defined by

$$\mathbf{W}_0 = 0, \quad \mathbf{W}_n = \sum_{k=1}^{n} \mathbf{X}_k \theta(\mathbf{S}_k),$$

where $\theta(x) = 1$ if $x > 0$, and 0 otherwise.

Prove that

$$\mathbf{P}[\mathbf{M}_n = x] = \mathbf{P}[\mathbf{Z}_n = x] = \mathbf{P}[\mathbf{W}_n = x]$$

for all non-negative integers n and x. The sequence $\mathbf{Z}_n$ has applications in *queueing theory* [17].

12. For an arbitrary *left-continuous* random walk, starting at 0, let $P(z) = \sum_{0}^{\infty} P(0, k-1) z^k$. If $\mathbf{T}^*$ is the time of the first visit to the left half line, then according to P17.5 and E17.2

$$\mathbf{E}[t^{\mathbf{T}^*}] = 1 - f_e(t; 1) = r(t),$$

and for $0 \leq t < 1$, $z = r(t)$ is the unique root of the equation $z = tP(z)$ in the disc $|z| \leq 1$. By a theorem of Lagrange this equation has the solution $z = z(t) = \sum_{1}^{\infty} a_k t^k$, where $k a_k$ is the coefficient of z^{k-1} in $[P(z)]^k$. Fill in the details and conclude that

$$\mathbf{P}[\mathbf{T}^* = k] = \frac{1}{k} \mathbf{P}[\mathbf{S}_k = -1], \quad \text{for} \quad k \geq 1.$$

For an application outside the theory of random walk consider a *simple branching process* $\mathbf{z}_n$ with $\mathbf{z}_0 = 1$. It is defined (see [40]) so that $\mathbf{z}_1 = k$ with given probabilities p_k for $k \geq 0$, and $\mathbf{z}_{n+1}$ is the sum of $\mathbf{z}_n$ independent random variables with the same distribution as $\mathbf{z}_1$. Let $\mathbf{N} = \sum_{0}^{\infty} \mathbf{z}_n \leq \infty$ denote the "total number of individuals in all generations" (which is finite if and only if $\mathbf{z}_n = 0$ for some n). If $r(t)$ is the generating function of $\mathbf{N}$, show that it satisfies the same equation as before, i.e., that

$$r(t) = t \sum_{0}^{\infty} p_k [r(t)]^k.$$

Finally, call $Q(i,j)$ the probability that $\mathbf{z}_{n+1} = j$, given that $\mathbf{z}_n = i$. Prove that

$$\mathbf{P}[\mathbf{N} = k] = \frac{1}{k} Q(k, k-1), \quad \text{for} \quad k \geq 1.$$

13. Suppose that $\mathbf{X}_1, \mathbf{X}_2, \ldots$, are independent real valued random variables (not integer valued!) with common distribution

$$F(y) = \mathbf{P}[\mathbf{X}_k \leq y], \quad -\infty < y < \infty.$$

Show, by replacing all Fourier series in the unit circle by Fourier transforms in the upper half plane, that every result of Chapter IV remains correct, with obvious modifications.

Now suppose in addition that $F(y)$ is continuous and symmetric, i.e., $F(y) = 1 - F(-y)$ for all real y. If

$$\phi(\theta) = \int_{-\infty}^{\infty} e^{i\theta y} dF(y), \quad -\infty < \theta < \infty,$$

show that

(a) $\quad 1 - t\phi(\theta) = |1 - \mathbf{E}[t^\mathbf{T} e^{i\theta \mathbf{Z}}]|^2, \quad 0 \leq t < 1,$

(b) $\quad \mathbf{P}[\mathbf{T} = k] = (-1)^{k+1} \binom{1/2}{k}, \quad k \geq 1,$

(c) $\quad \mathbf{P}[\mathbf{N}_k = 0] = (-1)^k \binom{-1/2}{k}, \quad k \geq 0,$

(d) $\quad \sum_{n=0}^{\infty} \mathbf{E}[e^{i\theta(2\mathbf{N}_n - n)}] t^n = |1 - e^{i\theta} t|^{-1}$

$$= \frac{1}{\sqrt{1 + t^2 - 2t \cos \theta}}, \quad 0 \leq t < 1,$$

so that

(e) $\quad \mathbf{E}[e^{i\theta(2\mathbf{N}_n - n)}] = P_n(\cos \theta),$

where $P_n(x)$ is the nth *Legendre polynomial*.

14. Imitating the proof of T20.2, conclude that

$$\lim_{n \to \infty} \mathbf{E}[e^{i\theta(2\mathbf{N}_n/n - 1)}] = \frac{1}{\pi} \int_{-1}^{1} \frac{e^{i\theta x}}{\sqrt{1 - x^2}} dx = J_0(\theta).$$

As pointed out by Rényi [85], $J_0(\theta)$ happens to be the *Bessel function of order zero*,

$$J_0(\theta) = \sum_{k=0}^{\infty} \frac{(-1)^k}{(k!)^2} \left(\frac{\theta}{2}\right)^{2k},$$

so that the arc-sine law yields a well-known theorem concerning the asymptotic behavior of Legendre polynomials:

$$\lim_{n \to \infty} P_n\left[\cos\left(\frac{\theta}{n}\right)\right] = J_0(\theta), \qquad -\infty < \theta < \infty.$$

15. *Coin tossing at random times* (Täcklind [98]). The *compound Poisson process* $\mathbf{x}(t)$ is defined as $\mathbf{x}(t) = \mathbf{x}_{N(t)}$, where $\mathbf{x}_n$ is simple one-dimensional random walk with $\mathbf{x}_0 = 0$. $N(t)$ is the simple Poisson process defined for $t \geq 0$ by $N(0) = 0$ and

$$\mathbf{P}[N(t_1) = n_1, N(t_2) = n_2, \ldots, N(t_k) = n_k]$$
$$= \prod_{j=1}^{k} \frac{e^{-a(t_j - t_{j-1})}}{(n_j - n_{j-1})!} [a(t_j - t_{j-1})]^{(n_j - n_{j-1})}$$

for real $0 = t_0 < t_1 < \cdots < t_k$ and integer valued $0 = n_0 \leq n_1 \leq \cdots \leq n_k$. Thus $\mathbf{x}(t)$ will be a step function; its jumps are of magnitude one, and the independent, exponentially distributed time intervals between jumps have mean $1/a$. Prove that

$$\mathbf{P}\left[\sup_{0 \leq \tau \leq t} \mathbf{x}(\tau) \geq n\right] = \sum_{k=0}^{\infty} e^{-at} \frac{(at)^k}{k!} \mathbf{P}_0[\max\{0, \mathbf{x}_1, \mathbf{x}_2, \ldots, \mathbf{x}_k\} \geq n]$$
$$= \begin{cases} 1 \text{ for } n = 0, \\ n \int_0^t e^{-ax} \frac{I_n(ax)}{x} dx \text{ for } n \geq 1. \end{cases}$$

Here $I_n(x) = i^{-n} J_n(ix)$, and $J_n(x)$ is the Bessel function of the first kind, of order n.

Chapter V

RANDOM WALK ON AN INTERVAL

21. SIMPLE RANDOM WALK

The purpose of this section is to review[1] certain aspects of the absorption problem for *simple random walk*. This problem has quite a long mathematical history, which is not surprising as we shall recognize it as a boundary value problem of the simplest possible type. It is discrete, and the transition function, which plays the role of a second-order difference operator, is symmetric. Therefore we shall be able to reduce the problem to the diagonalization of certain symmetric matrices.

The absorption problem makes just as good sense for arbitrary random walk as for the simple random walk. However, the methods used in this section for simple random walk will be seen to fail in the general case. The extent of this failure will then motivate the development of more powerful techniques in sections 22 and 23.

We begin with a few definitions which will also, in sections 22 and 23, serve for arbitrary one-dimensional random walk. We shall always exclude the degenerate case when $P(0,0) = 1$.

D1 *N is a non-negative integer, $[0,N]$ the interval consisting of the integers* $0, 1, 2, \ldots, N$.

$Q_N(x,y) = P(x,y)$ *for x,y in* $[0,N]$,

$Q_N^0(x,y) = I_N(x,y) = \delta(x,y), \quad Q_N^1(x,y) = Q_N(x,y), \quad x,y$ *in* $[0,N]$,

$Q_N^{n+1}(x,y) = \sum_{t=0}^{N} Q_N^n(x,t) Q_N(t,y)$ *for x,y in* $[0,N]$.

---

[1] For a more complete discussion, see Feller [31], Vol. I, Chs. III and XIV, Kac [51], or Rényi [84], Ch. VII, §11.

$$g_N(x,y) = \sum_{n=0}^{\infty} Q_N^n(x,y), \qquad x,y \in [0,N],$$

$$R_N(x,k) = \sum_{y=0}^{N} g_N(x,y) P(y,N+k) \text{ for } x \in [0,N], \qquad k \geq 1,$$

$$L_N(x,k) = \sum_{y=0}^{N} g_N(x,y) P(y,-k) \text{ for } x \in [0,N], \qquad k \geq 1,$$

$$R_N(x) = \sum_{k=1}^{\infty} R_N(x,k),$$

$$L_N(x) = \sum_{k=1}^{\infty} L_N(x,k) \text{ for } x \in [0,N].$$

Some of these definitions are old ones, such as that of Q_N^k. It is simply the kth iterate of the transition function $P(x,y)$ restricted to the set $[0,N]$. Similarly $g_N(x,y)$ was often encountered before, but in the terminology of D10.1 it would have to be written in the unappealingly complicated form

$$g_N(x,y) = g_{R-[0,N]}(x,y).$$

Going on to R_N and L_N, note that R stands for "right" and L for "left." These functions are simply hitting probabilities, and in the terminology of D10.1 we would have had to write

$$R_N(x,k) = H_{R-[0,N]}(x,N+k),$$
$$L_N(x,k) = H_{R-[0,N]}(x,-k),$$

for $x \in [0,N]$, $k \geq 1$.

To calculate these quantities for simple symmetric random walk is very easy. If we look at $Q_N(x,y)$ as an $N+1$ by $N+1$ square matrix (which it is!) then

$$Q_N = \tfrac{1}{2} \begin{bmatrix} 0 & 1 & 0 & \cdots & 0 & 0 \\ 1 & 0 & 1 & \cdots & 0 & 0 \\ \vdots & \vdots & & & \vdots & \vdots \\ 0 & 0 & 0 & \cdots & 0 & 1 \\ 0 & 0 & 0 & \cdots & 1 & 0 \end{bmatrix},$$

and it is natural to ask for its eigenvalues and eigenvectors. They are given by

P1 *The eigenvalues of Q_N are*

(a) $$\lambda_k = \cos\frac{k+1}{N+2}\pi, \qquad k = 0, 1, \ldots, N,$$

and the associated eigenvectors are

(b) $$v_k(x) = \sqrt{\frac{2}{N+2}} \sin\frac{k+1}{N+2}\pi(x+1), \qquad k = 0, 1, \ldots, N;$$
$$0 \le x \le N,$$

where for $k, m = 0, 1, \ldots, N$,

(c) $$Q_N v_k = \lambda_k v_k, \qquad (v_k, v_m) = \sum_{x=0}^{N} v_k(x) v_m(x) = \delta(k, m).$$

Proof: Let $\Delta_N = \Delta_N(\lambda)$ denote the determinant of $Q_N - \lambda I_N$. Then it is a simple exercise in expanding a determinant in terms of its minors to obtain the difference equation

$$\Delta_n = -\lambda \Delta_{n-1} - \tfrac{1}{4}\Delta_{n-2}, \qquad n = 2, 3, \ldots.$$

Direct computation of the two lowest-order determinants gives the initial conditions

$$\Delta_0 = -\lambda, \qquad \Delta_1 = \lambda^2 - \tfrac{1}{4}.$$

Under these conditions the difference equation has a unique solution. By standard methods (see E1.2) one gets

$$\Delta_n(\lambda) = A r_1^n + B r_2^n$$

where r_1 and r_2 are the roots of the quadratic equation

$$x^2 + \lambda x + \tfrac{1}{4} = 0.$$

Making the substitution $\lambda = -\cos t$, one finds that

$$x^2 - x \cos t + \tfrac{1}{4} = \left(x - \frac{e^{it}}{2}\right)\left(x - \frac{e^{-it}}{2}\right) = 0,$$

so that $2x = e^{\pm it}$, and the general solution of the difference equation becomes

$$\Delta_n(\lambda) = 2^{-n}[A \cos nt + B \sin nt].$$

In view of the initial conditions

$$A = -\lambda = \cos t, \qquad A \cos t + B \sin t = 2\lambda^2 - \tfrac{1}{2},$$

$$\Delta_n(\lambda) = 2^{-n} \frac{\sin 2t \cos nt + \cos 2t \sin nt}{2 \sin t} = 2^{-n} \frac{\sin(n+2)t}{2 \sin t}.$$

Hence $\Delta_n(\lambda) = 0$ when $(n + 2)t$ is a multiple of π, but not zero, so that one tries $(n + 2)t = (k + 1)\pi$, $k = 0, 1, \ldots, n$. That gives the eigenvalues in (a) of P1 with the sign reversed. But the sign does not matter since

$$\cos \frac{k + 1}{N + 2} \pi = -\cos \frac{(N - k) + 1}{N + 2} \pi$$

for $k = 0, 1, \ldots, N$. As we have found $N + 1$ eigenvalues, (a) is proved since a matrix of size $N + 1$ can have no more.

The task of verifying that the functions in (b) are eigenfunctions is quite straightforward. Indeed it suffices to show that

$$Q_N v_k = \lambda_k v_k \quad \text{and} \quad (v_k, v_k) = 1 \quad \text{for } 0 \leq k \leq N.$$

The fact that $(v_k, v_m) = 0$ when $k \neq m$ is a basic result in matrix theory; two eigenvectors belonging to distinct eigenvalues of a symmetric matrix are always orthogonal. To get $Q_N v_k = \lambda_k v_k$ one can use trigonometric identities, and finally the proof of P1 is completed by checking that

$$(v_k, v_k) = \frac{2}{N + 2} \sum_{x=0}^{N} \sin^2 \left[\frac{k + 1}{N + 2} (x + 1)\pi \right]$$

$$= 1 - \frac{1}{N + 2} \sum_{x=-1}^{N+1} \cos \left[2 \frac{k + 1}{N + 2} (x + 1)\pi \right] = 1.$$

The spectral theorem for symmetric matrices (see [38], Ch. III) now yields, in one stroke, the solution of all our problems. In other words, every one of the functions in D1 above can be expressed in terms of the eigenfunctions and eigenvalues of Q_N. In particular it is worth recording that

P2 (a) $\quad Q_N^n(x,y) = \sum_{k=0}^{N} \lambda_k^n v_k(x) v_k(y),$

(b) $\quad g_N(x,y) = \sum_{k=0}^{N} (1 - \lambda_k)^{-1} v_k(x) v_k(y).$

There is nothing to prove since these are exactly the spectral representations of Q_N and $(I - Q_N)^{-1}$, respectively. The matrix $I - Q_N$ of course has an inverse, since by P1 all its eigenvalues lie strictly between zero and one.

In anticipation of later results it is a good idea to look at the analogous eigenvalue problem in the continuous case. The matrix $Q_N - I_N$ is nothing but the second difference operator, acting on functions defined on the interval $[0,N]$. Thus one is tempted to regard

(1) $$\frac{d^2}{dx^2}f(x) + \mu f(x) = 0, \quad 0 \le x \le 1$$

as the proper analogue of

$$(Q_N - I_N)v = -\mu v \quad \text{or} \quad Q_N v = \lambda v, \quad \lambda = 1 - \mu.$$

The correct boundary condition to impose in (1) is

(2) $$f(0) = f(1) = 0.$$

This follows from the observation that the matrix equation $Q_N v = \lambda v$ is the same as the difference equation

$$\tfrac{1}{2}v(x+1) + \tfrac{1}{2}v(x-1) = \lambda v(x), \quad 0 \le x \le N,$$

with $v(-1) = v(N+1) = 0$. The well-known theory of the boundary value problem consisting of (1) and (2) may be summarized by

P3 *Equations (1) and (2) are equivalent to the integral equation*

(a) $$f(x) = \mu \int_0^1 R(x,y) f(y)\, dy, \quad 0 \le x \le 1,$$

with symmetric kernel

(b) $$R(x,y) = \min(x,y) - xy \quad \text{for } 0 \le x, y \le 1.$$

The eigenvalues are

(c) $$\mu_k = \pi^2 k^2, \quad k = 1, 2, \ldots,$$

and the eigenfunctions

(d) $$\phi_k(x) = \sqrt{2} \sin \pi k x, \quad 0 \le x \le 1, \quad k = 1, 2, \ldots,$$

form an orthonormal set, i.e.,

$$(\phi_k, \phi_m) = \int_0^1 \phi_k(x) \phi_m(x)\, dx = \delta(k,m).$$

The proof can be found in almost any book on eigenvalue problems. The theory of completely continuous symmetric operators[2] delimits the full extent to which matrix results carry over into a more general setting. One particularly important result, the continuous analogue

---

[2] Cf. Riesz-Nagy [87], Ch. IV, or Courant-Hilbert [16], Vol. I, Ch. III. In particular the Green function $R(x,y) = \min(x,y) - xy$ arises in the simplest mathematical description of the *vibrating string*, with fixed end points. It is often called the vibrating string kernel.

of part (b) of P2, is Mercer's theorem: the series

$$R(x,y) = \sum_{k=1}^{\infty} \frac{\phi_k(x)\phi_k(y)}{\mu_k},$$

converges uniformly in the square $0 \leq x, y \leq 1$, and so do the series expansions for the iterates of $R(x,y)$ which will be defined and used in the proof of T23.2.

At this point it should not be too surprising that one gets the following formulas for simple random walk

P4 (a) $\quad \dfrac{1}{2(N+2)} g_N(x,y) = R\left(\dfrac{x+1}{N+2}, \dfrac{y+1}{N+2}\right),$

(b) $\quad R_N(x) = \dfrac{x+1}{N+2}, \quad L_N(x) = R_N(N-x) = \dfrac{N-x+1}{N+2}.$

Here $R(x,y)$ is the kernel defined in equation (b) of P3.

Proof: The shortest proof of (a) is to verify, by matrix multiplication, that the matrix on the right in (a) is indeed the inverse of $2(N+2)(I_N - Q_N)$. This is left to the reader, and once that is done (b) follows from the definition of $R_N(x)$ in D1.

Now we shall show that there are problems where eigenfunction expansions, such as those in P2, are quite useful—in fact almost indispensable. Such problems concern the asymptotic behavior of hitting probabilities, Green functions, etc. We shall confine our attention to the random variable $\mathbf{T}_N$—the hitting time of the exterior of the interval $[0,N]$.

D2 $\qquad \mathbf{T}_N = \min \{k \mid k \geq 1, \mathbf{x}_k \in R - [0,N]\}.$

To simplify the problem a little, without losing the essential flavor, let us assume that the random walk has as its starting point $\mathbf{x}_0 = N$, the mid-point of the interval $[0,2N]$.

Quite evidently

$$\lim_{N \to \infty} \mathbf{P}_N[\mathbf{T}_{2N} > n] = 1$$

for each fixed integer n. Thus it is reasonable to look for a normalizing sequence $a(N)$ such that, if possible,

$$\lim_{N \to \infty} \mathbf{P}_N[\mathbf{T}_{2N} \leq xa(N)] = F(x), \qquad x > 0,$$

exists and is a nondegenerate probability distribution defined for

21. SIMPLE RANDOM WALK 243

$x > 0$, with $F(0) = 0$, $F(\infty) = 1$. It is very natural to try a sequence $a(N)$ such that

$$E_N[T_{2N}] \sim a(N), \qquad N \to \infty.$$

But P4 gives

$$E_N[T_{2N}] = \sum_{x=0}^{2N} g_{2N}(N,x)$$

$$= 2(2N + 2) \sum_{x=0}^{2N} R\left(\frac{N+1}{2N+2}, \frac{x+1}{2N+2}\right) = (N+1)^2,$$

so that we shall take $a(N) = N^2$.

Now the problem is to evaluate (and prove the existence of)

$$\lim_{N \to \infty} P_N[T_{2N} > N^2 x], \qquad x > 0.$$

The calculation of the distribution of T_{2N} is facilitated by the observation that

$$P_N[T_{2N} > n] = \sum_{y=0}^{2N} Q_{2N}{}^n(N,y).$$

Using part (a) of P2 one gets

$$P_N[T_{2N} > n] = \sum_{y=0}^{2N} \sum_{k=0}^{2N} \lambda_k{}^n v_k(N) v_k(y).$$

In view of P1,

$P_N[T_{2N} > n]$

$$= \sqrt{\frac{1}{N+1}} \sum_{k=0}^{2N} \left[\cos\left(\frac{k+1}{2N+2}\pi\right)\right]^n \sin\left(\frac{k+1}{2}\pi\right) \sum_{y=0}^{2N} v_k(y)$$

$$= \frac{1}{N+1} \sum_{j=0}^{N} \left[\cos\frac{2j+1}{2N+2}\pi\right]^n (-1)^j \sum_{y=0}^{2N} \sin\left[\frac{2j+1}{2N+2}\pi(y+1)\right]$$

$$= \frac{1}{N+1} \sum_{j=0}^{N} (-1)^j \left[\cos\left(\frac{2j+1}{2N+2}\pi\right)\right]^n \cot\left[\left(\frac{2j+1}{2N+2}\right)\frac{\pi}{2}\right].$$

Here the last step consisted of evaluating the sum on y, by converting it into a geometric progression.

It is easy to see that for every $\epsilon > 0$

$$\lim_{N \to \infty} \frac{1}{N+1} \sum_{N\epsilon \le j \le N} (-1)^j \left(\cos\frac{2j+1}{2N+2}\pi\right)^n \cot\left(\frac{2j+1}{2N+2}\frac{\pi}{2}\right) = 0,$$

uniformly in n. That serves as justification for using the approximations

$$\left(\cos\frac{2j+1}{2N+2}\pi\right)^{N^2x} \sim \left(1 - \frac{\pi^2}{8}\frac{(2j+1)^2}{(N+1)^2}\right)^{N^2x} \sim e^{-\frac{\pi^2(2j+1)^2x}{8}},$$

$$\cot\frac{2j+1}{2N+2}\frac{\pi}{2} \sim \frac{4}{\pi}\frac{N+1}{2j+1},$$

which apply when j is small compared to N. This way one obtains

P5 $\quad \lim_{N\to\infty} \mathbf{P}_N[\mathbf{T}_{2N} > N^2 x] = 1 - F(x)$

$$= \frac{4}{\pi}\sum_{j=0}^{\infty} \frac{(-1)^j}{2j+1} e^{-\frac{\pi^2}{8}(2j+1)^2 x}, \qquad x > 0.$$

In section 23 we shall develop methods which enable us to prove P5 for arbitrary one-dimensional random walk with mean zero and finite variance.

22. THE ABSORPTION PROBLEM WITH MEAN ZERO, FINITE VARIANCE

There are good reasons for suspecting strong similarity between simple random walk and any other random walk with mean zero and finite variance. Thus the Central Limit Theorem asserts that the asymptotic behavior of $P_n(x,y)$ is in a sense the same for every random walk with $\mu = 0$, $\sigma^2 < \infty$. But it is not clear how relevant the Central Limit Theorem is to the absorption problem where we are concerned with $Q_N(x,y)$ and its iterates rather than with the unrestricted transition function $P(x,y)$. For this reason we shall begin by investigating to what extent the operator $Q_N - I_N$ has a behavior resembling that of the second-order differential operator.

P1 *Let $P(x,y)$ be the transition function of a random walk with $\mu = 0$, $0 < \sigma^2 < \infty$, and let $Q_N(x,y)$ denote its restriction to the interval $[0,N]$. Let $f(x)$ be a function with a continuous second derivative on $0 \le x \le 1$. Then for every t in $0 < t < 1$*

$$\lim_{N\to\infty} \frac{2N^2}{\sigma^2} \sum_{k=0}^{N} [I_N - Q_N]([Nt],k) f\left(\frac{k}{N}\right) = -f''(t).$$

Proof: The assumptions on $f(x)$ imply that for each pair x,y in $[0,1]$

$$f(y) = f(x) + (y-x)f'(x) + \tfrac{1}{2}(y-x)^2 f''(x) + \rho(x,y),$$

where $(x-y)^{-2} \cdot \rho(x,y) \to 0$ uniformly in any interval $0 < a \leq x$, $y \leq b < 1$. Furthermore one can find a positive constant M such that

$$\max_{0 \leq x,\, y \leq 1} [f(x), f'(x), f''(x), \rho(x,y)] \leq M.$$

If we take $0 < t < 1$ and denote $[Nt]N^{-1} = z(N,t) = z$, then

$$\frac{2N^2}{\sigma^2} \sum_{k=0}^{N} [\delta([Nt],k) - Q_N([Nt],k)] f\left(\frac{k}{N}\right)$$

$$= \frac{2N^2}{\sigma^2} \sum_{k=0}^{N} \left[(N+1)^{-1} f(z) - P(zN,k) \left\{ f(z) + \left(\frac{k}{N} - z\right) f'(z) \right.\right.$$

$$\left.\left. + \frac{1}{2}\left(\frac{k}{N} - z\right)^2 f''(z) + \rho\left(z, \frac{k}{N}\right) \right\} \right]$$

$$= I_1 + I_2 + I_3 + I_4.$$

Here the terms have been decomposed in such a way that

$$I_1 = f(z) \frac{2N^2}{\sigma^2} \left[1 - \sum_{k=0}^{N} P(zN,k)\right],$$

$$I_2 = f'(z) \frac{2N}{\sigma^2} \sum_{k=0}^{N} P(zN,k)(zN-k),$$

$$I_3 = -f''(z) \frac{1}{\sigma^2} \sum_{k=0}^{N} P(zN,k)(zN-k)^2,$$

$$I_4 = -\frac{2N^2}{\sigma^2} \sum_{k=0}^{N} P(zN,k) \rho\left(z, \frac{k}{N}\right).$$

Since $f(t)$ and $f'(t)$ are bounded and $0 < z < 1$ one can conclude that $I_1 \to 0$ and $I_2 \to 0$ as $N \to \infty$. It should be clear how to do that, using in an essential way the assumption that $\mu = 0$, $\sigma^2 < \infty$. The third term, I_3, obviously converges to $-f''(t)$ as $N \to \infty$. To complete the proof of P1 it is necessary to break I_4 into two parts. Given $\epsilon > 0$ we choose A so that

$$\left|\rho\left(z, \frac{k}{N}\right)\right| \leq \epsilon \left(z \frac{k}{N}\right)^2,$$

when $|z - (k/N)| \leq A$, for all sufficiently large N. Then one has

$$|I_4| \leq \frac{2\epsilon}{\sigma^2} \sum_{k=-\infty}^{\infty} P(zN,k)(zN-k)^2 + \frac{2N^2}{\sigma^2} M \sum_{[k \mid z - \frac{k}{N}\mid > A]} P(zN,k).$$

246   RANDOM WALK ON AN INTERVAL

The first term is exactly 2ϵ and the second term tends to zero as $N \to \infty$ for the same (undisclosed, but obvious) reason that I_1 went to zero. Since ϵ is arbitrary, that completes the proof of P1.

According to P21.3, one has reason to suspect that the inverse $g_N(x,y)$ of $I_N - Q_N$ will approach, as $N \to \infty$, the kernel

$$R(s,t) = \min(s,t) - st,$$

in much the same way as $I_N - Q_N$ approaches the second derivative operator. One might guess the formally correct analogue of P1 to be

P2 *For random walk with mean $\mu = 0$ and variance $0 < \sigma^2 < \infty$, and for $f(x)$ bounded and Riemann integrable on $0 \leq x \leq 1$,*

$$\lim_{N \to \infty} \frac{\sigma^2}{2N^2} \sum_{k=0}^{N} g_N([Nt],k) f\left(\frac{k}{N}\right) = \int_0^1 f(s) R(s,t)\, ds,$$

uniformly for $0 \leq t \leq 1$.

Proposition P2, just as P1, is of interest, and of course makes perfect sense, without any reference whatsoever to probabilistic terms or ideas. But whereas the proof of P1 depended only on elementary calculus, P2 seems to be mathematically deeper. In the next section we shall prove it, and in fact considerably more—but only by making strong use of probabilistic results from Chapter IV.

Much too little is known about the asymptotic properties of eigenfunction expansions in general to attempt a proof of P2 along the lines of section 21. Even if we are willing to restrict attention to symmetric random walk so that $g_N(x,y)$ has a representation of the form

$$g_N(x,y) = \sum_{k=0}^{N} \frac{v_k(x) v_k(y)}{1 - \lambda_k},$$

the difficulties are formidable. Observe that in the above formula we should have written $v_k(x) = v_k(N;x)$, $\lambda_k = \lambda_k(N)$, since both the eigenvalues and the eigenfunctions depend on N. Consequently the asymptotic behavior of $g_N(x,y)$ as $N \to \infty$ depends on that of both $\lambda_k(N)$ and $v_k(N;x)$. Although a great deal is known about the distribution of the eigenvalues $\lambda_0(N), \lambda_1(N), \ldots, \lambda_N(N)$ for large N (see problems 8 and 9 for a small sample), one knows very little about the eigenfunctions. The blunt truth of the matter is that there are much better ways to approach our problem. Although spectral decomposition of an operator, which is what we have been contemplating, is a very general and very powerful mathematical tool, it is not always the best one. For one thing it is too general. Our present problem has a special feature which deserves to be taken advantage of: the operators $P(x,y)$ and $Q_N(x,y)$ are both *difference kernels* ($P(x,y) = P(0, y - x)$, and Q_N is the truncation of P).

In 1920 G. Szegö [97] succeeded in associating with symmetric difference operators a system of orthogonal polynomials. Although these polynomials are not eigenfunctions of the operator, they perform a very similar function. Referring the reader to the literature [36] (see also problems 5, 8, and 9) we shall be content to describe a small part of his theory which is directly relevant to symmetric random walk. This is a very special context from the point of view of Szegö's theory—his matrices need not have non-negative elements, as they do in the context of random walk.

P3 *Suppose that $P(x,y) = P(y,x)$ is the transition function of symmetric one-dimensional random walk with characteristic function*

$$\phi(\theta) = \sum_{x=-\infty}^{\infty} P(0,x) e^{ix\theta} = \overline{\phi(\theta)} = \phi(-\theta).$$

Then there exists a sequence of polynomials

(a) $$p_n(z) = \sum_{k=0}^{n} p_{n,k} z^k, \qquad n \geq 0$$

with the property that

(b) $$p_{n,n} > 0 \text{ for } n \geq 0,$$

(c) $$\frac{1}{2\pi} \int_{-\pi}^{\pi} p_k(e^{i\theta}) \overline{p_m(e^{i\theta})} [1 - \phi(\theta)] \, d\theta = \delta(k,m), \qquad k,m = 0, 1, 2, \ldots$$

These polynomials are uniquely determined by requirements (b) *and* (c). *All the coefficients $p_{n,k}$ are non-negative. The Green function $g_N(i,j)$ of D21.1 has the representation*

(d) $$g_N(i,j) = \sum_{k=\max(i,j)}^{N} p_{k,i} p_{k,j}$$

$$= \sum_{k=0}^{\min(i,j)} p_{N-k,N-i} p_{N-k,N-j} \text{ for } N \geq 0, \quad i \geq 0, \quad j \geq 0.$$

The limiting behavior of the orthogonal polynomials is described by

(e) $$\lim_{n \to \infty} p_{n,n-k} = u(k), \qquad k \geq 0,$$

where $u(k)$ is the sequence defined by

(f) $$\sum_{k=0}^{\infty} u(k) z^k = U(z) = V(z),$$

with $U(z)$ and $V(z)$ as defined in D18.2.

Proof: The existence of a unique system of polynomials $p_n(z)$ satisfying (b) and (c) follows from the well known Gram Schmidt orthogonalization process. The sequence $f_n(e^{i\theta}) = e^{in\theta}$, $n = 0, \pm 1, \pm 2, \ldots$, forms a complete orthonormal set for the Hilbert space $L_2(-\pi,\pi)$ of functions square integrable on the interval $[-\pi,\pi]$. Here orthogonality is defined in the sense that

$$\frac{1}{2\pi} \int_{-\pi}^{\pi} f_n(e^{i\theta}) \overline{f_m(e^{i\theta})} \, d\theta = \delta(m,n), \qquad m,n = 0, \pm 1, \ldots.$$

We are interested, however, in constructing a system of polynomials $p_n(z)$ which are orthogonal with respect to the inner product

$$(f,g) = \frac{1}{2\pi} \int_{-\pi}^{\pi} f(e^{i\theta}) \overline{g(e^{i\theta})} [1 - \phi(\theta)] \, d\theta.$$

It is easy to verify that the explicit formulas

$$p_0(z) = [1 - P(0,0)]^{-1/2},$$

(1) $$p_n(z) = (D_n D_{n-1})^{-1/2} \begin{vmatrix} (f_0,f_0) & (f_1,f_0) & \cdots & (f_n,f_0) \\ (f_0,f_1) & (f_1,f_1) & \cdots & (f_n,f_1) \\ \vdots & \vdots & & \vdots \\ (f_0,f_{n-1}) & (f_1,f_{n-1}) & \cdots & (f_n,f_{n-1}) \\ f_0(z) & f_1(z) & & f_n(z) \end{vmatrix}, \quad n \geq 1,$$

where $$D_n = \begin{vmatrix} (f_0,f_0) & (f_1,f_0) & \cdots & (f_n,f_0) \\ (f_0,f_1) & (f_1,f_1) & \cdots & (f_n,f_1) \\ \vdots & \vdots & & \vdots \\ (f_0,f_n) & (f_1,f_n) & \cdots & (f_n,f_n) \end{vmatrix}, \quad n \geq 0,$$

yield a system of polynomials such that

$$(p_n, p_m) = \delta(n,m), \qquad n,m = 0, 1, 2, \ldots,$$

and such that (b) and (c) are satisfied. Nor is it difficult to prove uniqueness.[3] The above representation of $p_n(z)$ as a ratio of determinants is of course quite general—it holds when $1 - \phi(\theta)$ is replaced by any non-negative weight function $w(\theta)$ whose integral over $[-\pi,\pi]$ is positive. In our case one has

$$(2) \quad (f_n, f_m) = \frac{1}{2\pi} \int_{-\pi}^{\pi} e^{in\theta} e^{-im\theta} [1 - \phi(\theta)] \, d\theta = \delta(m,n) - P(n,m),$$

in other words, *the matrix (f_n, f_m), $0 \le m,n \le N$, is exactly the $(N+1) \times (N+1)$ matrix which we called $I_N - Q_N$ in definition* D21.1.

The proof of (d) is based on the observation that $g_N(j,k)$ is the $(j,k)^{\text{th}}$ element of the inverse of the matrix $I_N - Q_N$. This was already pointed out in connection with P21.2. It is equally true here, for although we do not know the eigenvalues of Q_N it is easy to see that they are between 0 and 1. Hence

$$g_N(j,k) = \sum_{n=0}^{\infty} Q_N{}^n(j,k) = (I_N - Q_N)^{-1}(j,k).$$

Because of the symmetry of the random walk it is clear that

$$g_N(k,j) = g_N(N - k, N - j).$$

Therefore it suffices to prove the first identity in (d) which, in terms of generating functions, becomes

$$(3) \quad \sum_{k=0}^{N} \sum_{j=0}^{N} g_N(k,j) z^j w^k = \sum_{n=0}^{N} p_n(z) p_n(w)$$

for arbitrary complex z and w. To verify (3) one first derives, using (1) and (2), that

$$(4) \quad \sum_{n=0}^{N} p_n(z) \overline{p_n(w)} = -\frac{1}{D_N} \begin{vmatrix} 1 - c_0 & -c_1 & \cdots & -c_N & 1 \\ -c_1 & 1 - c_0 & \cdots & -c_{N-1} & \bar{w} \\ \vdots & \vdots & & & \\ -c_N & -c_{N-1} & \cdots & 1 - c_0 & \bar{w}^N \\ 1 & z & \cdots & z^N & 0 \end{vmatrix}$$

where $c_k = P(0,k)$ for $k \in R$. Equation (4) is known as the formula

---

[3] See [36], p. 14, for a proof that works in any Hilbert space.

for the *reproducing kernel*.[4] Inspection of the determinant in (4) shows that the coefficient of $\bar{w}^k z^j$ is simply

$$-(1/D_N)M_{k,j}(-1)^{k+j}, \quad 0 \leq k, j \leq N,$$

where $M_{k,j}$ is the minor of the $(k,j)^{\text{th}}$ element of the matrix $(I_N - Q_N)^{-1}$. Consequently we have

$$(5) \qquad \sum_{k=0}^{N} \sum_{j=0}^{N} g_N(k,j) z^j \bar{w}^k = \sum_{n=0}^{N} p_n(z)\overline{p_n(w)}.$$

From (5) we should be able to conclude (3), and hence (d), if we knew that all coefficients p_{nk} of $p_n(z)$ are real. That comes from (5) which gives

$$g_N(N,k) = g_N(k,N) = p_{N,N}\overline{p_{N,k}} = \overline{p_{N,N}p_{N,k}},$$

because $g_N = I_N - Q_N$ is a symmetric matrix. Furthermore $p_{N,N} = \overline{p_{N,N}}$ according to (b), so that $p_{N,k}$ is real for all $0 \leq k \leq N$.

For the proof of part (e) we shall rely on results from Chapter IV. If

$$g_B(x,y) = g(x,y), \quad B = [x \mid -\infty < x \leq -1]$$

is the Green function of the half-line in D19.3, we have

$$\lim_{N \to \infty} g_N(k,j) = g(k,j) \text{ for } k \geq 0, \quad j \geq 0,$$

since the sequence $g_N(k,j)$ is the expected number of visits from k to j before leaving $[0,N]$ and therefore increases monotonically to $g(k,j)$ which is a similar expectation for the half-line $[0,\infty)$. Since we are here in the symmetric case, P19.3 gives

$$(6) \qquad \lim_{N \to \infty} g_N(k,j) = \sum_{n=0}^{\min(k,j)} u(k-n)u(j-n).$$

Now we specialize (6) by setting $j = 0$, to obtain

$$(7) \qquad \lim_{N \to \infty} g_N(k,0) = \lim_{N \to \infty} g_N(N-k,N)$$
$$= \lim_{N \to \infty} p_{N,N}p_{N,N-k} = u(k)u(0).$$

---

[4] Call the left-hand side in (4) $K_N(z,w)$. This "kernel" owes its name to the property that

$$\frac{1}{2\pi} \int_{-\pi}^{\pi} K_N(z,e^{i\theta})g(e^{i\theta})[1 - \phi(\theta)] \, d\theta = g(z)$$

whenever $g(z)$ is a polynomial of degree at most N.

Setting $k = 0$ in (7), while observing that $p_{N,N} > 0$, yields

(8) $\qquad \lim_{N \to \infty} p_{N,N} = u(0), \qquad \lim_{N \to \infty} p_{N,N-k} = u(k).$

That completes the proof of P3.

In the sequel we shall actually never use P3 explicitly, but only the added conviction it gives us that one should pay attention to the algebraic aspects of the problem—i.e., to the fact that $P(x,y) = P(0, y - x)$, and its various manifestations. From Chapter IV we shall now copy certain results, to obviate the need for tedious references. They concern the Green function

$$g(x,y) = g_B(x,y), \qquad B = [x \mid -\infty < x \leq -1].$$

It is given by the amazingly simple formula (6) in the proof of P3, which is valid precisely because $P(x,y)$ is a difference kernel. We shall restrict ourselves to random walk with $\mu = 0$, $\sigma^2 < \infty$ so that all the asymptotic results concerning $g(x,y)$ from section 19 become available.

P4 *For aperiodic one-dimensional random walk with mean $\mu = 0$ and variance $\sigma^2 < \infty$,*

(a) $\qquad g(x,y) = \sum_{k=0}^{\min(x,y)} u(x - k)v(y - k), \qquad x \geq 0, \ y \geq 0,$

and

(b) $\qquad \lim_{x \to +\infty} u(x) = u > 0, \qquad \lim_{x \to +\infty} v(x) = v > 0,$

where

$$uv = \frac{2}{\sigma^2}.$$

Proof: We just look up P19.3, or equation (6) above, to get (a)—this part does not require the assumptions of P4—and then we use P18.8 to verify (b). This latter part depends crucially on aperiodicity, $\mu = 0$, and $\sigma^2 < \infty$. The exact values of the limits u and v will be of no interest. All that matters is their product. Using P4 it will now be possible to begin to extend the theory in section 21 to arbitrary *aperiodic one-dimensional random walk with $\mu = 0$, $\sigma^2 < \infty$*. (Certain auxiliary results, such as P5 and P6 below will in fact be valid under more general assumptions.)

P5 There is a constant $M > 0$ such that
$$g_N(x,y) \le M[1 + \min(x, y, N - x, N - y)] < \infty,$$
whenever $0 \le x, y \le N$.

Proof: In view of P4 we can choose M such that $0 \le u(x) \le M$, $0 \le v(x) \le M$, for all $x \ge 0$. But it is probabilistically obvious that
$$g_N(x,y) \le g(x,y) \le M \min(1 + x, 1 + y) = M[1 + \min(x,y)].$$
On the other hand, we have
$$g_N(x,y) = g_N(N - y, N - x).$$
This is not a mistake! To verify it one need not even introduce the reversed random walk, as this simple identity follows directly from the definition of $g_N(x,y)$ in D21.1. Therefore we also have
$$g_N(x,y) \le g(N - y, N - x) \le M \min[1 + N - y, 1 + N - x]$$
which completes the proof of P5.

Now we take up the study of the hitting probabilities $R_N(x,k)$ and $L_N(x,k)$ of the two components of the complement of the interval $[0,N]$.

P6 (a) $\sum_{k=1}^{\infty} R_N(x,k) + \sum_{k=1}^{\infty} L_N(x,k) = 1, \qquad 0 \le x \le N,$

(b) $\sum_{k=1}^{\infty} (N + k) R_N(x,k) + \sum_{k=1}^{\infty} (-k) L_N(x,k) = x,$
$$0 \le x \le N.$$

Proof: The argument is based on the identities
$$R_N(x,k) = \sum_{y=0}^{N} g_N(x,y) P(y, N + k),$$
$$L_N(x,k) = \sum_{y=0}^{N} g_N(x,y) P(y, -k),$$
valid for $0 \le x \le N$, $k \ge 0$. They are obvious from
$$g_N(x,y) P(y, N + k) = \sum_{j=0}^{\infty} Q_N^j(x,y) P(y, N + k)$$
$$= \sum_{j=0}^{\infty} \mathbf{P}_x[T_N = j + 1; \mathbf{x}_{T_N - 1} = y, \mathbf{x}_{T_N} = N + k]$$
$$= \mathbf{P}_x[\mathbf{x}_{T_N - 1} = y, \mathbf{x}_{T_N} = N + k],$$

22. THE ABSORPTION PROBLEM WITH MEAN ZERO

where $\mathbf{T}_N$ is the first exit time from the interval $[0,N]$ defined in D21.2. The identity for L_N is proved in the same way.

The left-hand side in (a) can then be written

$$\sum_{y=0}^{N} g_N(x,y)\left[1 - \sum_{s=0}^{N} P(y,s)\right].$$

Keeping in mind that

$$\sum_{y=0}^{N} g_N(x,y) P(y,s) = g_N(x,s) - \delta(x,s)$$

when x and s are both in $[0,N]$, one gets

$$\sum_{k=1}^{N} [R_N(x,k) + L_N(x,k)] = \sum_{y=0}^{N} g_N(x,y) - \sum_{s=0}^{N} g_N(x,s) + \sum_{s=0}^{N} \delta(x,s) = 1.$$

The proof of (b) is similar, but it depends in an essential way on the hypothesis that $m < \infty$, $\mu = 0$. The left-hand side in (b) is transformed into

$$\sum_{y=0}^{N} g_N(x,y)\left[\sum_{k=1}^{\infty} (N+k) P(y, N+k) - \sum_{k=1}^{\infty} k P(y, -k)\right]$$

$$= \sum_{y=0}^{N} g_N(x,y)\left[\sum_{j=-\infty}^{\infty} j P(y,j) - \sum_{j=0}^{N} j P(y,j)\right]$$

$$= \sum_{y=0}^{N} g_N(x,y)\left[y - \sum_{j=0}^{N} j P(y,j)\right]$$

$$= \sum_{y=0}^{N} y g_N(x,y) - \sum_{j=0}^{N} j g_N(x,j) + \sum_{j=0}^{N} j\, \delta(x,j) = x.$$

That completes the proof of this intuitively obvious proposition. Part (a) asserts that the random walk leaves $[0,N]$ either on the right or on the left. Part (b) is a statement characterizing the random walk with mean $\mu = 0$ as a "fair game": if $\mathbf{x}_0 = x$, and $\mathbf{T}_N$ the stopping time we are concerned with, part (b) states that

$$\mathbf{E}_x[\mathbf{x}_{\mathbf{T}_N}] = \mathbf{x}_0 = x.$$

This is of course true for every "reasonable" stopping time.

P7 $\quad R_N(x) = \dfrac{x}{N} + \dfrac{1}{N} \sum_{k=1}^{\infty} k L_N(x,k) - \dfrac{1}{N} \sum_{k=1}^{\infty} k R_N(x,k).$

Proof: First we transform

$$R_N(x) = \sum_{k=1}^{\infty} R_N(x,k) = \frac{1}{N} \sum_{k=1}^{\infty} (N+k) R_N(x,k) - \frac{1}{N} \sum_{k=1}^{\infty} k R_N(x,k).$$

The proof is completed by applying P6 (b) to the next to last term to get

$$R_N(x) = \frac{x}{N} + \frac{1}{N}\sum_{k=1}^{\infty} kL_N(x,k) - \frac{1}{N}\sum_{k=1}^{\infty} kR_N(x,k).$$

The main theorem of this section (T1 below) concerns the asymptotic behavior of $R_N(x)$ and $L_N(x)$ as $N \to \infty$. The proof will depend on P5 which was proved under the assumption that $\sigma^2 < \infty$. But that is deceptive—P5 can in fact be shown to be true under much more general conditions. However, essential use of the finiteness of the variance will be made in an estimate very much like the ones required to prove P1. In the proof of T1 we shall have occasion to define

(1) $$a(s) = \sum_{k=1}^{\infty} kP(0, s+k), \qquad s \geq 0,$$

and we shall need the conclusion that

(2) $$\lim_{n \to \infty} \frac{1}{n} \sum_{s=0}^{n} (1+s)a(s) = 0.$$

This is easily done, for (1) yields

$$\sum_{s=0}^{\infty} a(s) = \sum_{k=0}^{\infty} \sum_{s=0}^{\infty} kP(0,s+k) = \sum_{j=1}^{\infty} (1 + 2 + \cdots + j)P(0,j)$$

$$\leq \sum_{j=1}^{\infty} j^2 P(0,j) \leq \sigma^2 < \infty,$$

and by Kronecker's lemma the convergence of the series $\sum a(s)$ implies that

$$\lim_{n \to \infty} \frac{1}{n} \sum_{s=1}^{n} sa(s) = 0,$$

which shows that (2) holds.

T1 *For one-dimensional random walk with $\mu = 0$, $0 < \sigma^2 < \infty$, $R_N(x) - x/N \to 0$ uniformly for all $0 \leq x \leq N$, as $N \to \infty$. A similar statement holds for $L_N(x) - (1 - x/N)$.*

Proof: We give the proof for aperiodic random walk, but observe that T1 is then automatically true in the periodic case, assuming of course that $P(0,0) \neq 1$, a possibility which was excluded by assuming that $\sigma^2 > 0$. (To make the obvious extension to periodic random walk it suffices to observe that in the periodic case there is some integer

$d > 1$ such that the random walk is aperiodic on the state space consisting of the multiples of d.)

In view of P7 it suffices to show that

$$\lim_{N \to \infty} \frac{1}{N} \sum_{k=1}^{\infty} k R_N(x,k) = 0.$$

uniformly in $0 \le x \le N$, and to obtain a similar result for $L_N(x,k)$. Since both problems yield to the same method it suffices to treat only one of them. Now

$$\frac{1}{N} \sum_{k=1}^{\infty} k R_N(x,k) = \frac{1}{N} \sum_{k=1}^{\infty} k \sum_{y=0}^{N} g_N(x,y) P(y, N+k),$$

and using P5 one gets

$$\frac{1}{N} \sum_{k=1}^{\infty} k R_N(x,k) \le \frac{M}{N} \sum_{k=1}^{\infty} k \sum_{y=0}^{N} (1 + N - y) P(y, N+k)$$

$$= \frac{M}{N} \sum_{k=1}^{\infty} k \sum_{s=0}^{N} (1+s) P(0, k+s) = \frac{M}{N} \sum_{s=0}^{N} (1+s) a(s),$$

in the terminology of equation (1) preceding the statement of T1. This upper bound is independent of x, and it was shown to go to zero as $N \to \infty$ in (2) preceding T1. As one can obtain analogous estimates for the probabilities L_N of leaving on the left, the proof of T1 is complete.

With T1 we have solved "one-half" of the problem of extending the basic theory of the absorption problem from simple random walk to the case $\mu = 0$, $\sigma^2 < \infty$. Note that T1 corresponds to part (b) of P21.4, where the result is "exact" in the sense that

$$R_N(x) = \frac{x+1}{N+2}, \quad 0 \le x \le N.$$

Our next task will be to get a result for $g_N(x,y)$, which will correspond to part (a) of P21.4. That will be accomplished in T23.1, in the next section.

This section is concluded with some remarks concerning possible extensions of T1 to random walk with $\sigma^2 = \infty$, investigated by Kesten (1961) [58, 59]. The essential flavor of the problem is preserved if one writes T1 in the form

$$\lim_{N \to \infty} R_N([Nx]) = x, \quad 0 \le x \le 1.$$

In other words, the limit, when $\mu = 0$, $\sigma^2 < \infty$, is the uniform distribution. Just as in our brief discussion of the arc-sine law, and of occupation time problems in general, at the end of section 20, one can ask the following question. For what random walks does

(1) $$\lim_{N \to \infty} R_N([Nx]) = F(x), \qquad 0 \leq x \leq 1,$$

exist, with $F(x)$ a probability distribution, and what distributions $F(x)$ can arise as limits of this type? Just as in occupation time problems, the limit may fail to exist altogether. For symmetric random walk Kesten exhibits classes of random walks where the limiting distribution is the incomplete beta-function

(2) $$F_\alpha(x) = \frac{\Gamma(\alpha)}{\left[\Gamma\left(\frac{\alpha}{2}\right)\right]^2} \int_0^x t^{\frac{\alpha}{2}-1}(1-t)^{\frac{\alpha}{2}-1} dt, \qquad 0 < \alpha \leq 2.$$

Quite likely these are all the possibilities. Just as in the occupation time problems the parameter α is related to the asymptotic behavior of $P(x,y)$ as $|x-y| \to \infty$; in particular $\alpha = 2$ when $\mu = 0$ and $\sigma^2 < \infty$, giving the result $F_\alpha(x) = x$ of T1.

As an amusing special case consider the random walk of E8.3 in Chapter II. It was shown there that

(3) $$\phi(\theta) = \sum_{x=-\infty}^{\infty} P(0,x)e^{ix\theta} = 1 - \left|\sin\frac{\theta}{2}\right|$$

is the characteristic function of

(4) $$P(0,0) = 1 - \frac{2}{\pi}, \qquad P(0,x) = \frac{2}{\pi} \frac{1}{4x^2 - 1} \quad \text{for } x \neq 0.$$

Since Kesten showed that (2) holds with parameter α, $0 < \alpha < 2$ when

$$P(0,x) \sim \text{constant} \cdot |x|^{-(1+\alpha)},$$

it follows that the random walk in (3) and (4) satisfies

(5) $$\lim_{N \to \infty} R_N([Nx]) = F_1(x) = \frac{2}{\pi} \sin^{-1} \sqrt{x}, \qquad 0 \leq x \leq 1.$$

We proceed to sketch a proof of (5) along lines entirely different from the methods of this chapter.

22. THE ABSORPTION PROBLEM WITH MEAN ZERO

E1 Consider simple random walk $\mathbf{x}_n$ in the plane R. Given any positive integer N we decompose R into the three sets

$$A_N = [z \mid z \in R, z = k(1 + i) \text{ for some } k \geq N + 1],$$
$$B_N = [z \mid z \in R, z = k(1 + i) \text{ for some } k < 0],$$
$$C_N = R - (A_N \cup B_N).$$

Finally, let us call D the entire diagonal $[z \mid z = n(1 + i), -\infty < n < \infty]$. It was shown in E8.3 that $P(x,y)$, defined in equation (4) above, is the transition function of the random walk executed by $\mathbf{x}_n$ when it visits D (the imbedded random walk associated with D). But then it follows that for any real t, $0 < t < 1$, $R_N([Nt])$ of equation (5) has an interesting probability interpretation in terms of simple two-dimensional random walk $\mathbf{x}_n$. All we have to do is to take

$$\mathbf{x}_0 = [Nt](1 + i), \quad \mathbf{T}_N = \min [k \mid k \geq 1, \mathbf{x}_k \in A_N \cup B_N],$$

to be able to conclude that

(6) $$R_N([Nt]) = \mathbf{P}_{[Nt](1+i)}[\mathbf{x}_{\mathbf{T}_N} \in A_N].$$

To calculate $R_N([Nt])$ consider the function

$$f_N(z) = \mathbf{P}_z[\mathbf{x}_{\mathbf{T}_N} \in A_N], \quad z \in R,$$

and observe that it solves the exterior Dirichlet problem (see P13.2)

(7) $$\tfrac{1}{4}[f_N(z + 1) + f_N(z - 1) + f_N(z + i) + f_N(z - i)] = f_N(z)$$

for $z \in C_N$, with boundary conditions

(8) $$f_N(z) = 1 \text{ for } z \in A_N, \quad f_N(z) = 0 \text{ for } z \in B_N.$$

This boundary value problem is too complicated for us to attempt an explicit solution, but on the other hand the limiting case (as $N \to \infty$) of this problem is very simple. To obtain the correct limiting partial differential equation from (7) one must of course take a finer and finer grid as $N \to \infty$. If one takes a grid of mesh length N^{-1} at the Nth stage, then (7) should go over into Laplace's equation, and the boundary conditions in (8) of course remain unchanged. This argument, if correct, implies

$$\lim_{N \to \infty} f_N(z) = f(z) = u(x,y),$$

where $\Delta u = 0$ everywhere in the (x,y) plane except on $A = [(x,y) \mid x = y \geq 1]$, where $u(x,y) = 1$, and on $B = [(x,y) \mid x = y \leq 0]$, where $u(x,y) = 0$. If we can find such a function $u(x,y)$ we are finished, for then

(9) $$\lim_{N \to \infty} R_N([Nt]) = f[t(1 + i)] = u(t,t).$$

The methods required to justify the passage to the limit are quite well known. They were devised in 1928 in an important study of the approximation of boundary value problems by difference equations (Courant,

Friedrichs, and Lewy [15]). Even the probabilistic aspects of the problem were already then fairly well understood.

Finally we have to find the harmonic function $u(x,y)$, which can easily be done by use of a suitable conformal mapping (a mapping $w(z)$ of $R - (A \cup B)$ on a domain D such that Re $w(z) = 1$ when $z \in A$ and 0 when $z \in B$). In this way one obtains

$$u(x,y) = \frac{1}{\pi} \tan^{-1} \frac{2 \operatorname{Im}(w)}{w\bar{w} - 1},$$

where

$$\frac{2z}{1+i} - 1 = \frac{1 + w^2}{1 - w^2}, \qquad \operatorname{Im}(w) > 0.$$

After verifying that $u(x,y)$ is the appropriate harmonic function, set $x = y = t$ to obtain

$$w = i\sqrt{\frac{1-t}{t}}, \qquad 0 < t < 1,$$

(10) $$u(t,t) = \frac{1}{\pi} \tan^{-1} \left[\frac{2 \operatorname{Im}(w)}{w\bar{w} - 1} \right] = \frac{1}{\pi} \tan^{-1} \left[\frac{2\sqrt{t}\sqrt{1-t}}{1 - 2t} \right]$$

$$= \frac{2}{\pi} \sin^{-1} \sqrt{t}.$$

In view of equation (9), this completes our proof (apart from the justification of the passage to the limit in (9)) that the random walk in (4) has absorption probabilities given by (5). Qualitatively, since

$$\frac{2}{\pi} \sin^{-1} \sqrt{t} \begin{cases} > t \text{ for } t < \frac{1}{2} \\ < t \text{ for } t > \frac{1}{2} \end{cases}$$

we conclude that symmetric random walk with infinite variance differs from random walk with $\mu = 0$, $\sigma^2 < \infty$ as follows. *It is less likely to leave an interval on the same side as the starting point, and more likely to leave it on the opposite side.*

23. THE GREEN FUNCTION FOR THE ABSORPTION PROBLEM

The Green function $g_N(x,y)$ was defined in D21.1 for $N \geq 0$, x and y in the interval $[0,N]$. It represents the expected number of visits to y, starting at $\mathbf{x}_0 = x$, before the random walk leaves the interval $[0,N]$. Our immediate goal is the extension of P21.4(a) to arbitrary random walk with mean $\mu = 0$ and variance $0 < \sigma^2 < \infty$. This will be done in T1, and in the remainder of this chapter we shall consider a number of applications of T1.

23. THE GREEN FUNCTION FOR ABSORPTION PROBLEM

To get information about the asymptotic behavior of $g_N(x,y)$ as $N \to \infty$ we shall use T22.1 and the identity

P1 $\quad g_N(x,y) = g(x,y) - \sum_{k=1}^{\infty} R_N(x,k) g(N+k,y), \qquad x, y \in [0,N].$

Proof: Let $\mathbf{x}_n$ be the random walk with $\mathbf{x}_0 = x$, $\mathbf{T}_N$ the first time (the smallest k) such that $\mathbf{x}_k \in R - [0,N]$, and $\mathbf{T}$ the first time that $\mathbf{x}_k < 0$. Clearly $\mathbf{T}_N \le \mathbf{T}$, so that

$$g_N(x,y) = \mathbf{E}_x \sum_{k=0}^{\mathbf{T}_N - 1} \delta(y, \mathbf{x}_k) = \mathbf{E}_x \sum_{k=0}^{\mathbf{T}-1} \delta(y, \mathbf{x}_k) - \mathbf{E}_x \sum_{k=\mathbf{T}_N}^{\mathbf{T}-1} \delta(y, \mathbf{x}_k)$$

$$= g(x,y) - \mathbf{E}_x \left[\mathbf{E}_{\mathbf{x}_{\mathbf{T}_N}} \sum_{k=0}^{\mathbf{T}-1} \delta(y, \mathbf{x}_k) \right] = g(x,y) - \mathbf{E}_x g(\mathbf{x}_{\mathbf{T}_N}, y)$$

$$= g(x,y) - \sum_{k=1}^{\infty} R_N(x,k) g(N+k, y).$$

Combined with T22.1 and P22.4, P1 will yield

T1 *For aperiodic one-dimensional random walk, with $\mu = 0$, $0 < \sigma^2 < \infty$,*

$$\frac{\sigma^2}{2N} g_N(x,y) - R\left(\frac{x}{N}, \frac{y}{N}\right) \to 0$$

as $N \to \infty$, uniformly in $0 \le x, y \le N$. Here

$$R(s,t) = \min(s,t) - st, \qquad 0 \le s, t \le 1.$$

Proof: Without loss of generality one may assume that $y \le x$. The other case then can be obtained from the reversed random walk. Since $u(x) \to u > 0$, $v(x) \to v > 0$, we write

$$u(x) = u + \mu(x), \qquad v(x) = v + \nu(x),$$

where

$$\lim_{x \to +\infty} \mu(x) = \lim_{x \to +\infty} \nu(x) = 0.$$

Then, keeping in mind that $y \le x$ and that $uv = 2/\sigma^2$,

$$g(x,y) = u(x)v(y) + u(x-1)v(y-1) + \cdots + u(x-y)v(0)$$

$$= (y+1)\frac{2}{\sigma^2} + \sum_{k=0}^{y} [u\nu(k) + v\mu(x-y+k) + \mu(x-y+k)\nu(k)].$$

It is more convenient to write

$$g(x,y) = (y+1)\frac{2}{\sigma^2} + \frac{2}{\sigma^2} \rho(x,y).$$

where the error term $\rho(x,y)$, defined for $0 \le y \le x$, has the property that

$$|\rho(x,y)| \le \sum_{k=0}^{y} \epsilon(k), \qquad 0 \le y \le x,$$

$\epsilon(n)$ being a fixed non-negative null sequence ($\lim_{n \to \infty} \epsilon(n) = 0$).

Using P1 one obtains

$$\frac{\sigma^2}{2N} g_N(x,y) - R\left(\frac{x}{N}, \frac{y}{N}\right)$$

$$= \frac{\sigma^2}{2N} g_N(x,y) - \frac{y}{N}\left(1 - \frac{x}{N}\right)$$

$$= \frac{\sigma^2}{2N} g(x,y) - \frac{\sigma^2}{2N} \sum_{k=1}^{\infty} R_N(x,k) g(N+k,y) - \frac{y}{N}\left(1 - \frac{x}{N}\right)$$

$$= \frac{y+1}{N} + \frac{1}{N} \rho(x,y) - \frac{y+1}{N} R_N(x)$$

$$\qquad - \frac{1}{N} \sum_{k=1}^{\infty} R_N(x,k) \rho(N+k,y) - \frac{y}{N}\left(1 - \frac{x}{N}\right).$$

Now it remains only to estimate the error terms. First we observe that

$$\left|\frac{1}{N} \rho(x,y)\right| \le \frac{1}{N} \sum_{k=0}^{y} \epsilon(k) \le \frac{1}{N} \sum_{k=0}^{N} \epsilon(k),$$

$$\left|\frac{1}{N} \sum_{k=1}^{\infty} R_N(x,k) \rho(N+k,y)\right| \le \epsilon(y) \frac{y}{N} \le \frac{1}{N} \sum_{k=0}^{N} \epsilon(k),$$

so that these two terms tend to zero uniformly, as required. Let us call the sum of these two terms ϵ_N'. Furthermore we have, in view of T22.1,

$$\left| R_N(x) - \frac{x}{N} \right| \le \epsilon_N'', \qquad 0 \le x \le N,$$

where ϵ_N'' is another null sequence, independent of x. Collecting terms, and using the triangle inequality,

$$\left| \frac{\sigma^2}{2N} g_N(x,y) - R\left(\frac{x}{N}, \frac{y}{N}\right) \right| \le \epsilon_N' + \epsilon_N'' \frac{y+1}{N}$$

$$+ \left| \frac{y+1}{N}\left(1 - \frac{x}{N}\right) - \frac{y}{N}\left(1 - \frac{x}{N}\right) \right| \le \epsilon_N' + \epsilon_N'' + \frac{1}{N}$$

which tends to zero, uniformly in $0 \le y \le x \le N$ and proves T1.

23. THE GREEN FUNCTION FOR ABSORPTION PROBLEM

Now we are in a position to harvest several interesting consequences of T1. We shall be quite selective and choose only three (cf. problems 6 and 7 for one more application). First of all, we can now prove P22.2 in the beginning of the last section, which to some extent motivated the development of T22.1 and T1, but which was never proved. We have to show that

$$\lim_{N \to \infty} \frac{\sigma^2}{2N^2} \sum_{k=0}^{N} g_N([Nt],k) f\left(\frac{k}{N}\right) = \int_0^1 R(t,s) f(s) \, ds.$$

Since $f(x)$ is bounded and since the convergence in T1 is uniform, we have

$$\lim_{N \to \infty} \frac{\sigma^2}{2N^2} \sum_{k=0}^{N} g_N([Nt],k) f\left(\frac{k}{N}\right) = \lim_{N \to \infty} \frac{1}{N} \sum_{k=0}^{N} R\left(\frac{[Nt]}{N}, \frac{k}{N}\right) f\left(\frac{k}{N}\right),$$

and the last limit is of course the approximating sum of the desired Riemann integral.

For an application which is no more profound, but probabilistically a lot more interesting, we shall study an occupation time problem. This problem will appear a little more natural if we change the absorbing interval from $[0,N]$ to $[-N,N]$, so that the process can start at a fixed point, say at the middle of the interval. Thus we shall make the following definitions.

D1 *The random walk* $\mathbf{x}_n$ *starts at* $\mathbf{x}_0 = 0$. $\mathbf{T}_N$ *is the first time outside* $[-N,N]$, y *is a fixed point in* R, *and*

$$\mathbf{N}_N(y) = \sum_{k=0}^{T_N - 1} \delta(y, \mathbf{x}_k)$$

is the number of visits to y *before leaving* $[-N,N]$.

We shall prove

P2 *For aperiodic random walk, with* $\mu = 0$, $\sigma^2 < \infty$,

$$\lim_{N \to \infty} \mathbf{P}_0[\mathbf{N}_N(y) \geq Nx] = e^{-\sigma^2 x}, \qquad x \geq 0.$$

Proof: We shall consider only the case when $y \neq 0$ and leave to the reader the trivial modification required when $y = 0$. We let

$$p_N(y) = \mathbf{P}_0[\mathbf{N}_N(y) > 0],$$
$$r_N(y) = \mathbf{P}_y[\mathbf{x}_k = y \text{ for some } 0 < k < \mathbf{T}_N].$$

Then

$$\mathbf{P}_0[\mathbf{N}_N(y) = j] = p_N(y) r_N^{j-1}(y)[1 - r_N(y)], \qquad j \geq 1.$$

To make use of T1 in an efficient manner we translate the interval $[-N,N]$ to $[0,2N]$ so that the condition $\mathbf{x}_0 = 0$ becomes $\mathbf{x}_0 = N$ while the point y turns into $y + N$. In this setting a simple computation shows that

$$p_N(y) = \frac{g_{2N}(N, N+y)}{g_{2N}(N+y, N+y)},$$

$$r_N(y) = 1 - \frac{1}{g_{2N}(N+y, N+y)}.$$

Hence

$$\mathbf{P}_0[\mathbf{N}_N(y) > Nx] = p_N(y)[r_N(y)]^{[Nx]}.$$

By T1 it is clear that

$$\lim_{N \to \infty} p_N(y) = 1.$$

To estimate $r_N(y)$, observe that

$$\frac{\sigma^2}{4N} g_{2N}(N+y, N+y) = \frac{1}{4} - \frac{y^2}{4N^2} + \epsilon_N(y)$$

where $\epsilon_N(y) \to 0$ uniformly in y. Consequently

$$r_N(y) = 1 - \frac{\sigma^2}{N} + o\left(\frac{1}{N}\right),$$

where $No(1/N) \to 0$ as $N \to \infty$. But that suffices to conclude

$$\lim_{N \to \infty} [r_N(y)]^{[Nx]} = \lim_{N \to \infty} \left(1 - \frac{\sigma^2}{N}\right)^{Nx} = e^{-\sigma^2 x},$$

and proves P2.

Our next application of T1 concerns the asymptotic behavior of the first time $\mathbf{T}_N$ outside the interval $[-N,N]$. We shall show that P21.5 holds in a very general context, by proving

T2 *For arbitrary one-dimensional random walk with $\mu = 0$ and $0 < \sigma^2 < \infty$, $\mathbf{T}_N$ (defined in D1) has the limiting distribution*

$$\lim_{N \to \infty} \mathbf{P}_0\left[\mathbf{T}_N > x \frac{N^2}{\sigma^2}\right] = 1 - F(x), \qquad x \geq 0,$$

where

$$F(x) = 1 - \frac{4}{\pi} \sum_{k=0}^{\infty} \frac{(-1)^k}{2k+1} e^{-\frac{\pi^2}{8}(2k+1)^2 x}.$$

Proof: To facilitate the use of T1 we shall assume the random walk to be aperiodic, keeping in mind that this assumption may be removed by an argument similar to that in the proof of T22.1. The proof will require several preliminary lemmas, the first of which concerns the moments of the distribution $F(x)$ in T2.

P3 $$m_p = \int_0^\infty x^p f(x)\, dx = p!\, \frac{\pi}{2} \left(\frac{8}{\pi^2}\right)^{p+1} \sum_{k=0}^\infty (-1)^k \left(\frac{1}{2k+1}\right)^{2p+1}, \quad p \geq 0,$$

where $f(x)$ is the density $F'(x)$. Moreover, if $G(x)$ is any distribution function such that

$$\int_{-\infty}^\infty x^p\, dG(x) = m_p \text{ for all } p = 0, 1, 2, \ldots,$$

then $G(x) = F(x)$.

Proof: The first part of P3 involves only straightforward calculation of the moments m_p of the distribution $F(x)$ defined in T2. The second part, which amounts to saying that the *moment problem* for the particular moment sequence m_p has a *unique solution* may be deduced from a well known criterion. *If*

$$c_p = \int_{-\infty}^\infty x^p\, dH(x), \quad p \geq 0,$$

is a sequence of moments of a probability distribution $H(x)$, and if

$$\sum_{k=0}^\infty \frac{c_k}{k!} r^k < \infty \text{ for some } r > 0$$

then $H(x)$ is the only distribution with these moments (the moment problem for the sequence c_k has a unique solution). The moments m_p are easily seen to satisfy the above condition for all $r < \pi^2/8$.

We need one more result from the general theory of distribution functions.[5] *If $H_n(x)$ is a sequence of distribution functions, such that*

$$\int_{-\infty}^\infty x^p\, dH_n(x) = c_p(n), \quad p \geq 0, \quad n \geq 0,$$

$$\lim_{n \to \infty} c_p(n) = c_p, \quad p \geq 0,$$

---

[5] See Loève [73], §§11 and 14, for both this limit theorem and the uniqueness criterion for moments described above. The technique of using this theorem to find a limiting distribution by calculating the limits of moments is often referred to as *the method of moments*.

and if there is one unique distribution function $H(x)$ such that

$$c_p = \int_{-\infty}^{\infty} x^p \, dH(x), \qquad p \geq 0,$$

then

$$\lim_{n \to \infty} H_n(x) = H(x)$$

at every point of continuity of $H(x)$.

This theorem permits us to conduct the proof of T2 by showing that the moments of $\mathbf{T}_N$ have the proper limiting behavior. In view of P3 we may record this state of affairs as

P4 *If for each integer $p \geq 0$,*

$$\lim_{N \to \infty} \mathbf{E}_0\left[\left(\frac{\sigma^2 \mathbf{T}_N}{N^2}\right)^p\right] = m_p$$

with m_p given in P3, then T2 is true.

The analysis of the moments of $\mathbf{T}_N$ begins with a little trick. Keeping in mind that the random walk starts at $\mathbf{x}_0 = 0$, we define the random variables

$$\begin{aligned}
\psi_{N,0} &= 1, \qquad N \geq 1, \\
\psi_{N,k} &= 1 \text{ if } |\mathbf{x}_i| \leq N \text{ for } i = 1, 2, \ldots, k \\
&= 0 \text{ otherwise}, \qquad N \geq 1.
\end{aligned}$$

Then

$$\mathbf{T}_N = \sum_{k=0}^{\infty} \psi_{N,k},$$

$$\mathbf{T}_N^2 = \sum_{k_1=0}^{\infty} \sum_{k_2=0}^{\infty} \psi_{N,k_1} \psi_{N,k_2} = \sum_{k_1=0}^{\infty} \sum_{k_2=0}^{\infty} \psi_{N,j},$$

where $j = j(k_1,k_2) = \max(k_1,k_2)$. The number of distinct pairs (k_1,k_2) of non-negative integers whose maximum is j turns out to be $(j+1)^2 - j^2$, so that

$$\mathbf{T}_N^2 = \sum_{j=0}^{\infty} \psi_{N,j}[(j+1)^2 - j^2].$$

The trick consists in observing that this procedure generalizes very nicely. The number of distinct p-tuples $(k_1, k_2, \ldots, k_p)$ of non-negative integers, all of which are less than or equal to j, turns out to be $(j + 1)^p - j^p$ (the difference between the volumes of two cubes!) and one can write

P5 $\quad T_N{}^p = \sum_{j=0}^{\infty} \psi_{N,j}[(j + 1)^p - j^p], \qquad p \geq 1, \quad N \geq 1.$

This gives us a very natural setup for the calculation of moments. In the terminology of D21.1

$$E_0[\psi_{N,j}] = \sum_{i=0}^{2N} Q_{2N}{}^j(N,i).$$

If we define $g_{2N}{}^p(x,y)$ as the pth power of the matrix $g_{2N}(x,y)$, $0 \leq x, y \leq 2N$ then

$$g_{2N}{}^p(x,y) = (I_{2N} - Q_{2N})^{-p}(x,y) = \sum_{k=0}^{\infty} \binom{-p}{k}(-1)^k Q_{2N}{}^k(x,y).$$

Therefore

(1) $\quad \sum_{y=0}^{2N} g_{2N}{}^p(N,y)$

$\qquad = \sum_{k=0}^{\infty} \binom{-p}{k}(-1)^k E_0[\psi_{N,k}]$

$\qquad = \dfrac{1}{(p-1)!} \sum_{k=0}^{\infty} (k+1)(k+2)\cdots(k+p-1)E_0[\psi_{N,k}].$

On the other hand,

(2) $\quad E_0[T_N{}^p] = \sum_{k=0}^{\infty} [(k+1)^p - k^p]E_0[\psi_{N,k}].$

Now we are in a position to show that

P6 *If*

$$\lim_{N \to \infty} \left(\frac{\sigma^2}{N^2}\right)^p \sum_{y=0}^{2N} g_{2N}{}^p(N,y) = \frac{m_p}{p!}, \qquad p \geq 1,$$

then

$$\lim_{N \to \infty} E_0\left[\left(\frac{\sigma^2}{N^2} T_N\right)^p\right] = m_p.$$

Proof: If the hypothesis of P6 holds, one can combine equations (1) and (2) above to conclude that

$$\begin{aligned}\frac{m_p}{p!} &= \lim_{N\to\infty} \left(\frac{\sigma^2}{N^2}\right)^p \frac{1}{(p-1)!} \sum_{k=0}^{\infty} [(k+1)\cdots(k+p-1)]\mathbf{E}_0[\psi_{N,k}] \\ &= \lim_{N\to\infty} \left(\frac{\sigma^2}{N^2}\right)^p \frac{1}{(p-1)!} \sum_{k=0}^{\infty} k^{p-1}\mathbf{E}_0[\psi_{N,k}] \\ &= \lim_{N\to\infty} \left(\frac{\sigma^2}{N^2}\right)^p \frac{1}{p!} \sum_{k=0}^{\infty} pk^{p-1}\mathbf{E}_0[\psi_{N,k}] \\ &= \lim_{N\to\infty} \left(\frac{\sigma^2}{N^2}\right)^p \frac{1}{p!} \sum_{k=0}^{\infty} [(k+1)^p - k^p]\mathbf{E}_0[\psi_{N,k}] \\ &= \lim_{N\to\infty} \frac{1}{p!} \mathbf{E}_0\left[\left(\frac{\sigma^2}{N^2} T_N\right)^p\right].\end{aligned}$$

(At several steps we used the hypothesis of P6 to discard a finite number of terms of order k^{p-2} or lower.)

Thus the proof of P4, and hence of T2, will be complete if we can show that the first limit in P6 exists and has the desired value. To investigate this limit it is natural to define the iterates of the Green function

$$R(s,t) = R^1(s,t) = \min(s,t) - st$$

by

$$R^{p+1}(s,t) = \int_0^1 R(s,x)R^p(x,t)\,dx, \qquad p \geq 1.$$

Now we shall obtain a result that is actually a bit stronger than is needed.

P7 For $p = 1, 2, 3, \ldots$

$$\left(\frac{\sigma^2}{2}\right)^p N^{1-2p} g_N^p(x,y) - R^p\left(\frac{x}{N}, \frac{y}{N}\right) \to 0 \text{ as } N \to \infty$$

uniformly in $0 \leq x, y \leq N$ and also in $p \geq 1$.

Proof: Let

$$K^p\left(\frac{x}{N}, \frac{y}{N}\right) = N^{1-p} \sum R\left(\frac{x}{N}, \frac{i_1}{N}\right) \cdots R\left(\frac{i_{p-1}}{N}, \frac{y}{N}\right)$$

where the summation here, and in the rest of the proof, extends over all $(p-1)$-tuples $(i_1, i_2, \ldots, i_{p-1})$ with $0 \le i_\nu \le N$, $\nu = 1, 2, \ldots, p-1$. We shall use T1 in the form

$$\frac{\sigma^2}{2N} g_N(x,y) = R\left(\frac{x}{N}, \frac{y}{N}\right) + \epsilon_N(x,y),$$

$$|\epsilon_N(x,y)| \le \epsilon_N \to 0 \text{ as } N \to \infty.$$

Then

$$\left|\left(\frac{\sigma^2}{2}\right)^p N^{1-2p} g_N{}^p(x,y) - K^p\left(\frac{x}{N}, \frac{y}{N}\right)\right|$$

$$= N^{1-p} \left| \sum \left[R\left(\frac{x}{N}, \frac{i_1}{N}\right) + \epsilon_N(x, i_1)\right] \cdots \left[R\left(\frac{i_{p-1}}{N}, \frac{y}{N}\right) + \epsilon_N(i_{p-1}, y)\right] - \sum R\left(\frac{x}{N}, \frac{i_1}{N}\right) \cdots R\left(\frac{i_{p-1}}{N}, \frac{y}{N}\right) \right|.$$

Using the fact that

$$0 \le R(x,y) \le \tfrac{1}{4}, \qquad 0 \le x, y \le 1,$$

one obtains

$$\left|\left(\frac{\sigma^2}{2}\right)^p N^{1-2p} g_N{}^p(x,y) - K^p\left(\frac{x}{N}, \frac{y}{N}\right)\right|$$

$$\le N^{1-p}\left[\sum (\tfrac{1}{4} + \epsilon_N)^{p-1} - \sum (\tfrac{1}{4})^{p-1}\right]$$
$$= (\tfrac{1}{4} + \epsilon_N)^{p-1} - (\tfrac{1}{4})^{p-1} \to 0,$$

as $N \to \infty$, uniformly for $p \ge 1$. The last line used three simple observations in quick succession. First, given two sets of real p-tuples, namely $a_1, a_2, \ldots, a_p$ and $b_1, b_2, \ldots, b_p$ such that $0 \le a_k \le a$, $|b_k| \le b$, then

$$\left|\prod_{k=1}^p (a_k + b_k) - \prod_{k=1}^p a_k \right| \le (a+b)^p - a^p.$$

Secondly, the sum $\sum$, defined in the beginning of the proof, has the property that $\sum 1 = N^{p-1}$. Finally if ϵ_n is a null-sequence, and $0 < a < 1$, then $(a + \epsilon_n)^p - a^p \to 0$ uniformly in $p \ge 0$ as $n \to \infty$.

The proof of P7 may now be completed by showing that

$$R^p\left(\frac{x}{N}, \frac{y}{N}\right) - K^p\left(\frac{x}{N}, \frac{y}{N}\right) \to 0 \text{ as } N \to \infty$$

uniformly in the integers $x, y, p = 0, 1, 2, \ldots$. Given any $\epsilon > 0$ we choose p_0 such that $4^{-p_0} < \epsilon/2$. Then

$$\left| R^p\left(\frac{x}{N}, \frac{y}{N}\right) - K^p\left(\frac{x}{N}, \frac{y}{N}\right) \right| \le 4^{-p} + 4^{-p} < \epsilon$$

when $p \geq p_0$ for all x, y, N. Thus it suffices to exhibit an integer $K > 0$ such that $N > K$ implies

$$\left| R^p\left(\frac{x}{N}, \frac{y}{N}\right) - K^p\left(\frac{x}{N}, \frac{y}{N}\right) \right| < \epsilon$$

for all $0 \leq x, y \leq N$ and all $p = 1, 2, \ldots, p_0 - 1$. Since there are only a finite number of values of p to worry about, it suffices to do so for one value of p at a time (choosing afterward the largest value of K). So we are through if for each positive integer p

$$R^p\left(\frac{x}{N}, \frac{y}{N}\right) - K^p\left(\frac{x}{N}, \frac{y}{N}\right) \to 0$$

uniformly in $0 \leq x, y \leq N$. When $p = 1$ this difference is already zero. When $p = 2$,

$$K^p\left(\frac{x}{N}, \frac{y}{N}\right) = \frac{1}{N} \sum_{k=0}^{N} R\left(\frac{x}{N}, \frac{k}{N}\right) R\left(\frac{k}{N}, \frac{y}{N}\right)$$

is the Riemann approximating sum to the integral defining the second iterate $R^2(x/N, y/N)$. Such sums do converge uniformly provided only that the function involved, namely $R(x, y)$, is continuous on the square $0 \leq x, y \leq 1$. It is easy to verify this, and then also the case $p > 2$ presents no new difficulties.

Let us now show how P7 completes the proof of T2. In view of P7

$$\lim_{N \to \infty} \left(\frac{\sigma^2}{N^2}\right)^p \sum_{y=0}^{2N} g_{2N}{}^p(N, y) = \lim_{N \to \infty} 8^p \frac{1}{2N} \sum_{y=0}^{2N} R^p\left(\frac{1}{2}, \frac{y}{2N}\right)$$

$$= 8^p \int_0^1 R^p\left(\frac{1}{2}, x\right) dx.$$

To evaluate the last integral we use Mercer's theorem, referred to in the discussion following P21.3, to write

$$R^p(x, y) = \sum_{k=1}^{\infty} \frac{\phi_k(x) \phi_k(y)}{(\mu_k)^p},$$

where $\phi_k(x) = \sqrt{2} \sin \pi k x$, $\mu_k = \pi^2 k^2$. Now one can easily perform the integration to obtain

$$8^p \int_0^1 R^p\left(\frac{1}{2}, x\right) dx = 2 \cdot 8^p \sum_{k=1}^{\infty} (\pi k)^{-2p} \sin \frac{k\pi}{2} \int_0^1 \sin k\pi x \, dx$$

$$= \frac{\pi}{2} \left(\frac{8}{\pi^2}\right)^{p+1} \sum_{j=0}^{\infty} (-1)^j \left(\frac{1}{2j+1}\right)^{2p+1} = \frac{m_p}{p!}.$$

That concludes the evaluation of the limit in P6 and hence T2 is proved.

It is worth pointing out a slightly different formulation of T2, which is not without interest.

T3 *For one-dimensional random walk with $\mu = 0$, $0 < \sigma^2 < \infty$*

$$\lim_{n \to \infty} \mathbf{P}\left[\max_{1 \le k \le n} |\mathbf{S}_k| \le \sigma x \sqrt{n}\right] = 1 - F\left(\frac{1}{x^2}\right), \qquad x > 0,$$

with $F(x)$ as defined in T2.

Proof: Let us call $G_n(x)$ the distribution of $(\sigma^2 n)^{-1/2} \max_{1 \le k \le n} |\mathbf{S}_k|$. Then T3 amounts to saying that the sequence

$$G_n(x) = \sum_{y=0}^{2[\sigma x \sqrt{n}]} Q_{2[x\sigma\sqrt{n}]}{}^n([\sigma x \sqrt{n}], y), \qquad x > 0,$$

converges to the limit

$$G(x) = 1 - F(x^{-2}), \qquad x > 0,$$

as $n \to \infty$. It seems impossible to obtain sufficiently powerful estimates for the iterates of Q to verify this result directly. However, we shall see that most of the work was already done in the course of proving T2. The crucial observation concerns the two events

(i) $\max_{1 \le k \le n} |\mathbf{S}_k| \le \sigma x \sqrt{n}$, (ii) $\mathbf{T}_{[\sigma x \sqrt{n}]} > n$.

They are obviously the same event! In other words

$$G_n(x) = \mathbf{P}[\mathbf{T}_{[\sigma x \sqrt{n}]} > n].$$

If we fix $x > 0$, and let $m = m(n) = [\sigma x \sqrt{n}]$, then

$$\frac{m^2}{\sigma^2 x^2} \le n < \frac{(m+1)^2}{\sigma^2 x^2},$$

$$\mathbf{P}\left[\mathbf{T}_m > \frac{(m+1)^2}{\sigma^2 x^2}\right] \le G_n(x) \le \mathbf{P}\left[\mathbf{T}_m > \frac{m^2}{\sigma^2 x^2}\right].$$

As $n \to \infty$, also $m = m(n) \to \infty$. Thus

$$\overline{\lim_{n \to \infty}} G_n(x) \le \lim_{m \to \infty} \mathbf{P}\left[\mathbf{T}_m > \frac{m^2}{\sigma^2} \frac{1}{x^2}\right] = 1 - F\left(\frac{1}{x^2}\right)$$

by T2, and also

$$\lim_{n \to \infty} G_n(x) \geq \overline{\lim_{m \to \infty}} \mathbf{P}\left[\mathbf{T}_m > \frac{m^2}{\sigma^2} \frac{1}{x^2}\left(1 + \frac{1}{m}\right)^2\right]$$

$$\geq \lim_{m \to \infty} \mathbf{P}\left[\mathbf{T}_m > \frac{m^2}{\sigma^2} \frac{c}{x^2}\right] = 1 - F\left(\frac{c}{x^2}\right)$$

for every real $c > 1$. But $F(x)$ as well as $F(x^{-2})$ are continuous functions of x, and therefore the fact that c may be taken arbitrarily close to one yields

$$\lim_{n \to \infty} G_n(x) = 1 - F(x^{-2}).$$

Remark: The first rigorous proof of T2 and T3 is due to Erdös and Kac [29], 1946. Their proof is not only shorter, but in a way more interesting than ours, as it reduces the problem to the proof of T3 for the Brownian motion process, which is the limiting case of a sequence of random walks, if the time and space scales are changed appropriately. In the context of Brownian motion T3 had indeed been known already to Bachelier, in the beginning of the century, so that the crux of the proof in [29] was the justification of the limiting process, by means of a so-called *invariance principle* [22].

Problems

1. Derive the analogue of P21.1 for simple random walk in the plane, i.e., find the eigenfunctions and eigenvalues of the transition function restricted to the rectangle

$$[z \mid 0 \leq \text{Re}(z) \leq M; 0 \leq \text{Im}(z) \leq N] = R_{M,N}.$$

2. Continuation. When M and N are even integers and

$$\mathbf{T} = \min [k \mid k \geq 1, \ \mathbf{x}_n \in R - R_{M,N}],$$

calculate $\mathbf{E}_z[\mathbf{T}]$ for $z \in R_{M,N}$, and in particular when $z = \frac{1}{2}(M + iN)$.

3. Calculate the orthogonal polynomials $p_n(z)$ in P22.3 for simple random walk, and verify directly the truth of parts (d), (e), (f) of P22.3.

4. Calculate the orthogonal polynomials $p_n(z)$ in P22.3 for the two-sided geometric random walk in problem 3 of Chapter IV.

5. Show that the proper analogue of P22.3 for unsymmetrical random walk concerns a (uniquely determined) *biorthogonal system* of polynomials $q_n(z)$ and $r_n(z)$ such that

$$(2\pi)^{-1} \int_{-\pi}^{\pi} q_n(e^{i\theta}) \overline{r_m(e^{i\theta})} [1 - \phi(\theta)] \, d\theta = \delta_{m,n}.$$

Hint: Verify that these orthogonality relations hold if

$$q_{nk} = \sum_{\nu=0}^{\infty} \mathbf{P}_k[\mathbf{x}_\nu = n; \mathbf{x}_j < n \text{ for } j = 1, 2, \ldots, \nu - 1] \text{ for } n > k,$$

$$q_{nn} = 1,$$

and

$$r_{nk} = \sum_{\nu=0}^{\infty} \mathbf{P}_n[\mathbf{x}_\nu = k; \mathbf{x}_j \leq n \text{ for } j = 1, 2, \ldots, \nu] \text{ for } n \geq k,$$

are defined as the coefficients of the polynomials

$$q_n(z) = \sum_{k=0}^{n} q_{nk} z^k, \qquad r_n(z) = \sum_{k=0}^{n} r_{nk} z^k.$$

Using these polynomials one has [6]

$$g_N(i,j) = \sum_{\nu=\max(i,j)}^{N} q_{\nu i} r_{\nu j}.$$

6. Let $\mathbf{D}_N$ denote the number of distinct points visited before leaving the interval $[-N, N]$, by a one-dimensional random walk with mean 0 and variance $0 < \sigma^2 < \infty$. Calculate

$$\lim_{N \to \infty} \mathbf{E}_0\left[\frac{\mathbf{D}_N}{2N}\right],$$

the "fraction of points visited." (How should one expect that the answer will depend on σ^2?)

[6] This biorthogonal system has been investigated by Baxter (see [2] and [3]) in a much more general context. Indeed all results concerning the orthogonal polynomials in P22.3 may be extended to the present biorthogonal system. Not only that, but it is even possible to estimate the coefficients p_{nk} when both n and k get large simultaneously. Thus it was shown [95] for symmetric aperiodic random walk that the coefficients p_{nk} in P22.3 satisfy

$$p_{nk} - \frac{\sqrt{2}}{\sigma} \frac{k}{n} \to 0,$$

uniformly in k and n, as $n - k \to \infty$. Of course that is enough to obtain T23.1 as a simple corollary. Finally Watanabe [102] simplified the methods in [95] and extended the results to the coefficients of the biorthogonal system in problem 5.

7. As in problem 10 of Chapter I, the integers in R are assumed to be "red" with probability α and "green" with probability $(1 - \alpha)$. Let $E_{N,\alpha}$ denote the expected number of red points in the interval $[-N,N]$ and $P_{N,\alpha}$ the probability that the random walk of problem 6 above will visit a red point before leaving the interval $[-N,N]$. Prove that

$$\lim_{N\to\infty} \lim_{\alpha\to 0} \frac{P_{N,\alpha}}{E_{N,\alpha}} = \lim_{\alpha\to 0} \lim_{N\to\infty} \frac{P_{N,\alpha}}{E_{N,\alpha}} = \ln 2.$$

Does the double limit exist?

8. For symmetric one-dimensional random walk $\mathbf{x}_n$, Q_n is the $(n + 1)$ by $(n + 1)$ matrix defined in D21.1, and $\lambda_0(n) \le \lambda_1(n) \le \cdots \le \lambda_n(n)$ are its (real) eigenvalues. Then the trace of the kth power of Q_n is

$$\text{Tr}\,(Q_n^k) = \sum_{j=0}^{n} \lambda_j^k(n).$$

Let

$$\mathbf{M}_k = \max_{0\le j\le k} \mathbf{x}_j - \min_{0\le j\le k} \mathbf{x}_j,$$

and prove that

$$(n + 1)P_k(0,0) - \text{Tr}\,(Q_n^k)$$
$$= (n + 1)\mathbf{P}_0[\mathbf{x}_k = 0;\, \mathbf{M}_k > n] + \mathbf{E}_0[\mathbf{M}_k;\, \mathbf{x}_k = 0,\, \mathbf{M}_k \le n].$$

Without any assumptions on the moments of the random walk, prove that

$$\lim_{n\to\infty} \frac{1}{n+1} \sum_{j=0}^{n} \lambda_j^k(n) = P_k(0,0), \qquad n \ge 0.$$

9. Continuation. Using the theorem of Weierstrass to approximate continuous functions by polynomials, show that

$$\lim_{n\to\infty} \frac{1}{n+1} \sum_{j=0}^{n} f[\lambda_j(n)] = \frac{1}{2\pi} \int_{-\pi}^{\pi} f[\phi(\theta)]\, d\theta,$$

for every continuous function f on $[-\pi, \pi]$. Here $\phi(\theta)$ is the characteristic function of the random walk. M. Kac [49], who devised the proof in problem 8, showed that the same method applies to arbitrary matrices of the form

$$Q_n = \begin{bmatrix} c_0 & c_1 & c_2 & \cdots & c_n \\ c_{-1} & c_0 & c_1 & \cdots & c_{n-1} \\ \cdot & \cdot & \cdot & \cdot & \cdot \\ c_{-n} & c_{-n+1} & & \cdots & c_0 \end{bmatrix},$$

provided that

$$c_k = \overline{c_{-k}} \quad \text{and} \quad \sum_{k=-\infty}^{\infty} |c_k| < \infty.$$

Thus the results of problems 8 and 9 are valid for real functions

$$\phi(\theta) = \sum_{k=-\infty}^{\infty} c_k e^{ik\theta}$$

whose Fourier series converge absolutely. By approximating step functions by polynomials one may rephrase our results as follows: For any interval I, let $N_n(I)$ denote the number of eigenvalues of Q_n which fall in the interval I, and $L_\phi(I)$ the Lebesgue measure of that subset of $[-\pi,\pi]$ where $\phi(\theta) \in I$. Then

$$\lim_{n \to \infty} \frac{1}{n+1} N_n(I) = \frac{L_\phi(I)}{2\pi}.$$

10. Show that the limit distribution

$$F(x) = 1 - \frac{4}{\pi} \sum_{n=0}^{\infty} \frac{(-1)^n}{2n+1} e^{-\frac{\pi^2}{8}(2n+1)^2 x}$$

in P21.5, T23.2, and T23.3 has the Laplace transform

$$\int_0^\infty e^{-sx} \, dF(x) = \frac{\pi}{2} \sum_{n=0}^{\infty} (-1)^n \frac{2n+1}{s + \frac{\pi^2}{8}(2n+1)^2} = (\cosh \sqrt{2s})^{-1}$$

$$= 2 e^{-\sqrt{2s}} \sum_{k=0}^{\infty} (-1)^k e^{-2k\sqrt{2s}}.$$

Inverting the last series (term by term) show that the density of F is

$$f(x) = \frac{dF}{dx} = \sqrt{\frac{2}{\pi}} \sum_{n=0}^{\infty} (-1)^n \frac{2n+1}{x^{3/2}} e^{-\frac{(2n+1)^2}{2x}},$$

which converges fast for small (positive) x, while

$$f(x) = \frac{\pi}{2} \sum_{n=0}^{\infty} (-1)^n (2n+1) e^{-\frac{\pi^2}{8}(2n+1)^2 x}$$

converges fast for large x. These two representations of the density f constitute a special case of the transformation law of the *Theta function*.

Chapter VI

TRANSIENT RANDOM WALK

24. THE GREEN FUNCTION $G(x,y)$

One might try to develop a theory for transient random walk along the same lines as in Chapter III for recurrent random walk in the plane. The first order of business would then be the calculation of hitting probabilities $H_B(x,y)$ for finite subsets B of the state space R. That turns out to be extremely easy in the transient case, and will be accomplished in a few lines in the beginning of section 25 (P25.1). The following discussion will clarify our reasons for deferring this calculation.

In the recurrent case the explicit formula for $H_B(x,y)$ was very interesting. It contained all the essential information concerning the asymptotic behavior of $H_B(x,y)$ for large $|x|$. The reader may recall that this was so because $H_B(x,y)$ was expressed in terms of the recurrent potential kernel $A(x,y) = a(x-y)$, and we had to learn a great deal about $a(x)$ before we succeeded in expressing $H_B(x,y)$ in terms of $a(x)$.

In the transient case, however, it is so easy to express $H_B(x,y)$ in terms of the potential kernel $G(x,y)$ that nothing at all is learned in the process about the asymptotic behavior of $G(0,x)$ for large $|x|$. Therefore we shall at once proceed to the central problem of transient potential theory—that of investigating the behavior of $G(0,x)$ for large $|x|$. The basic result (T2) of this section will then have immediate interesting applications in section 25. One of these applications is P25.3, the core of which is as follows. For aperiodic transient random walk, and for any finite subset B of R, there are the following possibilities. If $d > 1$, or if $d = 1$ and $m = \sum |x| P(0,x) = \infty$, then

$$\lim_{|x| \to \infty} H_B(x,y) = 0 \text{ for each } y \in B.$$

If $d = 1$ and $m < \infty$, $\mu = \sum xP(0,x) > 0$, then

$$\lim_{x \to -\infty} H_B(x,y) = H_B(-\infty,y), \qquad y \in B$$

exists, and is positive for some y in B, whereas

$$\lim_{x \to +\infty} H_B(x,y) = 0, \qquad y \in B.$$

When $m < \infty$, $\mu < 0$, a similar statement holds, of course, with $+\infty$ and $-\infty$ reversed.

This then is a sample of typical results for transient random walk, and up to this point the theory (as it stands in 1963) is pleasingly complete. But new problems deserve attention in the transient case, which did not exist for recurrent random walk. The most striking one of these concerns the "total" hitting probability

$$H_B(x) = \sum_{y \in B} H_B(x,y).$$

In the recurrent case $H_B(x) = 1$ for all $x \in R$ for every nonempty set $B \subset R$, finite or infinite. In the transient case, however, there exist two kinds of sets: the sets B such that $H_B(x) \equiv 1$, and those for which $H_B(x) < 1$ at least for some x. There are sets of each type; that is clear. If $B = R$, the whole space, then it is clearly of the first type (a so-called *recurrent set*), and it seems reasonable to assume that many infinite sets B have the property that $H_B(x) \equiv 1$, i.e., that they are visited with probability one by the random walk with arbitrary starting point. On the other hand, the other type of set (the so-called *transient set*) must also be common enough. Every finite set is clearly a transient set—and it can be shown that every random walk also has infinite transient sets. This problem, of determining which sets are transient and which are recurrent for a given random walk, will also be dealt with in section 25, but only quite superficially. A really satisfactory answer will be seen to depend on more detailed aspects of the asymptotic behavior of $G(0,x)$ than we shall study in section 25. Section 26 will deal exclusively with three-dimensional random walk with mean zero and finite second moments. Due to these restrictive assumptions, a great deal of information is available about $G(x,y)$, and for these random walks we shall indeed obtain the desired characterization of transient and recurrent sets (in T26.1 and T26.2). Section 27, finally, will be devoted to an extension of potential theory. There we shall enlarge the scope of our investigation to include transition functions which are not difference kernels. The general potential theory so developed will be indispensable for a deeper understanding even of some very concrete random walk problems, and will be used again in Chapter VII which is devoted to recurrent random walk.

According to D13.1 a function $f(x)$ is *regular* with respect to the transition function $P(x,y)$ of a random walk, if

$$Pf(x) = \sum_{y \in R} P(x,y)f(y) = f(x), \quad x \in R,$$

and $f \geq 0$ on R. It was shown in Chapter III, P13.1, that the only regular functions are the constants, when P is aperiodic and recurrent. As shown in E13.2 the situation is more complicated in the transient case: aperiodic transient random walk may have other regular functions than the constants. Thus one might be tempted to proceed in one of two directions. First one might see if there is a restricted class of aperiodic transient transition functions $P(x,y)$ with the property that all the regular functions are constant. There are in fact such random walks (simple random walk in dimension $d \geq 3$ is an example). But the proof (see E27.2 or problem 8) even for this very special case is quite delicate, and therefore it is more expedient to restrict the class of regular functions, rather than the transition functions under consideration. The following important and elegant result has been known for ten years or more—but the first simple proof was given only in 1960, by Choquet and Deny [8]. Their theorem is valid in a slightly more general setting, but in the present context it reads

T1 *If $P(x,y)$ is aperiodic, and if f is regular relative to P and bounded on R, then f is constant on R.*

Proof: Given a transition function $P(x,y)$ and a function $f(x)$ satisfying the hypotheses of T1, we set

$$g(x) = f(x) - f(x - a)$$

where a is (for the moment at least) an arbitrary element of R^+. Clearly $g(x)$ is a bounded function, and one easily verifies that

$$Pg(x) = g(x), \quad x \in R.$$

(This is so because the translate of any regular function is regular, and g is the difference of two translates.) Now let

$$\sup_{x \in R} g(x) = M < \infty,$$

choose a sequence x_n of points in R such that

$$\lim_{n \to \infty} g(x_n) = M,$$

and let

$$g_n(x) = g(x + x_n).$$

Since $g(x)$ is bounded, one can select a subsequence x_n' from the sequence x_n such that, for a certain $x = x_0$

$$\lim_{n \to \infty} g(x_0 + x_n') \quad \text{exists}.$$

But one can do much better! One can take a subsequence x_n'' of this subsequence such that $g(x + x_n'')$ has a limit at $x = x_0$ and also at x_1. This process can be continued. It is famous as Cantor's diagonal process, the name being due to a convenient constructive procedure for getting a sequence which, from some point on, is a subsequence of each of a countable infinity of sequences $x_n^{(1)}, x_n^{(2)}, x_n^{(3)}, \ldots$ such that $x_n^{(k+1)}$ is a subsequence of $x_n^{(k)}$. Since R is countable, this is all that is needed for the following conclusion: there exists a subsequence n' of the positive integers such that

$$\lim_{n' \to \infty} g_{n'}(x) = g^*(x)$$

exists for every x in R. It is clear that

$$g^*(0) = M, \qquad g^*(x) \leq M, \quad x \in R.$$

It is also true (by dominated convergence) that

$$Pg^*(x) = g^*(x), \qquad x \in R.$$

What has been accomplished is this: we have exhibited a function $g^*(x)$ which is "essentially regular," i.e., it becomes regular if one adds to it a suitable constant, and it has the property that it assumes its maximum at the origin. It is the latter property, which together with the aperiodicity of $P(x,y)$, will lead to the conclusion that $f(x)$ is constant.

Iterating the operation by P on g^* one gets

$$\sum_{x \in R} P_n(0,x)g^*(x) = g^*(0) = M, \qquad n \geq 0.$$

According to the definition of R^+ this implies

$$g^*(x) = M \text{ on } R^+.$$

Given any positive integer r and any $\epsilon > 0$, one can therefore find an integer n large enough so that

$$g_n(a) > M - \epsilon, \qquad g_n(2a) > M - \epsilon, \ldots g_n(ra) > M - \epsilon.$$

Going back to the definition of $g_n(x)$ and adding these r inequalities one has

$$f(ra + x_n) - f(x_n) > r(M - \epsilon)$$

for all large enough n. If M were positive, then r could have been chosen so large that $r(M - \epsilon)$ exceeds the least upper bound of $f(x)$. This is impossible and hence $M \leq 0$. Hence $g(x) \leq 0$ or $f(x) \leq f(x - a)$ for all $x \in R$, $a \in R^+$. But exactly the same process of reasoning could have been applied to the function $-f(x)$ giving $f(x) \geq f(x - a)$ so that

$$f(x) = f(x - a) \text{ for } x \in R, \qquad a \in R^+.$$

That is all one can prove (no more is true) in the general case. But if the random walk is aperiodic, we can conclude that f is constant; for arbitrary $y \in R$, it is then possible to write $y = a - b$ with a and b in R^+ so that $f(y) = f(y - a) = f(-b) = f(0)$.

Now we shall embark on a long sequence of propositions (P1 through P8) each describing some different aspect of the asymptotic behavior of the Green function $G(x,y)$. Theorem T1 will be an indispensable tool at several points in our work. In fact after P3 or, at the latest after P6, it will be clear that T1 makes possible a simple and natural proof of the renewal theorem for positive random variables which was proved by Fourier analytical methods in Chapter II. Although T1 is sufficiently powerful so that, by taking full advantage of it, one could dispense entirely with the results of P9.2 and P9.3 of Chapter II, it will be much more convenient to use them. As those results really belong in this chapter, and to facilitate frequent reference to them, we list them first as P1 and P2.

P1 *For every aperiodic transient random walk*

$$\lim_{|x| \to \infty} [G(0,x) + G(x,0)] \quad \text{exists,}$$

and

P2 *For aperiodic random walk in one dimension, with $P(0,x) = 0$ for $x \leq 0$ and $\mu = \sum_{x=1}^{\infty} xP(0,x) \leq \infty$,*

$$\lim_{x \to +\infty} G(0,x) = \frac{1}{\mu} \quad (=0 \text{ if } \mu = +\infty).$$

Next we shall derive

P3 *For every aperiodic transient random walk*
$$\lim_{|x|\to\infty} [G(x,y) - G(x,0)] = 0, \qquad y \in R.$$

Proof: For convenience, call $G(x,0) = u(x)$. If P3 were false, then there would be some $\epsilon > 0$ and a sequence x_n in R with
$$\lim_{n\to\infty} |x_n| = \infty$$
such that
$$|u(x_n + t) - u(x_n)| > \epsilon$$
for all n at some point t in R. Suppose this is the case. Then, using the diagonal process, as in the proof of T1, one can extract a subsequence y_n from the sequence x_n such that
$$\lim_{n\to\infty} u(y_n + y) = v(y)$$
exists for every y in R, and such that $|v(t) - v(0)| \geq \epsilon$. The proof of P3 will be completed by showing that this is impossible, because $v(x)$ must be constant on R.

Since
$$G(x,0) = \sum_{y \in R} P(x,y)G(y,0) + \delta(x,0), \qquad x \in R,$$
$$u(y_n + x) = \sum_{y \in R} P(y_n + x, y)u(y) + \delta(y_n + x, 0)$$
$$= \sum_{s \in R} P(0,s)u(y_n + s + x) + \delta(y_n, -x),$$
for all $x \in R$ and all n. Letting $n \to \infty$ one obtains
$$v(x) = \sum_{s \in R} P(0,s)v(s + x) = \sum_{y \in R} P(x,y)v(y), \qquad x \in R.$$

Thus $v(x)$ is a regular function, by T1 it is constant, and this contradicts the earlier statement that $v(t) \neq v(0)$.

At this point we abandon, but only momentarily, this apparently fruitful approach. It will be resumed in P6 below when we try to extend P2 to a much larger class of one-dimensional random walk. First there is one more quite general property of $G(x,y)$ which can be obtained by elementary probabilistic arguments. Observe that periodicity is irrelevant to the truth of

P4 *For any transient random walk*
$$\lim_{|x|\to\infty} G(0,x)G(x,0) = 0.$$

Proof: We shall take $x \neq 0$, $\mathbf{T}_x = \min[k \mid k \geq 1; \mathbf{x}_k = x] \leq \infty$, and evaluate the sum

$$S(x) = \mathbf{E}_0\left[\sum_{k=\mathbf{T}_x}^{\infty} \delta(\mathbf{x}_k,0)\right]$$

in two different ways. First

$$S(x) = \mathbf{E}_0\left[\sum_{j=0}^{\infty} \delta(\mathbf{x}_{\mathbf{T}_x+j},0); \mathbf{T}_x < \infty\right]$$

and by the property of stopping times in P3.2 this becomes

$$S(x) = \mathbf{E}_0\left[\mathbf{E}_x \sum_{j=0}^{\infty} \delta(\mathbf{x}_j,0); \mathbf{T}_x < \infty\right]$$
$$= \mathbf{P}_0[\mathbf{T}_x < \infty]G(x,0) = \frac{G(0,x)G(x,0)}{G(0,0)}.$$

Since $G(0,0) > 0$ it will clearly suffice to prove that $S(x) \to 0$ as $|x| \to \infty$. This is verified by writing

$$S(x) = \mathbf{E}_0\left[\sum_{k=0}^{\infty} \delta(\mathbf{x}_k,0); \mathbf{T}_x \leq k\right] = \sum_{k=0}^{\infty} \mathbf{P}_0[\mathbf{x}_k = 0; \mathbf{T}_x \leq k]$$
$$\leq \sum_{k=0}^{n} \mathbf{P}_0[\mathbf{T}_x \leq k] + \sum_{k=n+1}^{\infty} \mathbf{P}_0[\mathbf{x}_k = 0],$$

for each $n \geq 0$. As the random walk is transient we may choose an arbitrary $\epsilon > 0$ and find an integer N such that

$$S(x) \leq \sum_{k=0}^{N} \mathbf{P}_0[\mathbf{T}_x \leq k] + \epsilon.$$

Hence $S(x) \to 0$, provided that

$$\lim_{|x|\to\infty} \mathbf{P}_0[\mathbf{T}_x \leq k] = 0$$

for each fixed k. But

$$\mathbf{P}_0[\mathbf{T}_x \leq k] \leq \sum_{j=0}^{k} P_j(0,x)$$

and obviously each term $P_j(0,x) \to 0$ as $|x| \to \infty$. Hence P4 is proved.[1]

---

[1] The results P1 through P4 can be obtained in a different order, without use of T1. To this end prove first P1, then a slightly strengthened form of P4 to the effect that $G(0,x)G(x,y) \to 0$ as $|x| \to \infty$ for each y. Together these two propositions immediately imply P4.

In the next proposition we complete our present study of $G(x,y)$ for transient random walk in dimension $d \geq 2$. We shall do so by using a very "weak" property of R with $d \geq 2$, but nevertheless one which very effectively distinguishes these additive Abelian groups from the one-dimensional group of integers. (This fact deserves emphasis since we shall get a result which according to P2 is false when $d = 1$.) The property of $R(d \geq 2)$ in question may be described as follows. Suppose that R is decomposed into three sets A, B, and C (disjoint sets whose union is R) such that A is finite or even empty, while B and C are both infinite sets. Then there exists a pair of neighbors x and y in R, i.e., a pair of points such that $|x - y| = 1$, with the property that x is in B, and y in C. It should be obvious that this is indeed a property of R with $d \geq 2$, but not of R with $d = 1$. Actually all that is essential for our purpose is that x and y are in a suitable sense not "too far apart"—the requirement that $|x - y| = 1$ merely serves to simplify the exposition.

P5 *For aperiodic transient random walk in dimension $d \geq 2$,*

$$\lim_{|x| \to \infty} G(0,x) = 0.$$

Proof: Suppose P5 is false, and call $G(x,0) = u(x)$. Then, in view of P1, we must have

$$\lim_{|x| \to \infty} [u(x) + u(-x)] = L > 0.$$

Now we choose an ϵ, $0 < \epsilon < L/3$, and decompose R into three sets A, B, C as follows. A is to be a sphere, i.e., $A = [x \mid |x| \leq r]$ with the radius r chosen so that

$$|u(x + t) - u(x)| < \epsilon \qquad \text{when } |t| = 1, \quad x \in R - A$$

and

$$\min [|u(x)|, |u(x) - L|] < \epsilon \qquad \text{when } x \in R - A.$$

That this can be done follows from P3 and from P4 combined with our hypothesis that $u(x) + u(-x)$ tends to L.

Now we shall decompose $R - A$ into B and C. B is the subset of $R - A$ where $|u(x)| < \epsilon$ and C the subset where $|u(x) - L| < \epsilon$. B and C are disjoint because $L > 3\epsilon$. Since $L > 0$, C must be an infinite set and P4 tells us that also B is infinite ($-x$ is in B when x is in C). But using the characteristic property of R with $d \geq 2$ which we discussed prior to P5, there is a pair x,y such that

$$|u(x)| < \epsilon, \qquad |u(y) - L| < \epsilon, \qquad |x - y| = 1, \quad x \in B, \quad y \in C.$$

Since $L > 3\epsilon$ the above inequalities give

$$|u(x) - u(y)| > L - 2\epsilon > \epsilon.$$

Now $y = x + t$ with $|t| = 1$, and according to the definition of the set A, x must be in A, which it is not. This contradiction proves P5. Observe that aperiodicity is not strictly necessary for P5. All we need is that R be a group with more than one generator, i.e., that the random walk does not take place on a one-dimensional subspace of R.

Now we are free to concentrate on one-dimensional random walk, which unfortunately presents much greater difficulties. First we deal with the comparatively straightforward case when the absolute first moment is finite. The results constitute a rather intuitive extension of the renewal theorem, which was obtained by Blackwell [4] and several other authors around 1952, but which had been conjectured and proved in special cases before 1940.[2]

P6 *Aperiodic transient random walk*, $d = 1$, $m = \sum_{x \in R} |x| P(0,x) < \infty$, $\mu = \sum_{x \in R} x P(0,x)$. *If $\mu > 0$, then*

$$\lim_{x \to +\infty} G(0,x) = \frac{1}{\mu}, \qquad \lim_{x \to -\infty} G(0,x) = 0.$$

If $\mu < 0$, the limits are reversed, and $(-\mu)^{-1}$ replaces μ^{-1}.

Proof: The proof uses in principle the same summation by parts as that of P2, which was given as P9.3 in Chapter II. Only far greater care is required in the present case. First we remark that the two limits

$$\lim_{x \to +\infty} G(0,x) = G(0, +\infty), \qquad \lim_{x \to -\infty} G(0,x) = G(0, -\infty)$$

certainly exist. This is a consequence of P1, P3, and P4, for P1 states that $G(0,x) + G(0,-x)$ has a limit as $x \to +\infty$, P4 tells us that one of these two limits is zero, and finally P3 prevents any possibility of oscillation whereby both limits might fail to exist. Thus we are concerned only with evaluating the limits. A close look at the statement of P6 shows that the identity

$$[G(0, +\infty) - G(0, -\infty)]\mu = 1$$

is equivalent to P6. If P6 is true, then the identity holds, whether $\mu > 0$ or $\mu < 0$ ($\mu = 0$ is impossible since the random walk would

---

[2] See [4], [52], and in particular [90] for a lucid account of this subject up to 1958.

then be recurrent). Conversely, if this identity holds, then P6 is true. For suppose $\mu > 0$ and it holds. Since we know from P4 that either $G(0, +\infty)$ or $G(0, -\infty)$ is zero, it follows that $G(0, +\infty) = \mu^{-1}$ and $G(0, -\infty) = 0$. If $\mu < 0$ one gets $G(0, -\infty) = (-\mu)^{-1}$ and $G(0, +\infty) = 0$.

Finally, let us introduce some simplifying notation. Let

$$f(a,b) = \sum_{x=a}^{b} P(0,x)$$

where a may be $-\infty$ and b may be $+\infty$;

$$\mu^+ = \sum_{x=1}^{\infty} xP(0,x) = \sum_{a=1}^{\infty} f(a, +\infty),$$

$$\mu^- = -\sum_{x=-\infty}^{-1} xP(0,x) = \sum_{b=-\infty}^{-1} f(-\infty, b),$$

$$\mu = \mu^+ - \mu^-.$$

Now we define $u(x) = G(0,x)$, so that

$$u(x) = \sum_{t=-\infty}^{\infty} u(t)P(0, x-t) + \delta(0,x), \qquad x \in R.$$

Taking M and N to be two positive integers, one obtains

$$\sum_{x=-M}^{N} u(x) = \sum_{t=-\infty}^{\infty} u(t)f(-M-t, N-t) + 1.$$

It is possible, by grouping terms properly, and changing variables of summation, to write this identity in the form

$$1 = \sum_{s=0}^{N} u(N-s)[f(-\infty, s-N-M-1) + f(s+1, \infty)]$$

$$+ \sum_{s=-M+1}^{0} u(-M-s)[f(-\infty, s-1) + f(M+N+s+1, \infty)]$$

$$- \sum_{s=1}^{\infty} u(N+s)f(-M-N-s, -s)$$

$$- \sum_{s=-\infty}^{-1} u(-M+s)f(-s, M+N-s).$$

The idea is now to let N and M get large, one at a time. Let us take $N \to +\infty$ first. The finiteness of the absolute moment is of course used in an essential way and gives dominated convergence that justifies limit processes such as

$$\lim_{N\to\infty} \sum_{s=0}^{N} u(N-s)f(s+1,\infty)$$
$$= \lim_{N\to\infty} u(N) \sum_{s=0}^{\infty} f(s+1,\infty) = G(0,+\infty)\mu^+.$$

In this way the above "decomposition of 1" becomes, as $N \to \infty$

$$1 = \sum_{s=0}^{\infty} u(s)f(-\infty, -M-s-1) + G(0,+\infty)\mu^+$$
$$+ \sum_{s=-M+1}^{0} u(-M-s)f(-\infty, s-1) + 0$$
$$- G(0,+\infty)\mu^- - \sum_{s=-\infty}^{-1} u(-M+s)f(-s,\infty).$$

(The terms have been kept in the original order to facilitate the verification.) Now let $M \to +\infty$. Then one gets

$$1 = 0 + G(0,+\infty)\mu^+ + G(0,-\infty)\mu^- + 0$$
$$- G(0,+\infty)\mu^- - G(0,-\infty)\mu^+$$
$$= [G(0,+\infty) - G(0,-\infty)][\mu^+ - \mu^-]$$
$$= [G(0,+\infty) - G(0,-\infty)]\mu.$$

This is the desired result which implies P6.

It now remains only to investigate transient one-dimensional random walk with $m = \infty$, and of course the desired result is

$$\lim_{|x|\to\infty} G(0,x) = 0.$$

Unfortunately this seems to be impossible to obtain from the methods developed so far in this chapter. Instead we make use of the rather more delicate apparatus of the "ladder" random variable **Z** introduced in Chapter IV, D18.1. It was, by the way, in connection with the renewal theorem that **Z** and the ladder random walk, consisting of partial sums of random variables with the same distribution as **Z**, were first studied by Blackwell [4]. But, whereas he required them in the case when the absolute moment m is finite, we need them only at this stage, when $m = \infty$, which seems to be the most difficult stage of the problem. It will be convenient to introduce appropriate notation, in

24. THE GREEN FUNCTION $G(x,y)$

D1 *The set A is the half-line $[x \mid x > 0]$, $\mathbf{T} = \min[n \mid n \geq 1, \mathbf{x}_n \in A]$ as in D17.1, $H_A(x,y) = \mathbf{P}_x[\mathbf{T} < \infty, \mathbf{x_T} = y]$ as in D10.1, and*

$$H_A(x) = \sum_{y \in A} H_A(x,y).$$

When $\mathbf{T} < \infty$ with probability one we call $\mathbf{x_T} = \mathbf{Z}$ as in D18.1. In this case $\hat{P}(x,y) = \mathbf{P}_0[\mathbf{Z} = y - x]$, and finally, $\hat{H}_A(x,y)$ is the hitting probability measure of the set A for the random walk whose transition function is $\hat{P}(x,y)$.

Obviously the hitting probabilities $H_A(x,y)$ are intimately connected with the Green function $G(x,y)$ of a transient random walk. For example, if $x \leq 0$ and $y > 0$, one has the identity P10.1(d)

$$G(x,y) = \sum_{t=1}^{\infty} H_A(x,t) G(t,y),$$

so that the asymptotic behavior of $H_A(x,t)$ as $x \to -\infty$ is related to that of $G(x,y)$. This connection seems sufficiently attractive to justify developing the basic facts about $H_A(x,y)$ in somewhat more detail than is absolutely necessary. These facts take the form of a simple classification. Obvious possibilities concerning the asymptotic behavior of $H_A(x,y)$ are given by

(i) $\qquad \lim_{x \to -\infty} H_A(x) = 0$

(ii) $\qquad H_A(x) = 1$ *for all x in R and*

$\qquad \lim_{x \to -\infty} H_A(x,y) = 0$ *for $y \in A$.*

(iii) $\qquad H_A(x) = 1$ *for all x in R,*

$\qquad \lim_{x \to -\infty} H_A(x,y) = \gamma_A(y)$ *for $y \in A$, and*

$$\sum_{y=1}^{\infty} \gamma_A(y) = 1.$$

We do not yet know, but we shall in fact show, that this is a complete classification, in the sense that (i), (ii), and (iii) exhaust all possibilities. Moreover this will be the case for recurrent as well as transient random walk. Once this fact is even suspected, it is not farfetched to ask if there is any connection with the following classification, which is quite familiar from Chapter IV.

(1) $\qquad \mathbf{P}_0[\mathbf{T} = \infty] > 0,$ i.e., $\mathbf{Z}$ *is undefined,*
(2) $\qquad \mathbf{P}_0[\mathbf{T} < \infty] = 1$ *and* $\mathbf{E}_0[\mathbf{Z}] = \infty,$
(3) $\qquad \mathbf{P}_0[\mathbf{T} < \infty] = 1$ *and* $\mathbf{E}_0[\mathbf{Z}] < \infty.$

This set of alternatives turns out to be exactly the same classification as (i) through (iii), as we shall now prove in

P7 *For arbitrary one-dimensional random walk (recurrent or transient), the two classifications are equivalent, that is* (i) *holds whenever* (1) *does, and so on. These are the only possibilities. Furthermore, in cases* (ii) = (2) *and* (iii) = (3), *one has*

$$\lim_{x \to -\infty} H_A(x,y) = \gamma_A(y) = \frac{1}{E_0[Z]} P_0[Z \geq y]$$

for $y \in A$. The limit is interpreted as zero in case (ii) = (2).

Proof: Clearly (1), (2), and (3) are mutually exclusive and exhaust all possibilities. Now (1) implies (i) by a very simple probability argument which was given in Chapter IV. For $x \leq 0$, $H_A(x) = P[M > x]$, where $M = \max_{k \geq 0} S_k$. But in P19.2, M was shown to be a random variable when (1) holds, and that proves (i).

Now suppose that either (2) or (3) holds. Then it is evident that $H_A(x) \equiv 1$ on R and also that

$$H_A(x,y) = \hat{H}_A(x,y) \text{ for } x \leq 0, \quad y > 0.$$

This relation makes it convenient to work with the random walk $\hat{x}_n$ defined by $\hat{P}(x,y)$ in D1 above. It is a transient random walk, whether $P(x,y)$ defines a transient random walk or not. Therefore it has a finite Green function which we may call

$$\hat{G}(x,y) = \sum_{n=0}^{\infty} \hat{P}_n(x,y).$$

When $x \leq 0$ and $y > 0$, $\hat{G}(x,y)$ is the probability that the random walk $\hat{x}_n$ defined by $\hat{P}$ visits y in a finite time, if it starts at $\hat{x}_0 = x$. Now one can write

$$\hat{G}(x,y) = \hat{H}_A(x,y) + \sum_{t=1}^{y-1} \hat{G}(x,t) P_0[Z = y - t],$$

by making the obvious decomposition of $\hat{G}(x,y)$ according to the last value of $\hat{x}_n$ in A before it visits the point y. A formal proof proceeds via

$$\hat{H}_A(x,y) = P_x[\hat{x}_k = y \text{ and } \hat{x}_{k-1} \leq 0 \text{ for some } k < \infty]$$

and

$$\hat{G}(x,t) P_0[Z = y - t] = P_x[\hat{x}_k = y \text{ and } \hat{x}_{k-1} = t \text{ for some } k < \infty].$$

Now we shall apply the renewal theorem for positive random variables (P2) to the Green function $\hat{G}(x,y)$. It gives

$$\lim_{x \to -\infty} \hat{G}(x,t) = \frac{1}{\mathbf{E}_0[Z]} \geq 0, \qquad t \in R.$$

Therefore

$$\lim_{x \to -\infty} H_A(x,y) = \lim_{x \to -\infty} \hat{H}_A(x,y)$$

$$= \lim_{x \to -\infty} \left\{ \hat{G}(x,y) - \sum_{t=1}^{y-1} \hat{G}(x,t) \mathbf{P}_0[Z = y - t] \right\}$$

$$= \frac{1}{\mathbf{E}_0[Z]} \left\{ 1 - \sum_{t=1}^{y-1} \mathbf{P}_0[Z = y - t] \right\} = \frac{\mathbf{P}_0[Z \geq y]}{\mathbf{E}_0[Z]} \geq 0$$

for $y \geq 1$.

But if (2) holds, then $\mathbf{E}_0[Z] = +\infty$ so that the above limit is to be interpreted as zero according to the renewal theorem P2. Hence (2) implies (ii), and for the same reason (3) implies (iii). The formula for $\gamma_A(y)$ in P7, is of course the one obtained above, as a by-product of the method of proof.

Only a small step remains to be taken to obtain

P8 *For transient random walk with $m = \infty$,*

$$\lim_{|x| \to \infty} G(0,x) = 0.$$

Proof: It will clearly suffice to show that

$$\lim_{x \to +\infty} G(0,x) = 0,$$

as there is nothing in the hypotheses of P8 to distinguish $+\infty$ from $-\infty$. The proof is based on the identity that motivated the development of P7, in the rather special form

(a) $$G(x,1) = \sum_{y=1}^{\infty} H_A(x,y) G(y,1).$$

Now we run through the three cases in the classification of P7. First suppose that (i) holds. Then (a) yields

$$\varlimsup_{x \to -\infty} G(x,1) \leq G(1,1) \lim_{x \to -\infty} H_A(x) = 0.$$

Next, let us take case (ii). Then the conclusion is the same, but the reasoning required to obtain it is a little more complicated. Since

$H_A(x,y)$ tends to zero for each $y > 0$ as $x \to -\infty$, but the total probability "mass" $H_A(x)$ is constant (one), it can be concluded from (a) that

$$\varlimsup_{x \to -\infty} G(x,1) \le \varlimsup_{y \to +\infty} G(y,1),$$

and since both limits are known to exist (this was pointed out at the start of the proof of P6)

$$\lim_{x \to -\infty} G(x,1) = G(0,+\infty) \le \lim_{y \to +\infty} G(y,1) = G(0,-\infty).$$

But if we had $G(0,+\infty) > 0$, then this inequality would entail $G(0,-\infty) > 0$ which contradicts P4. Hence P8 holds if we are in cases (i) or (ii).

Finally consider case (iii). There seems to be nothing here to prevent $G(-\infty,0) = G(0,+\infty)$ from being positive. So, there must be another way out of the dilemma, and that is the realization that transient random walk cannot possibly come under the classification (iii) when its absolute moment m is infinite! For when $m = \infty$, then either μ^+ or μ^- or both are infinite (they were defined in the proof of P6). If $\mu^+ < \infty$ and $\mu^- = \infty$, then we are in case (i) since $\mathbf{T} = \infty$ with positive probability (by the law of large numbers). Finally, if $\mu^+ = \infty$, we may still be in case (i) and then there is nothing left to prove. We may also be in case (ii) and that case has also been settled. But case (iii) is out of the question since, when $\mathbf{T} < \infty$ with probability one, then

$$\mathbf{E}_0[\mathbf{Z}] \ge \mathbf{E}_0[\mathbf{Z}; \mathbf{T} = 1] = \sum_{x=1}^{\infty} x P(0,x) = \mu^+.$$

That completes the proof of P8, and now it remains only to collect the results from P5, P6, and P8, and to combine them into a statement that is rather pleasing in its generality.[3]

T2 *The Green function of transient aperiodic random walk satisfies*

$$\lim_{|x| \to \infty} G(0,x) = 0$$

in all cases except when the dimension $d = 1$ and $m < \infty$. In this case

$$\lim_{x \to -\infty} G(0,x) = G(0,-\infty) \quad \text{and} \quad \lim_{x \to +\infty} G(0,x) = G(0,+\infty)$$

[3] For $d = 1$ this theorem is due to Feller and Orey [32]. Chung [10] was the first to discuss the renewal theorem in higher dimensions.

are distinct, with

$$G(0,+\infty) = \frac{1}{\mu} \quad \text{and} \quad G(0,-\infty) = 0 \quad \text{when} \quad \mu > 0,$$

$$G(0,-\infty) = \frac{1}{|\mu|} \quad \text{and} \quad G(0,+\infty) = 0 \quad \text{when} \quad \mu < 0.$$

The most important applications of T2 are associated with the study of hitting probabilities in the next section. But as a good illustration of its uses we can make a remark which improves and completes the fundamental result T2.1 of Chapter I. For transient random walk the probability $F(0,0)$ of a return to the starting point is less than one. However $F(0,x)$ presented a problem which could not be solved by earlier methods. Now we have

P9 *Aperiodic transient random walk has the property that $F(0,x) < 1$ for all $x \in R$, with one and only one exception. If the dimension $d = 1$, $m < \infty$, $\mu < 0$, and the random walk is left continuous (i.e., $P(0,x) = 0$ for $x \leq -2$), then $F(0,x) = 1$ for $x < 0$ while $F(0,x) < 1$ for $x > 0$. An analogous rule applies of course to the reversed, right continuous random walk with $m < \infty$, $\mu > 0$.*

Proof: Suppose that $F(0,a) = 1$ for a certain aperiodic transient random walk, where $a \neq 0$. Then $F(0,na) = 1$ for every integer $n \geq 1$. For if $\mathbf{T}_x$ is the first time of $\mathbf{x}_n$ at x, and if $\mathbf{x}_0 = 0$, then

$$F(0,a) = \mathbf{P}_0[\mathbf{T}_a < \infty],$$

$$F(0, na) = \mathbf{P}_0[\mathbf{T}_{na} < \infty] \geq \mathbf{P}_0[\mathbf{T}_a < \mathbf{T}_{2a} < \cdots < \mathbf{T}_{na} < \infty]$$
$$= [\mathbf{F}(0,a)]^n = 1.$$

But by P1.5

$$1 = F(0,na) = \frac{G(0,na)}{G(0,0)}.$$

Letting $n \to \infty$, we see from T2 that the right-hand side tends to zero, unless the random walk under consideration is one-dimensional with absolute moment $m < \infty$. Now it suffices to assume that $a < 0$, and then we have to show that $\mu < 0$ and that, in addition, the random walk must be left continuous. The case $a > 0$ can then receive analogous treatment. Observe that μ is negative, for if $\mu > 0$, then $F(0,x) < 1$ for $x < 0$ since $\mathbf{x}_n \to +\infty$ with probability one, and cannot be zero since the random walk is transient.

It is quite clear that left-continuous random walk with $\mu < 0$ has the desired property. Since $\mathbf{x}_n \to -\infty$ with probability one, and no point can be "skipped," $F(0,-1) = F(0,-n) = 1$ for $n \geq 1$.

Finally, if $\mu < 0$ and the random walk is not left continuous, one can show that for every $a < 0$ there is some "path" $0 = x_0, x_1, \ldots, x_n$ such that $x_n < a$,

$$P(0,x_1)P(x_1,x_2)\ldots P(x_{n-1},x_n) > 0,$$

and $x_k \neq a$ for $k = 1, 2, \ldots, n-1$. That may be done by a simple combinatorial argument (as for example in the proof of P18.8) which of course has to take account of the aperiodicity of the random walk. Because $\mu < 0$, the existence of such a path implies that $F(x_n,a) < 1$ and therefore also that

$$1 - F(0,a) \geq P(0,x_1)\ldots P(x_{n-1},x_n)[1 - F(x_n,a)] > 0.$$

This inequality shows that left continuity is a necessary property of an aperiodic random walk with $\mu < 0$ and $F(0,a) = 1$ for some $a < 0$. The proof of P9 is therefore complete.

25. HITTING PROBABILITIES

Let $P(x,y)$ be the transition function of a *transient* random walk. We shall also assume aperiodicity throughout this section, although it will of course usually be possible to rephrase the results in such a way as to remove this condition. The matter of calculating hitting probabilities is quite a bit easier than it was for recurrent random walk in Chapter III. Therefore we shall now attack the general situation where $A \subset R$ is any nonempty, *possibly infinite*, proper subset of R. The relevant definitions can be found in D10.1, but since we shall add to them two new ones (H_A and E_A) we summarize them as

D1 For $A \subset R$, $1 \leq |A| \leq \infty$, $R - A$ nonempty, $H_A(x,y)$ is defined as in D10.1, for $x \in R$, $y \in A$, and so is $\Pi_A(x,y)$ for $x \in A$ and $y \in A$. In addition, let

$$H_A(x) = \sum_{y \in A} H_A(x,y), \qquad x \in R,$$
$$E_A(x) = 1 - \sum_{y \in A} \Pi_A(x,y), \qquad x \in A.$$

H_A is called the entrance and E_A the escape probability function of the set A.

In general, one cannot say much about the probabilities $H_A(x)$ and $E_A(x)$. Obviously $H_A(x)$ is one on A and less than or equal to one on $R - A$. But to get more information is not easy. To appreciate the difficulties, it suffices to take for A the set consisting of the origin, and to look up P24.9 in the last section. According to P24.9, $H_{\{0\}}(x)$ is less than one for all $x \neq 0$ for "most" transient random walks—but there is the exception of the left- and right-continuous random walks with $m < \infty$.

Now consider the escape function. If

$$\mathbf{T} = \min [n \mid n \geq 1, \mathbf{x}_n \in A],$$

then the definition of $E_A(x)$ gives

$$E_A(x) = 1 - \mathbf{P}_x[\mathbf{T} < \infty] = \mathbf{P}_x[\mathbf{T} = +\infty], \qquad x \in A.$$

That is why we called it the escape function of A. The only general information one has from P10.4, or from the probability interpretation in terms of the hitting time $\mathbf{T}$, is that $E_A(x) \leq 1$ on A. Of course, if one wants an example of a set A such that $E_A(x) > 0$ on A, one should look for a "small" set—say a finite one. But here again P24.9 teaches one respect for the difficulty of the problem. If the random walk is left continuous, with $m < \infty$ and $\mu < 0$, and if A is the set $A = \{0, -1\}$, then $E_{\{0,-1\}}(0) = 0$. But for every other one-dimensional transient random walk $E_{\{0,-1\}}(0) > 0$. Still more difficult is the question of how "large" the set A should be so that $E_A(x) \equiv 0$ on A. We shall return to all these questions after getting some information concerning $H_A(x,y)$ and $\Pi_A(x,y)$.

P1 $H_A(x,y) = \sum_{t \in A} G(x,t)[\delta(t,y) - \Pi_A(t,y)]$, for $x \in R$, $y \in A$.

Proof: The proof goes back to part (a) of P10.1:

$$\sum_{t \in R} P(x,t) H_A(t,y) - H_A(x,y) = \Pi_A(x,y) - \delta(x,y) \text{ if } x \in A, y \in A,$$
$$= 0 \text{ if } x \in R - A, y \in A.$$

Operating by the transition function $P(x,y)$ on the left k times,

$$\sum_{t \in R} P_{k+1}(x,t) H_A(t,y) - \sum_{t \in R} P_k(x,t) H_A(t,y)$$
$$= \sum_{t \in A} P_k(x,t)[\Pi_A(t,y) - \delta(t,y)] \text{ for } x \in R, y \in A.$$

This procedure was used in the proof of P11.1 and, as was done there, we add the $n+1$ equations one gets by setting $k = 0, 1, 2, \ldots, n$. The result is

$$H_A(x,y) = \sum_{t \in A} G_n(x,t)[\delta(t,y) - \Pi_A(t,y)]$$
$$+ \sum_{t \in R} P_{n+1}(x,t)H_A(t,y), \qquad x \in R, \quad y \in A, \quad n \geq 1.$$

But from this point on the situation is drastically simpler for transient than it was for recurrent random walk. One simply obtains

$$H_A(x,y) = \lim_{n \to \infty} \sum_{t \in A} G_n(x,t)[\delta(t,y) - \Pi_A(t,y)], \qquad x \in R, \quad y \in A,$$

from the observation that

$$0 \leq \sum_{t \in R} P_{n+1}(x,t)H_A(t,y) \leq \sum_{t \in R} P_n(x,t)F(t,y)$$
$$\leq \sum_{k=n+1}^{\infty} P_k(x,y) = G(x,y) - G_n(x,y),$$

which tends to zero as $n \to +\infty$. To finish the proof of P1 it remains only to show that

$$\lim_{n \to \infty} \sum_{t \in A} G_n(x,t)[\delta(t,y) - \Pi_A(t,y)] = \sum_{t \in A} G(x,t)[\delta(t,y) - \Pi_A(t,y)].$$

This follows (even when A is an infinite set) from the inequalities

$$G_n(x,t) \leq G(x,t) \leq G(0,0)$$

together with

$$\sum_{t \in A} \Pi_A(t,y) \leq 1, \qquad y \in R,$$

proved in P10.4.

Several corollaries of P1, or rather of the method of proof of P1, are occasionally useful:

P2 (a) $\sum_{y \in R} P(x,y)H_A(y) - H_A(x) = -E_A(x)$ for $x \in A$,
$\qquad\qquad\qquad\qquad\qquad\qquad = 0 \qquad$ for $x \in R - A$.

(b) $\delta(x,y) = \sum_{t \in A} G(x,t)[\delta(t,y) - \Pi_A(t,y)]$
$\qquad\qquad = \sum_{t \in A} [\delta(x,t) - \Pi_A(x,t)]G(t,y)$ for $x \in A, \quad y \in A$.

Proof: Part (a) follows from the first identity in the proof of P1 by simply summing y over the set A. Part (b), on the other hand, is just the result of P1 when $x \in A$, for then

$$H_A(x,y) = \delta(x,y) = \sum_{t \in A} G(x,t)[\delta(t,y) - \Pi_A(t,y)].$$

To finish up, one looks at the reversed random walk with $G^*(x,y) = G(y,x)$, $\Pi^*(x,y) = \Pi(y,x)$, so that when $x \in A$, $y \in A$,

$$\begin{aligned}\delta(x,y) &= \sum_{t \in A} G^*(x,t)[\delta(t,y) - \Pi_A^*(t,y)] \\ &= \sum_{t \in A} [\delta(y,t) - \Pi_A(y,t)]G(t,x).\end{aligned}$$

Since $\delta(x,y) = \delta(y,x)$ the proof of P2 is complete.

Another corollary of P1, but a far deeper and more important one than P2, concerns the hitting probabilities $H_A(x,y)$ when A is a *finite* subset of R. Then one can assert

P3 *Consider aperiodic transient random walk, and a finite subset A of R. Then there are three possibilities.*

(a) *The dimension $d \geq 2$, or $d = 1$ but $m = \infty$, in which case*

$$\lim_{|x| \to \infty} H_A(x,y) = 0 \text{ for } y \in A.$$

(b) *$d = 1$ and $m < \infty$, $\mu > 0$, in which case*

$$\lim_{x \to -\infty} H_A(x,y) = \frac{1}{\mu}\left[1 - \sum_{t \in A} \Pi_A(t,y)\right], \qquad y \in A,$$

$$\lim_{x \to +\infty} H_A(x,y) = 0, \qquad y \in A.$$

(c) *$d = 1$, $m < \infty$, and $\mu < 0$, which is just like case (b) with the limits interchanged and μ replaced by $|\mu|$.*

Proof: P3 is obtained immediately by applying T24.2 to P1.

We shall have much more to say about the subject of the asymptotic behavior of hitting probabilities in Chapter VII. This is not surprising because we already found in Chapter III, T14.1, that

$$\lim_{|x| \to \infty} H_A(x,y), \qquad y \in A$$

exists for two-dimensional aperiodic recurrent random walk. The full story, extending the result of P3 to arbitrary recurrent random walk is to be told in T30.1 in Chapter VII.

Now we turn to a study of the *escape probabilities* $E_A(x)$ and their relation to the *entrance* probability function $H_A(x)$. The reader who has recognized part (a) of P2 above as the Poisson equation, discussed in Chapter III, D13.2, will not be surprised to see that we now proceed to solve this equation for $H_A(x)$ in terms of $E_A(x)$. The result is (remember that A is once again an *arbitrary nonempty proper subset of R, finite or infinite*)

P4 $$H_A(x) = h_A(x) + \sum_{t \in A} G(x,t) E_A(t), \qquad x \in R,$$

where the sum on the right converges, and where $h_A(x)$ is the limit

$$h_A(x) = \lim_{n \to \infty} \sum_{t \in R} P_n(x,t) H_A(t), \qquad x \in R.$$

Moreover, $h_A(x)$ is independent of x; it is a constant h_A which depends on the transition function $P(x,y)$ and on the set A.

Proof: We know from part (a) of P2 that $f(x) = H_A(x)$ satisfies the Poisson equation

$$f(x) - \sum_{y \in R} P(x,y) f(y) = \begin{cases} E_A(x) & \text{for } x \in A \\ 0 & \text{for } x \in R - A. \end{cases}$$

To prove P4 we therefore study the general Poisson equation

(a) $$f(x) - \sum_{y \in R} P(x,y) f(y) = \psi(x), \qquad x \in R,$$

where $\psi(x)$ is assumed to be a *given non-negative function* on R. We require the following three basic properties of equation (a) (which will be encountered in an even more general setting in section 27).

(i) *If $f(x)$ is a non-negative solution of* (a), *then it may be written*

$$f(x) = h(x) + \sum_{t \in R} G(x,t) \psi(t) < \infty, \qquad x \in R,$$

where

(ii) $$0 \le h(x) = \lim_{n \to \infty} \sum_{t \in R} P_n(x,t) f(t) < \infty,$$

and where

(iii) $Ph(x) = h(x)$, *so that $h(x)$ is a regular function*.

In order to prove (i), (ii), and (iii) we operate on equation (a) by the iterates $P_k(x,y)$ of $P(x,y)$. Using the convenient operator notation one gets

$$f - Pf = \psi, \quad Pf - P_2 f = P\psi, \ldots, P_n f - P_{n+1} f = P_n \psi.$$

At no step is there any question about convergence. The sum representing $P\psi(x)$ exists since $Pf(x)$ exists and $P\psi(x) \le Pf(x)$. Continuing this way, once $P_n f(x)$ exists, the non-negativity of f and ψ together with the inequality $P_n\psi(x) \le P_n f(x)$ implies that also $P_n\psi$ exists. Similarly $P_{n+1}f(x)$ exists as $P_{n+1}f(x) \le P_n f(x)$. Now we add the equations $P_k f - P_{k+1} f = P_k \psi$ over $k = 0, 1, \ldots, n$ to obtain

$$f(x) - P_{n+1}f(x) = G_n\psi(x).$$

Since

$$G_n\psi(x) = \sum_{y \in R} G_n(x,y)\psi(y) \le f(x) < \infty,$$

and since

$$G_n\psi(x) \le G_{n+1}\psi(x),$$

one has

$$\lim_{n \to \infty} G_n\psi(x) = \sum_{y \in R} G(x,y)\psi(y).$$

Hence also

$$\lim_{n \to \infty} \sum_{y \in R} P_{n+1}(x,y)f(y) < \infty,$$

and if we call the last limit $h(x)$ we have established (i) and (ii). Finally, one has

$$h(x) = \lim_{n \to \infty} \sum_{y \in R} P_{n+1}(x,y)f(y)$$
$$= \sum_{t \in R} P(x,t)\left[\lim_{n \to \infty} \sum_{y \in R} P_n(t,y)f(y)\right] = \sum_{t \in R} P(x,t)h(t),$$

so that $h(x)$ is regular and (iii) holds.

To complete the proof of P4, we set

$$\psi(x) = E_A(x) \text{ for } x \text{ in } A, \qquad \psi(x) = 0 \text{ for } x \in R - A.$$

All the conclusions of P4 are now contained in equations (i), (ii), and (iii) with the exception of the last statement of P4, to the effect that $h_A(x)$ is constant. But since, by (ii),

$$h_A(x) = \lim_{n \to \infty} \sum_{t \in R} P_n(x,t)H_A(t),$$

it is non-negative and bounded. From (iii) it follows that $h_A(x)$ is a bounded regular function. Therefore we can apply T24.1 to conclude that $h_A(x) = h_A$ is constant.

To illuminate the result and the proof of P4 from a nonprobabilistic point of view, let us look at a simple special case. The equation

(b) $$f(x) - \sum_{y \in R} P(x,y)f(y) = \delta(x,0), \qquad x \in R,$$

has the solution $f(x) = G(x,0)$. We have used this observation many times, most recently in several proofs in section 24. It would be interesting to be able to characterize the function $G(x,0)$ by the property that it satisfies (b) together with as few other natural requirements as possible. Now (i) through (iii) in the proof of P4 imply that the only non-negative solutions of (b) are the functions

$$f(x) = h(x) + G(x,0),$$

where $h(x)$ is any regular function. Although we have no information as yet concerning the possible unbounded regular functions (the bounded ones are constant) we are in a position to assert

P5 *The Green function $G(x,0)$ is the minimal non-negative solution of* (b). *Alternatively, it is the only bounded non-negative solution of* (b) *such that*

$$\lim_{|x| \to \infty} f(x) = 0.$$

Proof: Since $f(x) = h(x) + G(x,0)$, $h(x) \geq 0$, $G(x,0)$ is minimal, i.e., every non-negative solution of (b) is greater than, or equal to, $G(x,0)$ everywhere. By T24.1 the only bounded non-negative solutions of (b) are $G(x,0) +$ constant, and in view of T24.2, such a solution will have lower limit zero as $|x| \to \infty$ if, and only if, the constant is zero.

Our next topic is the classification of subsets A of R (only infinite subsets will turn out to be of real interest) according to whether they are *transient* or *recurrent*. To avoid any possible misunderstanding we re-emphasize that the random walk is assumed to be *transient* and *aperiodic*. It seems tempting to call a set A recurrent if the random walk is certain to visit the set A, no matter where it starts. That would be the case if $H_A(x) = 1$ for every x in R. Alternatively, if $H_A(x) < 1$ for some $x \in R$ one might call the set A transient. On the basis of this idea we introduce the formal classification

D2 *A proper subset $A \subset R$, $1 \leq |A| \leq \infty$, is called recurrent if $H_A(x) = 1$ for all x in R, and transient if it is not recurrent.*

But let us temporarily complicate matters, if only to question the wisdom of definition D2. An equally reasonable classification of sets might have been based on the escape probabilities $E_A(x)$. If $E_A(x) = 0$ for all x in A, then we conclude that the set A is in some sense "large enough" so that it is impossible for the random walk to leave it forever. Such a set might be called recurrent. Conversely, a set A such that $E_A(x) > 0$ for some x in A would then be transient. Fortunately *this classification turns out to be the same as that in* D2. From P4 one has

$$H_A(x) = h_A + \sum_{t \in A} G(x,t) E_A(t), \qquad x \in R.$$

If $E_A(t) = 0$ for all t in A, than $H_A(x) \equiv h_A$. But the constant h_A must have the value one, as $H_A(x) = 1$ when x is in A. Therefore a set A on which E_A vanishes identically is necessarily a recurrent set. Conversely, let us suppose that A is a recurrent set, so that $H_A(x) \equiv 1$. Then P4 shows that

$$h_A = \lim_{n \to \infty} \sum_{t \in R} P_n(x,t) H_A(t) = 1.$$

Hence

$$1 = 1 + \sum_{t \in A} G(x,t) E_A(t), \qquad x \in R.$$

But this equation justifies the conclusion that $E_A(t) \equiv 0$ on A, for if we had $E_A(t_0) > 0$ for some t_0 in A the result would be $0 = G(x,t_0)$ for all x in R, which is impossible. Therefore we have proved

P6 *If A is recurrent $E_A(x) = 0$ on A, but if A is transient, then $E_A(x) > 0$ for some x in A.*

That is not quite the end of the story, however. In the proof of P6 we saw that the constant $h_A = 1$ for a recurrent set, but no information was obtained concerning its possible values for a transient set A. In fact it is only now, in studying h_A, that we shall fully use all the information available. The result of P6 remains correct for the far larger class of transition functions to be studied in section 27. By using the information that we are dealing with random walk (that $P(x,y) = P(0, y - x)$) in a much more essential way than before, we can show that h_A *can assume only the value zero or one.*

First we clarify the probability interpretation of h_A.

P7 *For our random walk (transient and aperiodic) let A_n denote the event that $\mathbf{x}_n \in A$. Then*

$$H_A(x) = \mathbf{P}_x \left[\bigcup_{n=0}^{\infty} A_n \right], \qquad x \in R,$$

$$h_A = h_A(x) = \mathbf{P}_x \left[\bigcap_{n=1}^{\infty} \bigcup_{k=n}^{\infty} A_k \right], \qquad x \in R.$$

Proof: Nothing more is involved than a simple rephrasing of definitions. When $x \in A$, $\mathbf{P}_x[A_0] = 1$, so that $H_A(x) = 1$, and when $x \in R - A$ the statement of P7 concerning $H_A(x)$ is the definition of $H_A(x)$. Since the events $B_n = \bigcup_{k=n}^{\infty} A_k$ form a monotone sequence,

$$\mathbf{P}_x\left[\bigcap_{n=1}^{\infty} \bigcup_{k=n}^{\infty} A_k\right] = \lim_{n \to \infty} \mathbf{P}_x[B_n].$$

Here B_n is the event of a visit to the set A at or after time n, but in a finite time. It follows from the probability interpretation of B_n that

$$\mathbf{P}_x[B_n] = \sum_{y \in R} P_n(x,y) H_A(y),$$

and the definition of $h_A(x)$ in P4 completes the proof of P7.

The stage is now set for the last step—the somewhat delicate argument required to show that *any set A is either visited infinitely often with probability one or with probability zero*.[4] As one might hope, this dichotomy corresponds in a natural way to the classification in D2.

P8 *If A is recurrent, $h_A(x) \equiv h_A = 1$, but if A is transient, then $h_A(x) \equiv h_A = 0$.*

Proof: The first statement has already been verified, in the course of the proof of P6. We also know, from P4, that $h_A(x)$ is a constant, called h_A. Therefore we may assume that A is transient and concentrate on proving that $h_A = 0$. But we shall actually forget about whether or not A is transient and prove an apparently stronger result, namely

(1) $\qquad h_A[1 - H_A(x)] = 0 \ \textit{for all } x \textit{ in } R.$

Equation (1) will finish the proof of P8, for if A is transient, then $1 - H_A(x) > 0$ for some $x \in R - A$, so that $h_A = 0$.

Because $h_A(x)$ is independent of x, one can write (1) in the form

(2) $\qquad h_A(x) = \sum_{y \in A} H_A(x,y) h_A(y).$

---

[4] A much more general theorem was proved by Hewitt and Savage [42]. Let $(\Omega, \mathscr{F}, \mathbf{P})$ be the probability space in definition D3.1 and suppose that $S \in \mathscr{F}$ is an event invariant under all point transformations $T: \Omega \leftrightarrow \Omega$ such that $T(\omega)$ simply *permutes a finite number of the coordinates of* $\omega = (\omega_1, \omega_2, \ldots)$. Then, according to [42], $\mathbf{P}[S]$ is either zero or one. The event S that a given set $A \in R$ is visited infinitely often clearly possesses the property of being invariant under all finite permutations of coordinates. Two far more direct proofs of this so-called *zero-one law* are given in [42], one by Doob, the other by Halmos.

This statement is probabilistically very plausible: $h_A(x)$ gives the probability of infinitely many visits to A, starting at x. This event is decomposed according to the point y of the first visit to A. After this first visit, the random walk, now starting at y, is still required to pay an infinite number of visits to A. The resulting sum of probabilities is precisely the right-hand side in (2).

In apparently similar cases in the past we have often "declared" a theorem proved after such a heuristic outline of what must obviously be done. Nevertheless, the present argument being by far the most complicated example of its kind we have encountered, we insist on presenting a complete proof. The events A_n will be the same as in the proof of P7, as will $B_n = \bigcup_{k=n}^{\infty} A_k$, and C_n will be the event that $\mathbf{T} = n$, where $\mathbf{T}$ is the stopping time $\mathbf{T} = \min [n \mid n \geq 1, \mathbf{x}_n \in A]$.

The right-hand side in (2) is

$$\sum_{y \in A} H_A(x,y) h_A(y) = \lim_{n \to \infty} \sum_{y \in A} H_A(x,y) \sum_{t \in R} P_n(y,t) H_A(t)$$

$$= \lim_{n \to \infty} \mathbf{P}_x \left[\bigcup_{k=1}^{\infty} \left(C_k \cap \bigcup_{j=k+n}^{\infty} A_j \right) \right] = \mathbf{P}_x \left[\bigcup_{k=1}^{\infty} \left(C_k \cap \bigcap_{n=1}^{\infty} B_{k+n} \right) \right].$$

Observing that the sets B_k are decreasing,

$$\bigcap_{n=1}^{\infty} B_{k+n} = \bigcap_{n=1}^{\infty} B_n$$

so that the last probability is

$$\mathbf{P}_x \left[\bigcup_{k=1}^{\infty} \left(C_k \cap \bigcap_{n=1}^{\infty} B_n \right) \right] = \mathbf{P}_x \left[\left(\bigcup_{k=1}^{\infty} C_k \right) \cap \left(\bigcap_{n=1}^{\infty} B_n \right) \right].$$

However, for obvious reasons,

$$\bigcup_{k=1}^{\infty} C_k \supset \bigcap_{n=1}^{\infty} B_n.$$

Hence

$$\sum_{y \in A} H_A(x,y) h_A(y) = \mathbf{P}_x \left[\bigcap_{n=1}^{\infty} B_n \right]$$

and that is $h_A(x)$ according to P7. Thus we have proved equation (2), which implies (1), which in turn proves P8.

The results of P4, P6, and P8 may be combined into

T1 Let A be a nonempty proper subset of the state space R of an aperiodic transient random walk. Then the set A is either recurrent, in which case

$$H_A(x) \equiv 1, \qquad E_A(x) \equiv 0, \qquad h_A = 1,$$

or it is transient, and in that case

$$H_A(x) < 1 \text{ for some } x \in R, \quad E_A(x) > 0 \text{ for some } x \in A, \quad h_A = 0, \text{ and}$$

$$H_A(x) = \sum_{y \in A} G(x,y) E_A(y), \qquad x \in R.$$

Remark: Every finite set A is transient, for every transient random walk. The proof of this remark may be reduced to the case when $|A| = 1$ by the following simple argument. A single point is a transient set, according to the definition of transient random walk in D1.5, if and only if the random walk is transient. Suppose now that a finite set A were recurrent. Then at least one of its points would be visited infinitely often with positive probability. This particular point would then be a recurrent set in the sense of D2, but that would make the random walk recurrent, which it is not.

Now we turn briefly to several other aspects of the potential theory associated with transient random walk. The idea is not to develop fully the analogy with classical Newtonian potential theory, nor even to prove theorems in analysis, just because they are "provable." Rather we want to prepare some of the machinery which is required to carry the probability theory of random walk a little further. Thus we would like to find effective methods for deciding whether a specific subset of the state space is recurrent or transient, and this objective will be attained in the next section for a regrettably small class of random walks.

Despite our limited objective, we need a few of the notions which belong to the standard repertoire of Newtonian potential theory.

D3 If A is a transient subset of R, $E_A(x)$ is called the equilibrium charge of A;

$$H_A(x) = \sum_{y \in A} G(x,y) E_A(y)$$

is called the equilibrium potential (or capacitory potential) of A, and

$$C(A) = \sum_{y \in A} E_A(y) \leq \infty,$$

which may be finite or infinite, is the total equilibrium charge, or capacity of A.

E1 The capacity of a set consisting of a single point is of course $1 - F(0,0) = [G(0,0)]^{-1}$. The capacity of any finite set can be calculated as follows: Let $A = \{x_1, x_2, \ldots, x_n\}$, the x_i being distinct points of R. According to D1, $E_A(x)$ is defined in terms of $\Pi_A(x,y)$. By P2, part (b), $\delta(x,y) - \Pi_A(x,y)$, for x and y in A, is the matrix inverse of the n by n matrix

$$G = G(x_i, x_j), \quad i, j = 1, 2, \ldots, n.$$

It follows from D3 that $C(A)$ is the sum of the elements in the n by n matrix G^{-1}.

When $n = |A| = 2$, $A = \{0, x\}$ one finds that

$$C(A) = \frac{2G(0,0) - G(x,0) - G(0,x)}{G^2(0,0) - G(0,x)G(x,0)}, \text{ if } x \neq 0.$$

One remarkable conclusion is that $C(\{0,x\}) = C(\{0,-x\})$, i.e., the capacity of a two-point set is invariant under reflection about zero. That generalizes nicely to arbitrary finite sets A, and in addition, of course, the capacity of any set *is invariant under translation* of the set.

Another interesting property of capacity concerns the limit of $C(\{0,x\})$ as $|x| \to \infty$. Let us exclude the case of one-dimensional random walk with finite mean, which will be treated in T2 and E2 following it at the end of this section. In all other cases, T24.2 gives

$$\lim_{|x| \to \infty} C(\{0,x\}) = \frac{2}{G(0,0)}.$$

This generalizes to

$$\lim C(\{x_1, x_2, \ldots, x_n\}) = \frac{n}{G(0,0)}$$

as $|x_k| \to \infty$ for $k = 1, 2, \ldots, n$, in such a way that $|x_i - x_j| \to \infty$ for each pair $i \neq j$, $i,j = 1, 2, \ldots, n$. (Under this limiting process the off diagonal elements in the matrix $G = G(x_i, x_j)$, $i,j = 1, 2, \ldots, n$, tend to zero, so that the inverse matrix G^{-1} tends to $[G(0,0)]^{-1}I$.) Furthermore, as we shall see presently (as a corollary of P11), every set of n points has a capacity smaller than $n[G(0,0)]^{-1}$. This phenomenon may be called the *principle of the lightning rod*. A "long, thin" set, such as a lightning conductor has a far larger capacity to absorb charge than "rounder" bodies of the same volume (cardinality). Nor is this situation startling, viewed in terms of its probability interpretation. The capacity is the sum of the escape probabilities, and a thin or sparse set provides a better opportunity for the random walk to "escape" than other sets of the same cardinality.

Now we shall derive some general properties of *potentials*, i.e., of *functions of the form*

$$f(x) = \sum_{y \in R} G(x,y)\psi(y), \quad x \in R$$

with non-negative charge $\psi(y) \geq 0$ on R. They will be useful in getting a new definition of capacity different from that in D3, and also in studying the properties of capacity as a set function. Our results take the form of a *maximum principle*.

P9 (a) *If $f(x) = G\psi(x)$ is a potential, whose charge ψ has support A, i.e. $\psi = 0$ on $R - A$, then*

$$f(x) \leq \sup_{t \in A} f(t).$$

(b) *If $f_1 = G\psi_1$, and $f_2 = G\psi_2$ are two potentials such that $\psi_1(x) = \psi_2(x) = 0$ for $x \in R - A$, and if $f_1(x) \leq f_2(x)$ for all x in A, then*

$$f_1(x) \leq f_2(x) \text{ for all } x \text{ in } R.$$

(c) *Under the same hypotheses as in (b) one can conclude that*

$$\sum_{x \in A} \psi_1(x) \leq \sum_{x \in A} \psi_2(x).$$

Here the right sum, or even both, may be infinite.

Proof: Whenever $x \in R$ and $y \in A$

$$\sum_{t \in A} H_A(x,t) G(t,y) = G(x,y),$$

which was a useful identity in the proof of P24.8. Here too it is of fundamental importance. Applying $H_A(x,t)$ as an operator to the identity $f = G\psi$ one gets

(1) $$\sum_{t \in A} H_A(x,t) f(t) = f(x), \qquad x \in R.$$

Hence

$$f(x) \leq \left[\sup_{t \in A} f(t)\right] H_A(x) \leq \sup_{t \in A} f(t).$$

That was the proof of part (a), and part (b) also follows from (1) which gives

$$f_2(x) - f_1(x) = \sum_{t \in A} H_A(x,t) [f_2(t) - f_1(t)] \geq 0, \qquad x \in R.$$

In connection with part (c) we may assume that

$$\sum_{x \in A} \psi_2(x) = M < \infty$$

for otherwise there is nothing to prove. Since the set A in P9 may not be transient we select a sequence of finite (and hence transient) sets A_n which increase to A as $n \to \infty$. Applying T1 to the reversed random walk we have for each of these sets

$$\sum_{t \in A_n} E^*_{A_n}(t) G(t,y) = H^*_{A_n}(y), \qquad y \in R.$$

Therefore

$$0 \le \sum_{t \in A_n} E^*_{A_n}(t)[f_2(t) - f_1(t)] = \sum_{y \in A_n} [\psi_2(y) - \psi_1(y)]$$
$$+ \sum_{y \in A - A_n} H^*_{A_n}(y)[\psi_2(y) - \psi_1(y)]$$
$$\le \sum_{y \in A_n} [\psi_2(y) - \psi_1(y)] + \sum_{y \in A - A_n} \psi_2(y),$$

so that for each of the sets $A_n \subset A$

$$\sum_{y \in A_n} \psi_1(y) \le \sum_{y \in A} \psi_2(y).$$

Letting $n \to \infty$ one obtains the desired conclusion of part (c) of P9.

P10 *The capacity of a transient set may also be defined as*

$$C(A) = \sup_\psi \sum_{x \in A} \psi(x),$$

the supremum being taken over all ψ such that

$$\psi(x) \ge 0 \text{ on } A, \qquad \sum_{y \in A} G(x,y)\psi(y) \le 1 \text{ on } A.$$

Proof: If $C(A) < \infty$, in the sense of D3, then the total equilibrium charge is finite, and its potential

$$H_A(x) = \sum_{y \in A} G(x,y) E_A(y) = 1 \text{ for } x \in A.$$

By part (c) of P9

$$\sum_{x \in A} \psi(x) \le \sum_{x \in A} E_A(x)$$

for every charge $\psi \ge 0$ on A such that

$$\sum_{y \in A} G(x,y)\psi(y) \le \sum_{y \in A} G(x,y) E_A(y) = 1.$$

Thus not only is $C(A)$ the supremum in P9, but the supremum is actually attained by the charge $\psi(x) = E_A(x)$.

If, on the other hand, $C(A) = +\infty$ in the sense of D3, then E_A is a charge ψ, satisfying the conditions in P10, such that $\sum_{x \in A} \psi(x) = +\infty$. That completes the proof.

P11 *If A_1 and A_2 are any two transient subsets of R, then*

(a) $\qquad C(A_1 \cup A_2) + C(A_1 \cap A_2) \le C(A_1) + C(A_2).$

Here the capacity of the set $A_1 \cap A_2$ is defined as zero if A_1 and A_2 happen to be disjoint.

(b) *If $A_1 \subset A_2$, then $C(A_1) \le C(A_2)$.*

Proof: For any transient subset $A \subset R$ let

$$\mathbf{T}_A = \min\,[k \mid 0 \le k \le \infty, \mathbf{x}_k \in A], \text{ if } 1 \le |A| \le \infty$$

and $\mathbf{T}_A = +\infty$ if A is the empty set. Then one argues probabilistically that, for each x in R,

$$\mathbf{P}_x[\mathbf{T}_{A_1 \cup A_2} < \infty] - \mathbf{P}_x[\mathbf{T}_{A_2} < \infty]$$
$$= \mathbf{P}_x[\mathbf{T}_{A_1 \cup A_2} < \infty, \mathbf{T}_{A_2} = \infty] \le \mathbf{P}_x[\mathbf{T}_{A_1} < \infty, \mathbf{T}_{A_1 \cap A_2} = +\infty]$$
$$= \mathbf{P}_x[\mathbf{T}_{A_1} < \infty] - \mathbf{P}_x[\mathbf{T}_{A_1 \cap A_2} < \infty].$$

Now observe that

$$\mathbf{P}_x[\mathbf{T}_A < \infty] = H_A(x),$$

unless A is empty, but in that case it is correct if we define $H_A(x)$ to be identically zero. Thus we have shown

$$H_{A_1 \cup A_2}(x) + H_{A_1 \cap A_2}(x) \le H_{A_1}(x) + H_{A_2}(x).$$

This is an inequality between two potentials. Their charges are, respectively,

$$E_{A_1 \cup A_2}(x) + E_{A_1 \cap A_2}(x) \text{ and } E_{A_1}(x) + E_{A_2}(x)$$

($E_A(x)$ being defined as zero if A is empty, and also if $x \in R - A$). By part (c) of P9 the total charges satisfy the same inequality as the potentials, if we consider all charges as charges on the same set $A_1 \cup A_2$. This proves part (a), if we consult the definition of capacity in D3.

The proof of part (b) is even simpler, since $T_{A_2} \leq T_{A_1}$ with probability one, so that

$$\mathbf{P}_x[T_{A_2} < \infty] - \mathbf{P}_x[T_{A_1} < \infty] = H_{A_2}(x) - H_{A_1}(x) \geq 0$$

which again gives the desired result by proper use of P9, part (c).

One of the simplest corollaries of P9, mentioned in E1 above, is that

$$C(A) \leq \frac{n}{G(0,0)} \quad \text{when} \quad |A| = n.$$

When $|A| = 1$, the capacity is exactly the reciprocal of $G(0,0)$, and part (a) of P11 gives the inequality for all n by induction on n.

Finally we present a theorem which shows that one is "almost never" interested in the capacity of an infinite transient set. The reason is that, if one excludes one important class of random walks, every infinite transient set has infinite capacity. Perhaps the main interest of T2 below lies in its simplicity, and in the elegant way it uses the renewal theorem, in its strongest form given by T24.2.

T2 *Let A be an infinite subset of the state space R of an aperiodic transient random walk, and suppose that it is transient in the sense of D2. Then there are two possibilities. If the dimension $d = 1$ and the random walk has its first absolute moment m finite, then $C(A) = |\mu|$. In all other cases $C(A) = +\infty$.*

Proof: Suppose first that $d = 1$, $m < \infty$. Since the problem is invariant under reflection about 0 (reversal of the random walk), we may assume that $\mu < 0$. This being the case, we shall show that the set A can contain at most a finite number of points to the left of 0; for suppose that we had an infinite sequence of points y_n in A, with $n = 1, 2, \ldots$,

$$\lim_{n \to \infty} y_n = -\infty.$$

Let A_n be the event that the point y_n is visited at some finite time. Clearly $h_A(x) = h_A$, the probability of an infinite number of visits to A, satisfies

$$h_A \geq \mathbf{P}_x\left[\bigcap_{n=1}^{\infty} \bigcup_{k=n}^{\infty} A_k\right],$$

and by results from elementary measure theory

$$\mathbf{P}_x\left[\bigcap_{n=1}^{\infty} \bigcup_{k=n}^{\infty} A_k\right] = \lim_{n \to \infty} \mathbf{P}_x\left[\bigcup_{k=n}^{\infty} A_k\right] \geq \overline{\lim_{n \to \infty}} \, \mathbf{P}_x[A_n].$$

By a simple calculation

$$\mathbf{P}_x[A_n] = \frac{G(x,y_n)}{G(0,0)},$$

and by the renewal theorem, P24.6 or T24.2,

$$\lim_{n\to\infty} \mathbf{P}_x[A_n] = [G(0,0)]^{-1} \lim_{y\to-\infty} G(x,y) = [G(0,0)|\mu|]^{-1} > 0.$$

Combining the inequalities, one finds that $h_A > 0$, which is impossible as A is a transient set.

At this point we know that A contains infinitely many points to the right of the origin (since $|A| = \infty$, and A contains only finitely many points to the left of the origin). Thus we may select a sequence z_n in A, with

$$\lim_{n\to\infty} z_n = +\infty,$$

and write, using T1,

$$H_A(z_n) = 1 = \sum_{t\in A} G(z_n,t)E_A(t).$$

For every integer $N > 0$,

$$1 \geq \lim_{n\to\infty} \sum_{[t|t\in A,\, t\leq N]} G(z_n,t)E_A(t) = \frac{1}{|\mu|} \sum_{[t|t\in A,\, t\leq N]} E_A(t).$$

(The interchange of limits was justified since A has only finitely many points to the left of N.) Thus we have

$$C(A) = \sum_{t\in A} E_A(t) \leq |\mu|,$$

and now one can repeat the limiting process using the facts that $C(A) < \infty$ and $G(z_n,t) \leq G(0,0)$ to conclude, by the argument of dominated convergence, that $C(A) = |\mu|$.

There remains only the case when $d = 1$ and $m = \infty$, or $d \geq 2$. Again one can base the proof on

$$H_A(z_n) = 1 = \sum_{t\in A} G(z_n,t)E_A(t)$$

where now z_n is any sequence of points in A such that

$$\lim_{n\to\infty} |z_n| = \infty.$$

It follows from T24.2 that $C(A) = +\infty$, for if that were not so, then a dominated convergence argument would give

$$1 = \lim_{n \to \infty} \sum_{t \in A} G(z_n, t) E_A(t) = \sum_{t \in A} \lim_{n \to \infty} G(z_n, t) E_A(t) = 0.$$

The last step used the full strength of T24.2 and completes the proof of T2.

E2 What are the transient sets for a one-dimensional aperiodic random walk with $m < \infty$ and $\mu < 0$? This problem was actually completely solved in the course of the above proof of T2. Obviously every finite set A is transient, and if A is infinite and transient we saw that it must contain only a finite number of points to the left of 0. Further it is easy to see (use the strong law of large numbers!) that every set A of this type is transient. Thus *the transient sets are exactly those which are bounded on the left.*

A natural extension of this question concerns the transient sets when $m = \infty$, $\mu^+ < \infty$, $\mu^- = \infty$ (with μ^+ and μ^- as defined in the proof of P24.6). Then it turns out that there are transient sets which are unbounded on the left. (See problem 3.)

26. RANDOM WALK IN THREE-SPACE WITH MEAN ZERO AND FINITE SECOND MOMENTS

Here we are concerned with somewhat deeper problems than in the last section, in the following sense. By now nearly everything has been said about transient random walk which depends only on the generalized renewal theorem T24.2. When $d \geq 2$, T24.2 tells us only that $G(0,x) \to 0$ as $|x| \to \infty$, and although the number zero is a "rather disarming constant"[4] we have come to appreciate that this result is far from trivial. Nevertheless, it is hardly surprising to find that certain interesting aspects of the behavior of a transient random walk depend on the asymptotic behavior of the Green function $G(x,y)$ in a much more elaborate way. Therefore we shall now confine our attention to a class of random walk for which one can easily characterize the rate at which $G(0,x)$ tends to zero as $|x| \to \infty$. This class is chosen so that it contains simple random walk in three-space. The ordinary Newtonian potential kernel is the reciprocal of the distance between two points, and we shall show that the Green function for simple random walk has the same behavior: $G(x,y)$ behaves like a constant times $|x-y|^{-1}$ as $|x-y| \to \infty$. Indeed we shall show that a large class of three-dimensional random walks share this property.

[4] Cf. Chung [10], p. 188.

P1 *If a random walk satisfies conditions*

(a) $P(x,y)$ *is three dimensional and aperiodic,*

(b) $$\mu = \sum_{x \in R} x P(0,x) = 0,$$

(c) $$m_2 = \sum_{x \in R} |x|^2 P(0,x) < \infty,$$

then its Green function has the asymptotic behavior

$$G(0,x) \sim \frac{1}{2\pi} |Q|^{-1/2} (x \cdot Q^{-1} x)^{-1/2},$$

as $|x| \to \infty$. *Here Q is the covariance matrix of the second moment quadratic form*

$$Q(\theta) = \sum_{x \in R} (x \cdot \theta)^2 P(0,x),$$

Q^{-1} *is its inverse, and $|Q|$ is the determinant of Q. In the particular (isotropic) case when Q is a multiple of the identity, i.e., when*

(d) $$Q(\theta) = |\theta|^2 \sigma^2$$

one has

$$\lim_{|x| \to \infty} |x| G(0,x) = \frac{1}{2\pi\sigma^2}.$$

(That applies to simple random walk in three-space, which has $Q(\theta) = |\theta|^2/3$.)

Proof: The proof depends on rather delicate Fourier analysis as we shall have to utilize the full strength of the Local Central Limit Theorems P7.9 and P7.10. They apply only to strongly aperiodic random walk, a restriction which will be removed at the last step of the proof. Under this restriction

(1) $|x| P_n(0,x) = |x|(2\pi n)^{-\frac{d}{2}} |Q|^{-\frac{1}{2}} e^{-\frac{1}{2n}(x \cdot Q^{-1} x)} + |x| n^{-\frac{d}{2}} E_1(n,x),$

(2) $|x| P_n(0,x) = |x|(2\pi n)^{-\frac{d}{2}} |Q|^{-\frac{1}{2}} e^{-\frac{1}{2n}(x \cdot Q^{-1} x)} + |x|^{-1} n^{1-\frac{d}{2}} E_2(n,x).$

Equation (1) is P7.9, (2) is P7.10, and both of the error terms $E_1(n,x)$ and $E_2(n,x)$ have the property of tending to zero as $n \to \infty$, uniformly in x. Of course we must assume that $x \neq 0$ for (2) to be valid.

Let us begin by showing that the principal terms in (1) and (2) give the correct answer for the symptotic behavior of

$$|x| G(0,x) = \sum_{n=0}^{\infty} |x| P_n(0,x).$$

When $x \neq 0$ and $d = 3$ they yield the series

$$S(x) = (2\pi)^{-3/2}|Q|^{-1/2}|x| \sum_{n=1}^{\infty} n^{-3/2} e^{-\frac{1}{2n}(x \cdot Q^{-1}x)}.$$

As $|x| \to \infty$ it will now be easy to replace this sum by an asymptotically equal Riemann integral. Let us call $(x \cdot Q^{-1}x)^{-1} = \Delta$, observing that $\Delta \to 0$ as $|x| \to \infty$ (the quadratic form Q^{-1} is positive definite!). Then

$$S(x) = \frac{(2\pi)^{-3/2}|Q|^{-1/2}|x|}{(x \cdot Q^{-1}x)^{1/2}} \sum_{n=1}^{\infty} (n\Delta)^{-3/2} e^{-(2n\Delta)^{-1}} \Delta,$$

and as $\Delta \to 0$ the sum on the right tends to the convergent improper Riemann integral

$$\int_0^{\infty} t^{-3/2} e^{-1/2t} dt = \sqrt{2\pi}.$$

Therefore

(3) $\qquad S(x) \sim (2\pi)^{-1}|Q|^{-1/2}(x \cdot Q^{-1}x)^{-1/2}|x|$ as $|x| \to \infty$.

This is the desired result, so that we now only have to explain why the error terms do not contribute. It is here that the need for the two different types of error terms in (1) and (2) becomes manifest. We shall use (1) for the range $[|x|^2] < n < \infty$, and (2) for the range $1 \leq n \leq [|x|^2]$. Here $[y]$ denotes the greatest integer in y. Since the contribution of the principal terms in (3) is positive, we have to show that

(4) $\qquad \lim_{|x| \to \infty} |x|^{-1} \sum_{n=1}^{[|x|^2]} n^{-1/2} |E_2(n,x)|$

$$+ \lim_{|x| \to \infty} |x| \sum_{n=[|x|^2]+1}^{\infty} n^{-3/2} |E_1(n,x)| = 0.$$

The limit of any finite number of terms in the first sum is automatically zero, since the finiteness of the second moments assures us that

$$\lim_{|x| \to \infty} |x| P_n(0,x) = 0$$

for each fixed n. Therefore we choose M so large that $|E_2(n,x)| < \epsilon$ whenever $n \geq M$. Then

$$|x|^{-1} \sum_{n=M}^{[|x|^2]} n^{-1/2} |E_2(n,x)| \leq \epsilon |x|^{-1} \sum_{n=M}^{[|x|^2]} n^{-1/2}$$

$$\leq \epsilon |x|^{-1} \sum_{n=1}^{[|x|^2]} n^{-1/2} \leq \epsilon k_1$$

for some positive k_1 which is independent of ϵ and x. Since ϵ is arbitrary, the first limit in (4) is zero. The second limit is also zero since

$$|x| \sum_{n=[|x|^2]+1}^{\infty} n^{-3/2} |E_1(n,x)| \leq |x| \sup_{n>[|x|^2]} |E_1(n,x)| \sum_{n=[|x|^2]}^{\infty} n^{-3/2}$$
$$\leq k_2 \sup_{n>[|x|^2]} |E_1(n,x)|,$$

which tends to zero as $|x| \to \infty$ (k_2 is a positive constant, independent of x).

The proof of P1 is now complete for strongly aperiodic random walk, but this requirement is superfluous and may be removed by a trick of remarkable simplicity (a version of which was used in the proof of P5.4). If $P(x,y)$ is aperiodic, but not strongly aperiodic, define the transition function

$$P'(x,y) = (1 - \alpha)\delta(x,y) + \alpha P(x,y),$$

with $0 < \alpha < 1$. P' will now be strongly aperiodic, and of course if P satisfies conditions (a), (b), (c) in P1, then so does P'. The Green function G' of P', as a trivial calculation shows, is given by

$$G'(x,y) = \frac{1}{\alpha} G(x,y).$$

In particular,

(5) $$|x|G(0,x) = \alpha |x| G'(0,x), \qquad x \in R.$$

As we have proved P1 for strongly aperiodic random walk, the asymptotic behavior of $|x|G'(0,x)$ is known, and that of $|x|G(0,x)$ may be inferred from it. It remains only to check that we obtain the right constant. The asymptotic behavior of $|x|G'(0,x)$ is governed by the second moments of the P' random walk. These are *continuous functions* of α. Since equation (5) is valid for all α between 0 and 1 we may let $\alpha \to 1$, and a continuity argument concludes the proof of P1.

Remark: Since the matrices Q and Q^{-1} are positive definite (see P7.4 and P7.5) we may conclude from P1 that

$$0 < \varliminf_{|x| \to \infty} |x|G(0,x) \leq \varlimsup_{|x| \to \infty} |x|G(0,x) < \infty,$$

whenever conditions (a), (b), (c) are satisfied. Hence every sufficiently distant point has positive probability of being visited at some time.

It follows that the semigroup R^+ must be all of $\bar{R}$. By the assumption of aperiodicity $R^+ = \bar{R} = R$; hence every point has positive probability of being visited at some time. This implies the almost obvious but exceedingly useful corollary to P1 that *for every aperiodic three-dimensional random walk with mean 0 and finite second moments the Green function $G(x,y)$ is everywhere positive.*

As the first application of P1 we consider the hitting probabilities of a finite set A. For any three-dimensional random walk

$$\lim_{|x| \to \infty} H_A(x,y) = 0.$$

This was P25.3. But one can refine the question and expect a more interesting conclusion by imposing the condition that A is visited in a finite time. Let

$$\mathbf{T}_A = \min [n \mid 0 \leq n \leq \infty, \mathbf{x}_n \in A].$$

Then the conditional probability of hitting A at y, given that $\mathbf{T}_A < \infty$, is

$$\frac{\mathbf{P}_x[\mathbf{x}_{\mathbf{T}_A} = y; \mathbf{T}_A < \infty]}{\mathbf{P}_x[\mathbf{T}_A < \infty]} = \frac{H_A(x,y)}{H_A(x)}.$$

To calculate the limit as $|x| \to \infty$, for a random walk which satisfies the conditions for P1, observe that by P25.1

$$H_A(x,y) = \sum_{t \in A} G(x,t)[\delta(t,y) - \Pi_A(t,y)],$$

so that

$$\lim_{|x| \to \infty} |x| H_A(x,y) = \lim_{|x| \to \infty} |x| G(x,0) \sum_{t \in A} [\delta(t,y) - \Pi_A(t,y)].$$

Let E_A^* be the equilibrium charge of the reversed random walk. Then, using P1 and D25.1

$$\lim_{|x| \to \infty} |x| H_A(x,y) = \frac{1}{2\pi\sigma^2} E_A^*(y).$$

Since $|A| < \infty$,

$$\lim_{|x| \to \infty} |x| H_A(x) = \lim_{|x| \to \infty} |x| \sum_{y \in A} H_A(x,y)$$
$$= \frac{1}{2\pi\sigma^2} \sum_{y \in A} E_A^*(y) = \frac{C^*(A)}{2\pi\sigma^2} = \frac{C(A)}{2\pi\sigma^2},$$

because $C^*(A)$, the capacity of the set A with respect to the reversed random walk, is easily shown to be the same as $C(A)$. Therefore we have proved

P2 *If a random walk satisfies conditions* (a) *through* (d) *of* P1, *then for any finite subset A of R*

(1) $\quad \lim_{|x|\to\infty} |x| H_A(x,y) = \dfrac{1}{2\pi\sigma^2} E_A^*(y), \quad \lim_{|x|\to\infty} |x| H_A(x) = \dfrac{C(A)}{2\pi\sigma^2},$

and

(2) $\quad \lim_{|x|\to\infty} \dfrac{H_A(x,y)}{H_A(x)} = \dfrac{E_A^*(y)}{C(A)}.$

Observe that part (2) is a far weaker statement than (1). It does not depend at all on the full strength of P1, but will hold whenever the Green function has the property

$$\lim_{|x|\to\infty} \frac{G(x,y)}{G(x,0)} = 1, \quad y \in R.$$

E1 It was observed by Ciesielski and Taylor ([14], 1962) that two interesting random variables associated with the Brownian motion process have the same probability distribution: the time until d-dimensional Brownian motion first leaves the unit sphere has the same distribution as the total time spent by $(d+2)$-dimensional Brownian motion in the unit sphere. (This makes sense, Brownian motion in dimension $d + 2 \geq 3$ being transient. Incidentally both Brownian motions must start at the origin, for the theorem to hold.)

We shall take the case $d = 1$, and use P1 to prove an analogous theorem for random walk. For one-dimensional random walk, let $\mathbf{T}_n$ denote the *first time outside the interval* $[-n,n]$. When the random walk has mean 0 and finite variance σ^2, and starts at the origin, it was shown in T23.2 that

(1) $\quad \lim_{n\to\infty} \mathbf{P}_0\left[\mathbf{T}_n > \dfrac{n^2 x}{\sigma^2}\right] = 1 - F(x) = \dfrac{4}{\pi} \sum_{k=0}^{\infty} \dfrac{(-1)^k}{2k+1} e^{-\frac{\pi^2}{8}(2k+1)^2 x}.$

Now consider *three-dimensional random walk, satisfying hypotheses* (a) *through* (d) *of* P1, and define

$$\chi_r(x) = \begin{cases} 1 & \text{if } |x| \leq r \\ 0 & \text{if } |x| > r, \end{cases}$$

$$\mathbf{N}_r = \sum_{k=0}^{\infty} \chi_r(\mathbf{x}_k),$$

so that $\mathbf{N}_r$ is the *total time spent in the sphere* $[x \mid |x| \leq r]$. ($\mathbf{N}_r$ is finite, since the random walk is transient.) We shall show that

(2) $$\lim_{r \to \infty} \mathbf{P}_0 \left[\mathbf{N}_r > \frac{r^2 x}{\sigma^2} \right] = 1 - F(x),$$

with σ^2 as defined in P1, and $F(x)$ the *same distribution function* as in equation (1) above.

The proof of (1) in section 23 was carried out by the method of moments. Referring to the justification given there, it will therefore suffice to prove that for $p \geq 1$

(3) $$\lim_{r \to \infty} \mathbf{E}_0 \left[\left(\frac{\sigma^2}{r^2} \mathbf{N}_r \right)^p \right] = p! \frac{\pi}{2} \left(\frac{8}{\pi^2} \right)^{p+1} \sum_{k=0}^{\infty} \frac{(-1)^k}{(2k+1)^{2p+1}},$$

these limits being the moments of $F(x)$, given in P23.3. The main idea of the proof of (3) will be clear if we consider the cases $p = 1$ and $p = 2$.

Letting $p = 1$,

(4) $$\mathbf{E}_0[\mathbf{N}_r] = \mathbf{E}_0 \left[\sum_{k=0}^{\infty} \chi_r(\mathbf{x}_k) \right] = \mathbf{E}_0 \left[\sum_{[x \mid |x| \leq r]} \sum_{k=0}^{\infty} \delta(x, \mathbf{x}_k) \right]$$
$$= \sum_{[x \mid |x| \leq r]} \sum_{k=0}^{\infty} P_k(0, x) = \sum_{[x \mid |x| \leq r]} G(0, x).$$

Using P1 we obtain, asymptotically as $r \to \infty$,

(5) $$\mathbf{E}_0[\mathbf{N}_r] \sim \frac{1}{2\pi\sigma^2} \sum_{[x \mid |x| \leq r,\, x \neq 0]} |x|^{-1}$$
$$\sim \frac{r^2}{2\pi\sigma^2} \int_{|\alpha| \leq 1} \frac{d\alpha}{|\alpha|} = \frac{2r^2}{\sigma^2} \int_0^1 t\, dt = \frac{r^2}{\sigma^2}.$$

Here the first integral was taken over the unit sphere, $|\alpha| \leq 1$, in E, $d\alpha$ as usual denoting the volume element.

When $p = 2$,

(6) $$\mathbf{E}_0[\mathbf{N}_r^2] = \mathbf{E}_0 \sum_{[x \mid |x| \leq r]} \sum_{[y \mid |y| \leq r]} \sum_{j=0}^{\infty} \sum_{k=0}^{\infty} \delta(x, \mathbf{x}_j)\, \delta(y, \mathbf{x}_k)$$
$$= -\mathbf{E}_0 \sum_{[x \mid |x| \leq r]} \sum_{k=0}^{\infty} \delta(x, \mathbf{x}_k) + 2\mathbf{E}_0 \left[\sum_{[x \mid |x| \leq r]} \sum_{[y \mid |y| \leq r]} \right.$$
$$\left. \sum_{j=0}^{\infty} \delta(x, \mathbf{x}_j) \sum_{k=j}^{\infty} \delta(y, \mathbf{x}_k) \right]$$
$$= -\mathbf{E}_0[\mathbf{N}_r] + 2 \sum_{[x \mid |x| \leq r]} \sum_{[y \mid |y| \leq r]} G(0, x) G(x, y)$$
$$\sim \frac{r^2}{2\pi\sigma^2} \int_{|\alpha| \leq 1} \frac{d\alpha}{|\alpha|} + 2 \frac{r^4}{(2\pi\sigma^2)^2} \int_{|\alpha| \leq 1} \int_{|\beta| \leq 1} \frac{d\alpha\, d\beta}{|\alpha||\beta - \alpha|}.$$

The dominant term in (6) is of course the one of order r^4, and the method by which it was derived generalizes, to give for arbitrary $p \geq 1$,

$$(7) \quad \mathbf{E}_0[(\mathbf{N}_r)^p] \sim p! \sum_{[x_1||x_1|\leq r]} \cdots \sum_{[x_p||x_p|\leq r]} G(0,x_1)G(x_1,x_2)\ldots G(x_{p-1},x_p)$$

$$\sim p!\left(\frac{r^2}{2\pi\sigma^2}\right)^p \int_{|\alpha_1|\leq 1} \cdots \int_{|\alpha_p|\leq 1} \frac{d\alpha_1}{|\alpha_1|}\frac{d\alpha_2}{|\alpha_2-\alpha_1|}\cdots\frac{d\alpha_p}{|\alpha_{p-1}-\alpha_p|}.$$

Let us denote the Newtonian potential kernel by $K(\alpha,\beta)$, for α,β in the unit sphere of three-dimensional Euclidean space E. Then

$$|\alpha-\beta|^{-1} = K(\alpha,\beta) = K_1(\alpha,\beta), \quad K_{p+1}(\alpha,\beta) = \int_{|\gamma|\leq 1} K_p(\alpha,\gamma)K(\gamma,\beta)\,d\gamma,$$

are the iterates of the kernel K, restricted to the unit sphere, and equation (7) may be written in the form

$$(8) \quad \mathbf{E}_0[\mathbf{N}_r^p] \sim p!\left(\frac{r^2}{2\pi\sigma^2}\right)^p \int_{|\alpha|\leq 1} K_p(0,\alpha)\,d\alpha.$$

Comparing equation (8) with equation (3), which we are striving to prove, it is seen that (3) is equivalent to

$$(9) \quad \int_{|\alpha|\leq 1} K_p(0,\alpha)\,d\alpha = \frac{4}{\pi}\left(\frac{16}{\pi}\right)^p \sum_{n=0}^{\infty} \frac{(-1)^n}{(2n+1)^{2p+1}}.$$

Equation (9) looks as though it should follow from an eigenfunction expansion of the kernel K_p. Indeed that is so, but first we shall simplify the problem considerably by observing that $K_p(0,\alpha)$ is *spherically symmetric*, i.e., depends only on $|\alpha|$. Let us therefore study the action of $K(\alpha,\beta) = |\alpha-\beta|^{-1}$ on the class of spherically symmetric functions. The integral

$$Kf(\alpha) = \int_{|\beta|\leq 1} K(\alpha,\beta)f(\beta)\,d\beta = \int_{|\beta|\leq 1} \frac{1}{|\alpha-\beta|}f(\beta)\,d\beta$$

is simply the *Newtonian potential* due to the charge f on the unit ball $|\alpha| \leq 1$. We shall compute it by using two familiar corollaries of Gauss' mean value theorem ([54], p. 83). The potential on or outside a sphere with a spherically symmetric charge is the same as if the total charge were concentrated at the origin. The potential inside a spherical shell on which there is a symmetric charge is constant. Therefore one can decompose the potential $Kf(\alpha)$ into its two contributions from the region $|\beta| \leq |\alpha| = r$

and from $|\beta| > |\alpha| = r$. The second part is conveniently evaluated at $|\alpha| = 0$ (since it is constant!) and in this way one gets

$$
(10) \qquad Kf(\alpha) = \int_{|\beta| \leq 1} \frac{f(\beta)}{|\alpha - \beta|} d\beta
$$

$$
= \frac{4\pi}{r} \int_0^r \tilde{f}(\rho)\rho^2 \, d\rho + 4\pi \int_r^1 \tilde{f}(\rho)\rho \, d\rho,
$$

if $|\alpha| = r$ and $f(\beta) = \tilde{f}(\rho)$ when $|\beta| = \rho$.

Suppose now that f is a spherically symmetric eigenfunction of K, i.e., that $f = \lambda Kf$ on the unit ball, for some $\lambda \neq 0$. It is convenient to write $r\tilde{f}(r) = \phi(r)$, for then it follows from equation (10) that $f = \lambda Kf$ assumes the simple form

$$
(11) \qquad \phi(r) = 4\pi\lambda \int_0^1 \phi(\rho) \min(\rho, r) \, d\rho, \qquad 0 \leq r \leq 1.
$$

In other words we are looking for the eigenfunctions of the integral operator $4\pi \min(\rho, r)$ which has every desirable property (it is symmetric, positive, and completely continuous; hence its eigenfunctions form a complete orthonormal set, and Mercer's theorem applies). Differentiating (11) one gets

$$
(12) \qquad \phi''(r) + 4\pi\lambda\phi(r) = 0, \qquad 0 \leq r \leq 1,
$$

and the boundary conditions are easily seen to be

$$
(13) \qquad \phi(0) = \phi'(1) = 0.
$$

The solutions of (12) and (13) are

$$
(14) \qquad \lambda_n = \frac{\pi}{16}(2n+1)^2, \qquad \phi_n(r) = \sqrt{2} \sin\left[\frac{\pi}{2}(2n+1)r\right],
$$

$$
\int_0^1 \phi_n(r)\phi_m(r) \, dr = \delta(m, n), \qquad n \geq 0, \, m \geq 0.
$$

The proof of (9) is now easily completed. When $p = 1$, (9) is obvious. When $p \geq 2$,

$$
K_p(0, \alpha) = \int_{|\beta| \leq 1} K(\alpha, \beta) K_{p-1}(0, \beta) \, d\beta
$$

so that $K_p(0, \alpha)$ is the result of applying K to a spherically symmetric function. Let $|\alpha| = r$, $|\alpha|K_p(0, \alpha) = \psi_p(r)$. Then

$$
\psi_p(r) = 4\pi \int_0^1 \psi_{p-1}(\rho) \min(\rho, r) \, d\rho, \qquad p \geq 2.
$$

Hence ψ_p is the result of applying the $(p-1)$st iterate of $4\pi \min(\rho,r)$ to the function $\psi_1(r) \equiv 1$. By Mercer's theorem (see P21.3)

$$\psi_p(r) = \sum_{n=0}^{\infty} \int_0^1 \frac{\phi_n(r)\phi_n(\rho)}{(\lambda_n)^{p-1}} \, d\rho.$$

Finally,

$$\int_{|\alpha| \le 1} K_p(0,\alpha) \, d\alpha = 4\pi \int_0^1 \psi_p(r) r \, dr$$

$$= \sum_{n=0}^{\infty} \frac{4\pi}{(\lambda_n)^{p-1}} \int_0^1 \phi_n(r) r \, dr \int_0^1 \phi_n(\rho) \, d\rho,$$

and straightforward computation using (14) shows that (9) is true. That proves the limit theorem in equation (2) for the time $\mathbf{N}_r$ spent in a sphere of radius r by the three-dimensional random walk.

A still deeper application of P1 yields a characterization of the infinite transient sets (all finite sets are of course transient). The following criterion (for simple random walk) was recently discovered (1959) by Itô and McKean [46].

T1 *Suppose a three-dimensional random walk satisfies conditions (a) through (c) in P1. Given an infinite set A, let A_n denote the intersection of A with the spherical shell of points x such that $2^n \le |x| < 2^{n+1}$, and let $C(A_n)$ be the capacity of A_n, defined in D25.3. Then A is recurrent if and only if*

$$\sum_{n=1}^{\infty} \frac{C(A_n)}{2^n} = \infty.$$

Proof: To emphasize the essential features of the proof—which are the results of section 25, the estimate which comes from P1, and the way they are applied to the problem—we shall be a little casual about the measure theoretical aspects. Suffice it to say that the following statements are equivalent.

(1) A is a recurrent set, i.e., it is visited infinitely often;
(2) the random walk $\mathbf{x}_n$, with $\mathbf{x}_0 = 0$ visits infinitely many of the sets A_n with probability one.

Clearly (2) implies (1), but (1) also implies (2) since $h_A(x)$ is independent of x and each A_n is finite and therefore only visited finitely often.

Let E_n be the event that A_n is visited in a finite time. Recurrence means that

$$\mathbf{P}_0 \left[\bigcap_{n=1}^{\infty} \bigcup_{k=n}^{\infty} E_k \right] = 1.$$

It seems reasonable to estimate the probabilities $\mathbf{P}_0[E_n]$ from T25.1 which gives

$$\mathbf{P}_0[E_n] = H_{A_n}(0) = \sum_{x \in A_n} G(0,x) E_{A_n}(x).$$

Letting $c_1, c_2, \ldots$ denote positive constants, one has

$$c_1 |x|^{-1} \leq G(0,x) \leq c_2 |x|^{-1} \text{ for all } x \neq 0$$

according to P1, and so

$$c_3 \sum_{x \in A_n} 2^{-n} E_{A_n}(x) \leq \mathbf{P}_0[E_n] \leq c_4 \sum_{x \in A_n} 2^{-n} E_{A_n}(x),$$

or

$$c_3 2^{-n} C(A_n) \leq \mathbf{P}_0[E_n] \leq c_4 2^{-n} C(A_n).$$

This inequality makes it plain what T1 is about, and what is left to prove. We have to show that

$$\mathbf{P}_0 \left[\bigcap_{n=1}^{\infty} \bigcup_{k=n}^{\infty} E_k \right] = 1$$

if and only if $\sum \mathbf{P}_0[E_n] = \infty$. This conclusion can be obtained from a well-known form of the Borel-Cantelli Lemma.[6]

P3 *Let E_n be any sequence of events (sets) in a probability space (a Borel field of subsets of a space on which there is a countably additive measure μ defined, of total mass one).*

(a) *If* $\quad \sum \mu(E_n) < \infty \quad$ *then* $\quad \mu \left[\bigcap_{n=1}^{\infty} \bigcup_{k=n}^{\infty} E_k \right] = 0.$

(b) *If* $\quad \sum \mu(E_n) = \infty \quad$ *and if for some* $\quad c > 0$

$$\lim_{n \to \infty} \frac{\sum_{k=1}^{n} \sum_{m=1}^{n} \mu(E_k \cap E_m)}{\left[\sum_{k=1}^{n} \mu(E_k) \right]^2} \leq c,$$

then

$$\mu \left[\bigcap_{n=1}^{\infty} \bigcup_{k=n}^{\infty} E_k \right] \geq \frac{1}{c}.$$

---

[6] The familiar form of this lemma consists of part (a) of P3, and of its converse when the events E_n are mutually independent. The more sophisticated converse in P3(b) seems to be rather recent; it has been discovered independently by a number of authors. Cf. [65], [84], and in particular [70], where Lamperti showed how a form of P3 may be used to prove T1 for a large class of transient Markov processes.

We shall prove P3 after completing the proof of T1. Part (a) of P3 states that the series in T1 diverges if the set A is recurrent. From part (b) we want to conclude the converse. So we shall suppose that the series $\sum \mathbf{P}_0[E_n]$ diverges. If we can then verify that the limit inferior in P3 is finite we shall know that the set A is visited infinitely often with *positive probability*. But by T25.1 this probability is $h_A(0) = h_A$ which is either *zero* or *one*. If we prove that it is positive, then it *is* one!

Now let $\mathbf{T}_k$ be the first time the random walk is in the set A_k. Thus $\mathbf{P}_0[E_k] = \mathbf{P}_0[\mathbf{T}_k < \infty]$ and

$$\mathbf{P}_0[E_k \cap E_m] = \mathbf{P}_0[\mathbf{T}_k < \mathbf{T}_m < \infty] + \mathbf{P}_0[\mathbf{T}_m < \mathbf{T}_k < \infty]$$
$$\leq \mathbf{P}_0[\mathbf{T}_k < \infty] \max_{x \in A_k} \mathbf{P}_x[\mathbf{T}_m < \infty] + \mathbf{P}_0[\mathbf{T}_m < \infty] \max_{x \in A_m} \mathbf{P}_x[\mathbf{T}_k < \infty]$$
$$= \mathbf{P}_0[E_k] \max_{x \in A_k} \sum_{y \in A_m} G(x,y) E_{A_m}(y) + \mathbf{P}_0[E_m] \max_{x \in A_m} \sum_{y \in A_k} G(x,y) E_{A_k}(y).$$

Then one uses P1 in the form $G(x,y) \leq c_2|x-y|^{-1}$ to obtain the estimate

$$\mathbf{P}_0[E_k \cap E_m] \leq c_2 \mathbf{P}_0[E_k] \max_{x \in A_k} \sum_{y \in A_m} |x-y|^{-1} E_{A_m}(y)$$
$$+ c_2 \mathbf{P}_0[E_m] \max_{x \in A_m} \sum_{y \in A_k} |x-y|^{-1} E_{A_k}(y).$$

If $k < m - 1$ it follows from the geometry of the sets A_n that

$$\mathbf{P}_0[E_k \cap E_m]$$
$$\leq c_2 \mathbf{P}_0[E_k]|2^{k+1} - 2^m|^{-1} C(A_m) + c_2 \mathbf{P}_0[E_m]|2^{k+1} - 2^m|^{-1} C(A_k).$$

The next step again depends on P1, in the form $C(A_n) \leq c_3^{-1} 2^n \mathbf{P}_0[E_n]$. Therefore,

$$\mathbf{P}_0[E_k \cap E_m] \leq \frac{c_2}{c_3} \mathbf{P}_0[E_k] \mathbf{P}_0[E_m] \left(\frac{2^m + 2^k}{2^m - 2^{k+1}} \right)$$
$$\leq c \mathbf{P}_0[E_k] \mathbf{P}_0[E_m] \text{ for } k < m - 1,$$

where $c = 4c_2/c_3$ is a constant, independent of k and m. Thus one obtains

$$\varliminf_{n \to \infty} \frac{\sum_{k=1}^{n} \sum_{m=1}^{n} \mathbf{P}_0[E_k \cap E_m]}{\sum_{k=1}^{n} \sum_{m=1}^{n} \mathbf{P}_0[E_k] \mathbf{P}_0[E_m]} \leq c < \infty,$$

because the estimate we derived holds also when $m < k - 1$, and the terms with $|k - m| \leq 1$ clearly are "too few" to affect the value of the limit superior.

Now it remains only to prove the Borel-Cantelli Lemma. Part (a) comes from

$$\mu\left[\bigcap_{n=1}^{\infty} \bigcup_{k=n}^{\infty} E_k\right] = \lim_{n \to \infty} \mu\left[\bigcup_{k=n}^{\infty} E_k\right] \leq \lim_{n \to \infty} \sum_{k=n}^{\infty} \mu(E_k) = 0.$$

To prove part (b), let $\phi_k = 1$ if E_k occurs, and zero otherwise. Let

$$\mathbf{N}_n = \sum_{k=1}^{n} \phi_k, \qquad \mathbf{E}(\mathbf{N}_n) = \sum_{k=1}^{n} \mu(E_k).$$

For any positive ϵ, let $B_{n,\epsilon}$ denote the measurable set defined by the statement that

$$\mathbf{N}_k \geq \epsilon \mathbf{E}(\mathbf{N}_k) \text{ for some } k \geq n.$$

The Schwarz inequality reads

$$[\mathbf{E}(\mathbf{fg})]^2 \leq \mathbf{E}(\mathbf{f}^2)\mathbf{E}(\mathbf{g}^2),$$

and choosing the measurable functions

$\mathbf{f}$ = characteristic function of the set $B_{n,\epsilon}$
$\mathbf{g} = \mathbf{fN}_n$,

$$\mu[B_{n,\epsilon}] = \mathbf{E}(\mathbf{f}^2) \geq \frac{[\mathbf{E}(\mathbf{fg})]^2}{\mathbf{E}(\mathbf{g}^2)} \geq \frac{[\mathbf{E}(\mathbf{fg})]^2}{\mathbf{E}(\mathbf{N}_n^2)} \geq \frac{\{\mathbf{E}(\mathbf{N}_n) - \mathbf{E}[\mathbf{N}_n(1 - \mathbf{f})]\}^2}{\mathbf{E}(\mathbf{N}_n^2)}.$$

But

$$\mathbf{E}[\mathbf{N}_n(1 - \mathbf{f})] \leq \epsilon \mathbf{E}(\mathbf{N}_n)$$

so that

$$\mu[B_{n,\epsilon}] \geq (1 - \epsilon)^2 \frac{[\mathbf{E}(\mathbf{N}_n)]^2}{\mathbf{E}(\mathbf{N}_n^2)}$$

for every $n \geq 1$. Because $\mathbf{E}(\mathbf{N}_n) \to \infty$ as $n \to \infty$

$$\mu\left[\bigcap_{n=1}^{\infty} \bigcup_{k=n}^{\infty} E_k\right] \geq \lim_{n \to \infty} \mu[B_{n,\epsilon}].$$

This is true for every positive ϵ and therefore

$$\mu\left[\bigcap_{n=1}^{\infty} \bigcup_{k=n}^{\infty} E_k\right] \geq \varlimsup_{n \to \infty} \frac{[\mathbf{E}(\mathbf{N}_n)]^2}{\mathbf{E}(\mathbf{N}_n^2)} = \varlimsup_{n \to \infty} \frac{\left[\sum_{k=1}^{n} \mu(E_k)\right]^2}{\sum_{k=1}^{n} \sum_{m=1}^{n} \mu(E_k \cap E_m)}.$$

Thus P3 is proved, and hence T1.

Wiener's test (that is the name of T1 because it imitates a test due to N. Wiener for singular points of Newtonian potential)[7] can occasionally be used to determine whether or not a particular set is recurrent. In general the computations are prohibitively difficult.

E2 Let A be a set of the form $A = \bigcup_{n \geq 1} (0,0,a(n))$ where $a(n)$ is a monotone sequence of positive integers. If $a(n) = n$ it is quite obvious on other grounds that A is recurrent. To apply Wiener's test, when A is any monotone sequence of points along the positive x-axis, observe that

$$C(A_n) \leq \frac{|A_n|}{G(0,0)} = k_1 |A_n|,$$

by P25.11, since single points have capacity $[G(0,0)]^{-1}$.

To get an inequality going the other way, observe that, in view of P1,

$$\sum_{y \in A_n} G(x,y) \leq k_2 \sum_{y \in A_n} |x - y|^{-1}$$

for some $k_2 > 0$. Suppose now that we are dealing with a sequence $a(n)$ for which

(1) $$\max_{x \in A_n} \sum_{y \in A_n} |x - y|^{-1} \leq k_3 < \infty,$$

where k_3 is independent of n. Then we have

$$f(x) = \sum_{y \in A_n} G(x,y) \leq k_3, \qquad x \in A_n,$$

which has the following potential theoretical interpretation: the charge which is identically one on A_n gives rise to a potential $f(x)$ whose boundary values are at most k_3. In view of P25.10 the total equilibrium charge of A_n (i.e., the capacity of A_n) is at least as great as k_3^{-1} times the number of points in A_n, so that

$$C(A_n) \geq k_3^{-1} |A_n|.$$

It follows that under condition (1)

$$k_4 \sum_1^\infty \frac{|A_n|}{2^n} \leq \sum_1^\infty \frac{C(A_n)}{2^n} \leq k_5 \sum_1^\infty \frac{|A_n|}{2^n}.$$

Therefore we have shown that *a set which satisfies* (1) *is recurrent if and only if the series* $\sum 2^{-n}|A_n|$ *diverges*.

---

[7] See [16], Vol. 2, p. 306, or [54], p. 331, for Wiener's characterization of singular points; see [47], Ch. VII, for a discussion in terms of Brownian motion, which explains the connection of Wiener's test with P1.

It remains to exhibit examples to show that (1) is ever satisfied. In fact, equation (1) turns out to be a relatively mild regularity condition concerning the sequence $a(n)$. It may be shown to hold when

$$a(n+1) - a(n) \geq \ln n$$

for large enough values of n. This is not enough (as pointed out by Itô and McKean [46]) to decide whether or not the set of primes on the positive x-axis is visited infinitely often (in fact it is, cf. [28] and [77]). However our result permits the conclusion that the set

$$A = \bigcup_{n \geq 1} (0,0,a(n)), \qquad a(n) = [n \cdot \ln n \cdot \ln_2 n \cdots (\ln_k n)^\alpha],$$

where $\ln_{k+1} n = \ln(\ln_k n)$, $k \geq 1$, $\ln_1 n = \ln n$, is visited infinitely often if and only if $\alpha \leq 1$.

Wiener's test (T1) was phrased in such a way that the criterion for whether a particular set is transient or not seemed to depend on the random walk in question. We shall now use the maximum principle of potential theory, in the form given by P25.10, to show that this is not the case.

T2 *An infinite subset A of three-space R is either recurrent (visited infinitely often with probability one) for each aperiodic three-dimensional random walk with mean 0 and finite second moments, or it is transient for each of these random walks.*

Proof: If P and P' are the transition functions of two aperiodic three-dimensional random walks with mean 0 and finite second moments, then their Green functions G and G' are everywhere positive (by the remark following P1). Hence P1 implies that there exists a pair of positive constants c_1 and c_2 such that

$$c_1 G(x,y) \leq G'(x,y) \leq c_2 G(x,y), \qquad x,y \in R.$$

Given a finite subset $B \subset R$, let $C(B)$ and $C'(B)$ denote the capacities induced by the two random walks. The characterization of capacity in P25.10 implies that

$$C'(B) = \sup_{[\psi \mid G'\psi \leq 1 \text{ on } B]} \sum_{x \in B} \psi(x) \leq \sup_{[\psi \mid c_1 G\psi \leq 1 \text{ on } B]} \sum_{x \in B} \psi(x)$$

$$= \sup_{[\varphi \mid G\varphi \leq 1 \text{ on } B]} \sum_{x \in B} c^{-1}\varphi(x) = \frac{1}{c_1} C(B).$$

There is a similar lower estimate so that

$$\frac{1}{c_2} C(B) \leq C'(B) \leq \frac{1}{c_1} C(B).$$

The two constants are independent of the set B. Therefore

$$\frac{1}{c_2} \sum_{n=1}^{\infty} \frac{C(A_n)}{2^n} \leq \sum_{n=1}^{\infty} \frac{C'(A_n)}{2^n} \leq \frac{1}{c_1} \sum_{n=1}^{\infty} \frac{C(A_n)}{2^n},$$

where the sets A_n are the spherical shells in Wiener's test T1. It follows that either both infinite series converge, or both diverge. Hence the set A is either transient for both random walks or recurrent for both, and the proof of T2 is complete.

27. APPLICATIONS TO ANALYSIS

Let S denote an arbitrary countably infinite or finite set. In all our applications S will actually be R or a subset of R, but for the moment it would only obscure matters to be specific about the nature of S. We shall be concerned with a function $Q(x,y)$ from the product of S with itself to the reals with the following properties

D1
$$Q(x,y) \geq 0 \text{ for } x,y \in S$$
$$\sum_{y \in S} Q(x,y) \leq 1 \text{ for } x \in S$$
$$Q_0(x,y) = \delta(x,y), \qquad Q_1(x,y) = Q(x,y),$$
$$Q_{n+1}(x,y) = \sum_{t \in S} Q_n(x,t) Q(t,y), \qquad x,y \in S, \quad n \geq 0.$$
$$\sum_{n=0}^{\infty} Q_n(x,y) = g(x,y) < \infty, \qquad x,y \in S.$$

Such a function Q is called a transient kernel.

We avoid the term transition function for $Q(x,y)$ since it need not be a difference kernel; indeed, no operation of addition or subtraction is defined on S. Nevertheless $Q(x,y)$ and its iterates $Q_n(x,y)$ behave much like transition functions. It is not hard to see that one can associate a stochastic process with $Q(x,y)$. Such a construction lies at the heart of the theory of Markov chains,[8] and random walk may be thought of as a special Markov chain. If one does construct a Markov chain with state space S such that $Q(x,y)$ determines its transition probabilities, then this chain will be transient according to the usual terminology of the theory of Markov chains.

[8] Cf. Chung [9], §I.2.

However it is not our purpose to carry out such an elaborate construction. Rather, it will meet our needs to derive some simple analytic properties of transient kernels defined by D1. Just as was done for transient random walk, we shall define a class of functions which are non-negative and satisfy the equation

$$Qf(x) = \sum_{y \in S} Q(x,y)f(y) = f(x), \qquad x \in S.$$

They will be called *Q-regular*. Two more classes of functions are defined, the *Q-potentials* and also a convenient class of functions which contains all *Q*-potentials. The latter are called *Q-excessive* and form an analogue of the superharmonic functions in classical analysis.

D2 *If Q is a transient kernel on $S \times S$, then a non-negative function $f(x)$ on S is called* (a) *Q-regular,* (b) *Q-excessive,* (c) *a Q-potential, if*

(a) $\qquad\qquad Qf(x) = f(x), \qquad x \in S,$

(b) $\qquad\qquad Qf(x) \leq f(x), \qquad x \in S,$

(c) $\qquad\qquad f(x) = \sum_{y \in S} g(x,y)\psi(y) < \infty, \qquad x \in S$

where $\psi(x) \geq 0$ for $x \in S$. In case (c) *f is called the potential of the charge ψ.*

In what follows we call *Q*-regular functions simply regular, and so forth, whenever there is no risk of confusion. A few elementary observations concerning these function classes are so frequently useful that we assemble them under

P1 (1) *Potentials are excessive.*
(2) *The minimum of two excessive functions is excessive.*
(3) *If $f(x)$ is a potential, then*

$$\lim_{n \to \infty} Q_n f(x) = 0, \qquad x \in S.$$

Proof: Part (1) follows from

$$\sum_{t \in S} Q(x,t)g(t,y) - g(x,y) = -\delta(x,y), \qquad x,y \in S,$$

which shows that $g(x,y)$ is excessive as a function of x, for each fixed y. But if f is a potential, then according to (c) in D2, it is a convex combination of excessive functions. Hence it is excessive, the inequality (b) of D2 being preserved under the process of taking (convergent) linear combinations with non-negative coefficients.

To get (2) note that, if $Pf \le f$, $Pg \le g$, then

$$\min [f(x), g(x)] \ge \min \left[\sum_{y \in S} P(x,y) f(y), \sum_{y \in S} P(x,y) g(y) \right]$$
$$\ge \sum_{y \in S} P(x,y) \min [f(y), g(y)].$$

Finally, suppose that $f(x)$ is a potential. If

$$f(x) = \sum_{y \in S} g(x,y) \psi(y) = g\psi(x),$$

then

$$Q_n f(x) = \sum_{y \in S} \psi(y) \left[\sum_{k=n}^{\infty} Q_k(x,y) \right],$$

which tends to zero for each x as $n \to \infty$.

A fundamental relation between the three classes of functions defined in D2 was discovered by F. Riesz ([86], p. 337). He showed that every excessive function may be written as the sum of a regular function and of a potential. This decomposition is unique, i.e., there is only one such decomposition for each excessive function. It is given explicitly by

T1 *If Q is a transient kernel and f is a Q-excessive function, then*

$$f(x) = h(x) + u(x),$$

where

$$h(x) = \lim_{n \to \infty} \sum_{y \in S} Q_n(x,y) f(y) < \infty, \qquad x \in S,$$

is Q-regular, and where

$$u(x) = \sum_{t \in S} g(x,t)[f(t) - Qf(t)], \qquad x \in S,$$

is a Q-potential. Moreover, this decomposition is unique.

Proof: Let $f(x) - Qf(x) = \psi(x)$. Since f is excessive, ψ is non-negative on S. Furthermore

$$Q_k f(x) - Q_{k+1} f(x) = Q_k \psi(x) \ge 0, \qquad x \in S, \quad k \ge 0.$$

Adding the first $n+1$ of these equations

$$f(x) - Q_{n+1} f(x) = g_n \psi(x),$$

where

$$g_n \psi(x) = \sum_{y \in S} g_n(x,y) \psi(y), \qquad g_n(x,y) = \sum_{k=0}^{n} Q_k(x,y).$$

Now $g_n(x,y)$ increases to $g(x,y)$ as $n \to \infty$, and in view of the inequality

$$g_n\psi(x) \le f(x) < \infty, \qquad x \in S,$$
$$\lim_{n \to \infty} g_n\psi(x) = g\psi(x) < \infty, \qquad x \in S.$$

The function $g\psi(x)$ is the potential $u(x)$ in T1, and the argument which gave the existence of the limit of $g_n\psi(x)$ also implies that

$$\lim_{n \to \infty} Q_{n+1}f(x) < \infty, \qquad x \in S,$$

exists. This is clearly a regular function (as $Q_n f$ and $Q_n(Qf)$ have the same limit) and so we identify it with $h(x)$ in T1. Thus the Riesz decomposition was obtained by letting $n \to \infty$ in $f = Q_{n+1}f + g_n\psi$.

To prove uniqueness, suppose that $f = h' + u'$, h' regular, u' a potential, in addition to $f = u + h$ with u and h as defined in T1. Then

$$v(x) = u'(x) - u(x) = h(x) - h'(x), \qquad x \in S,$$

and since

$$Qh(x) - Qh'(x) = h(x) - h'(x), \qquad x \in S,$$

we have

$$Qv(x) = Q_n v(x) = v(x), \qquad x \in S, \quad n \ge 0.$$

Because $v(x)$ is the difference of two potentials, part (c) of P1 gives

$$\lim_{n \to \infty} Q_n v(x) = 0 = v(x), \qquad x \in S.$$

This implies that $u(x) \equiv u'(x)$ and therefore also that $h(x) \equiv h'(x)$, establishing the uniqueness of the decomposition.

This is all we shall need in the sequel, but it is tempting to pursue the subject a little further. Remarkably enough, the potential theory of an arbitrary transient kernel exhibits many features of the potential theory we have developed for random walk in section 25.

E1 *The maximum principle* (the analogue of P25.9). *Let $u(x) = \sum_{y \in S} g(x,y)\psi(y)$ be a Q-potential, with the property that $\psi(x) = 0$ for all x in $S - A$, where A is an arbitrary subset of S. Then*

$$u(x) \le \sup_{t \in A} u(t), \qquad x \in S.$$

Proof: We define $Q'(x,y) = Q(x,y)$ when x and y are in the set $S - A$. And as usual, let $Q_n'(x,y)$ be the iterates of Q' over $S - A$, and

$$g'(x,y) = \sum_{n=0}^{\infty} Q_n'(x,y), \qquad x,y \in S - A.$$

Since $Q_n'(x,y) \le Q_n(x,y)$ for x,y in $S - A$, we have $g'(x,y) \le g(x,y)$ for x,y in $S - A$, so that $g'(x,y) < \infty$, which means that Q' is a *transient kernel* over the set $S - A$. Next we define

(1) $$H_A(x,y) = \sum_{t \in S-A} g'(x,t) Q(t,y), \qquad x \in S - A, \quad y \in A.$$

Observe that if we think of $Q(x,y)$ as defining a random process on the set S, then $H_A(x,y)$ would have the same probability interpretation as in D10.1. It would be the probability that the first visit to A occurs at the point y, if the process starts at the point x in $S - A$. It suggests the identity

(2) $$g(x,y) = \sum_{t \in A} H_A(x,t) g(t,y), \qquad x \in S - A, \quad y \in A,$$

and the inequality

(3) $$\sum_{t \in A} H_A(x,t) \le 1, \qquad x \in S - A.$$

For random walk (2) and (3) were obtained in P10.1. Assuming, for the moment, that they are valid in this more general context, equation (2) implies

$$u(x) = \sum_{y \in A} g(x,y) \psi(y) = \sum_{t \in A} H_A(x,t) u(t), \qquad x \in S - A,$$

so that

$$u(x) \le \left[\sum_{t \in A} H_A(x,t) \right] \sup_{t \in A} u(t), \qquad x \in S - A.$$

Finally, one uses (3) to complete the proof of the maximum principle.

For the proof of (2) and (3) the use of measure theory would indeed be a convenience, as it makes the intuitive use of stopping times legitimate. But an elementary proof of (2) and (3) based on the definition of $H_A(x,y)$ in (1) is not at all difficult. It requires induction or counting arguments such as those in the similar nonprobabilistic proof of P1.2 in Chapter I, which are cheerfully left as an exercise for the reader.

The simplest special case of the maximum principle deserves special mention. When $\psi(x) = 1$ at the point $x = y$ and 0 elsewhere on S, the maximum principle asserts

(4) $$g(x,y) \le g(y,y) \text{ for all } x \in S.$$

This statement again is obvious—if one admits as obvious the probability interpretation of $g(x,y)$ as the expected number of visits of a process, starting at x, to the point y. For then $g(x,y)$ becomes $g(y,y)$ times the probability (which is at most one) of at least one visit to y in finite time, starting at x.

Now we shall turn to a few applications of probability theory to analysis. Using the powerful Riesz decomposition theorem (T1), one can study the regular functions of a transient kernel. More specifically, it will be seen that there is an intimate connection between the asymptotic behavior of the ratio

$$\frac{g(x_1,y)}{g(x_2,y)} \quad \text{for large } |y|$$

and the various Q-regular functions (if there are any). This connection was first systematically used by R. S. Martin [75], 1941, in his work with harmonic functions in arbitrary domains in the plane. All we shall attempt to do is to present useful special cases of a method for finding and representing Q-regular functions. In recent work in probability theory of Doob, Hunt, and others,[9] this fruitful and, in its full generality, rather deep method is called the construction of the *Martin boundary*.

Our excuse for attempting this foray into abstract potential theory is that probabilistic results from earlier chapters will find, rather unexpectedly, elegant applications in a new setting. For example, in E3 at the end of this section, we shall return to random walk on a half-line. Thus R will be the set of integers, and it will be natural to take for S the set $[0,\infty)$ of non-negative integers. If $P(x,y)$ is the transition function of an aperiodic random walk, and if $Q(x,y)$ is the restriction of $P(x,y)$ to S, then Q *becomes a transient kernel on S*, in the sense of definition D1. If $g(x,y)$ is the Green function of Q in the sense of D1, it is clear that *it is simultaneously the Green function* $g_B(x,y) = g(x,y)$ in the sense of definition D19.3. About this particular Green function we know a good deal. Thus, according to P19.3

$$g(x,y) = \sum_{n=0}^{\min(x,y)} u(x-n)v(y-n), \quad x \geq 0, \quad y \geq 0,$$

and in section 18 of Chapter IV we studied the functions $u(x)$ and $v(x)$ quite thoroughly. In particular the asymptotic properties of $g(x,y)$ should present no formidable difficulties, at least under reasonable auxiliary hypotheses. Using this information about $g(x,y)$ we shall be able to prove that, under certain restrictions, there is *one and only one Q-regular function*. This is of probabilistic interest, the equation $Qf(x) = f(x)$ for $x \in S$ being exactly the Wiener-Hopf equation

$$\sum_{y=0}^{\infty} P(x,y)f(y) = f(x), \quad x \geq 0,$$

[9] Cf. Doob [24], Hunt [45], and Watanabe [100].

of D19.4. In P19.5 this equation was shown to have the solution

$$f(x) = u(0) + u(1) + \cdots + u(x), \qquad x \geq 0.$$

Thus E3 will show that for recurrent aperiodic random walk the Wiener-Hopf equation has a unique non-negative solution, or, if one prefers a different viewpoint, that the function $u(x)$, which plays such a crucial role in fluctuation theory, has a simple potential theoretical characterization.

Similarly, a strong case can be made for another look at aperiodic recurrent random walk in the plane. If $P(x,y)$ is the transition function on R, let $S = R - \{0\}$, $Q(x,y) = P(x,y)$ for x,y in $R - \{0\}$. Then Q is a transient kernel. According to P13.3

$$\sum_{y \in R} P(x,y)a(y) - a(x) = \delta(x,0), \qquad x \in R,$$

where $a(x)$ is the potential kernel in P12.1. Since $a(0) = 0$, the above equation implies

$$\sum_{y \in S} Q(x,y)a(y) = a(x), \qquad x \in S,$$

so that $a(x)$ is a Q-regular function on S. Is it (apart from a multiplicative constant) the only Q-regular function? It will follow, from P3 below, that this is so, and the reason we can prove it is that, according to P11.6 and P12.2, we know that

$$g(x,y) = a(x) + a(-y) - a(x - y),$$

and

$$\lim_{|y| \to \infty} [a(x + y) - a(y)] = 0.$$

The details will be given in E4. The last application, in E5 to analysis proper, will concern the Hausdorff moment problem.

Two innovations of terminology are needed. Given any countable set S, and a real valued function $f(x)$ on S

D3 $\lim_{|x| \to \infty} f(x) = a$, *if, given any $\epsilon > 0$, $|f(x) - a| < \epsilon$ for all but a finite number of points x in S.*

Clearly this is of interest only when S is infinite (otherwise $f(x)$ has every possible real number as a limit), but S will be infinite in all cases of concern to us.

Much more essential is

D4 *If Q is a transient kernel on the set S, and if $h(x)$ is a positive Q-regular function, then*

$$Q^h(x,y) = \frac{Q(x,y)h(y)}{h(x)}, \qquad x,y \in S,$$

and

$$g^h(x,y) = \frac{g(x,y)h(y)}{h(x)}, \qquad x,y \in S.$$

D4 was motivated by

P2 $Q^h(x,y)$ *is a transient kernel on S, and if we call its iterates $Q_n{}^h(x,y)$ and its Green function*

$$g^h(x,y) = \sum_{n=0}^{\infty} Q_n{}^h(x,y), \quad x,y \in S,$$

then this is the same $g^h(x,y)$ as that in D4. *The kernel Q^h has the property that the constant function $e(x) \equiv 1$ is Q^h-regular.*

The proof is omitted, being completely obvious. Now we are ready to derive a set of sufficient conditions for a transient kernel Q to have a unique regular function (up to a multiplicative constant; this will always be understood).

P3 *Let $Q(x,y)$ be a transient kernel on S such that $g(x,y) > 0$ for x,y in S, and suppose that for some point ξ in S*

$$\lim_{|y|\to\infty} \frac{g(x,y)}{g(\xi,y)} = f(x) < \infty$$

exists for all x in S. Suppose further that $f(x)$ is Q-regular.[10] *Then the positive multiples of $f(x)$ are the only Q-regular function.*

Proof: Suppose that $h(x)$ is Q-regular, and not identically zero. We have to show that for some $c > 0$, $h(x) = cf(x)$ on S. First observe that $h(x) > 0$ on S, in view of

$$\sum_{y \in S} Q_n(x,y)h(y) = h(x), \qquad x \in S, \quad n \geq 0.$$

If we had $h(x_0) = 0$ for some $x_0 \in S$, we could choose y_0 such that $h(y_0) > 0$, and conclude that $Q_n(x_0,y_0) = 0$ for all $n \geq 0$. This

---

[10] It is easy to show that this hypothesis may be weakened: it suffices to require that there exists at least one nontrivial Q-regular function. Show that one cannot do better, however, by constructing a transient kernel Q which *has no regular functions*, but such that the limit in P3 nevertheless exists.

contradicts our assumption that $g(x,y) > 0$ for all x, y in S. Therefore every Q-regular function $h(x)$ is strictly positive on all of S.

Using the positive Q-regular function $h(x)$ we now define Q^h and $g^h(x,y)$ according to D3 and P2. The function $e(x) \equiv 1$ on S is Q^h regular, and will now be approximated from below by the sequence

$$v_n(x) = \min [e(x), ng^h(x,\eta)].$$

Here the point η is an arbitrary fixed point of S. It is easy to see that $v_n(x) \leq e(x)$ and $v_n(x) \to e(x)$ as $n \to \infty$.

Now we use T1 to show that each $v_n(x)$ is a Q^h-potential. Being the minimum of two excessive functions, $v_n(x)$ is excessive by P1, part (2). Hence, according to T1

$$v_n(x) = k(x) + u(x),$$

where $k(x)$ is Q^h-regular and $u(x)$ is a Q^h-potential. But

$$k(x) \leq ng^h(x,\eta), \qquad x \in S,$$

which shows, by P1(3), that $k(x) \equiv 0$. Hence we have succeeded in exhibiting a sequence of Q^h-potentials $v_n(x)$ which converge to $e(x)$. All other properties of this sequence will be quite irrelevant from now on.

According to the definition of potentials, there is a sequence of charges $\mu_n(x)$, such that

(1) $$v_n(x) = \sum_{y \in S} g^h(x,y)\mu_n(y).$$

With ξ denoting the point in the hypothesis of P3,

(2) $$v_n(x) = \sum_{y \in S} \frac{g^h(x,y)}{g^h(\xi,y)} \gamma_n(y),$$

where $\gamma_n(y) = g^h(\xi,y)\mu_n(y)$. It follows from (1) and (2) that

(3) $$0 \leq \sum_{y \in S} \gamma_n(y) = v_n(\xi) \leq 1.$$

It is also known that

(4) $$e(x) = 1 = \lim_{n \to \infty} \sum_{y \in S} \frac{g^h(x,y)}{g^h(\xi,y)} \gamma_n(y).$$

Now we choose a subsequence n' of the integers (by the diagonal process) such that

(5) $$\lim_{n' \to \infty} \gamma_{n'}(y) = \gamma(y)$$

APPLICATIONS TO ANALYSIS

exists for each y in S, and observe, using the hypothesis of P3, that

(6) $$\lim_{|y|\to\infty}\frac{g^h(x,y)}{g^h(\xi,y)} = \lim_{|y|\to\infty}\frac{g(x,y)}{g(\xi,y)}\frac{h(\xi)}{h(x)} = f(x)\frac{h(\xi)}{h(x)}.$$

Choosing M so large that

$$\left|\frac{g^h(x,y)}{g^h(\xi,y)} - f(x)\frac{h(\xi)}{h(x)}\right| < \epsilon \quad \text{when} \quad |y| > M,$$

it follows from (4), (5), and (6) that

$$1 = \sum_{|y|\le M}\frac{g^h(x,y)}{g^h(\xi,y)}\gamma(y) + \lim_{n'\to\infty}\frac{f(x)h(\xi)}{h(x)}\sum_{|y|>M}\gamma_{n'}(y) + R(M),$$

where $R(M)$ is an error term whose magnitude does not exceed ϵ. Now $R(M)$ tends to zero as $M \to \infty$, and if we call

$$\lim_{M\to\infty}\lim_{n'\to\infty}\sum_{|y|>M}\gamma_{n'}(y) = \gamma_\infty,$$

then it follows that

(7) $$1 = \sum_{y\in S}\frac{g^h(x,y)}{g^h(\xi,y)}\gamma(y) + \gamma_\infty f(x)\frac{h(\xi)}{h(x)}, \quad x \in S.$$

In (7), the constant γ_∞ has a value between zero and one, and the measure $\gamma(y)$ has total mass at most one. That is all one can conclude from (3). But now we shall show that $\gamma(y)$ is identically zero, by appealing to the Riesz decomposition theorem. In equation (7), the constant on the left is Q^h-regular. On the right the function

(8) $$\sum_{y\in S}\frac{g^h(x,y)}{g^h(\xi,y)}\gamma(y), \quad y \in S,$$

is clearly a Q^h-potential. Finally $f(x)/h(x)$ is Q^h-regular since

$$\sum_{y\in S}\frac{Q^h(x,y)f(y)}{h(y)} = \sum_{y\in S}\frac{Q(x,y)}{h(x)}f(y) = \frac{f(x)}{h(x)}.$$

Using the uniqueness part of T1 one sees that the potential in equation (8) is zero, while

(9) $$1 = \gamma_\infty h(\xi)\frac{f(x)}{h(x)}, \quad x \in S.$$

Hence $h(x)$ is a constant times $f(x)$, which proves P3.

Our first application of P3 concerns the transient random walk treated in the last section.

E2 *Consider aperiodic random walk in three dimensions, satisfying the hypotheses* (a) *through* (c) *of* P26.1. Then P26.1 gives

$$\lim_{|y|\to\infty} \frac{G(x,y)}{G(0,y)} = 1, \qquad x \in R.$$

In order to apply P3 to the problem of finding the P-regular functions, we make the obvious identifications,

$$R = S, \qquad P(x,y) = Q(x,y), \qquad x, y \in S,$$
$$g(x,y) = G(x,y), \qquad \xi = 0.$$

One must also note that

$$f(x) = \lim_{|y|\to\infty} \frac{g(x,y)}{g(0,y)} = 1, \qquad x \in S,$$

is P-regular (the constant function is P-regular for every random walk) and hence Q-regular. Now P3 applies, so that *aperiodic three-dimensional random walk with mean zero, and finite second moments has only the constant regular function*. The hypotheses are of course much too stringent—see problem 6. Nevertheless there is no "easy" proof, even in the special case of simple random walk. (A short but sophisticated proof for simple random walk is outlined in problem 8.)

E3 Theorem: *If $P(x,y)$ is recurrent aperiodic random walk in one dimension, then the Wiener-Hopf equation*

$$\sum_{y=0}^{\infty} P(x,y) f(y) = f(x), \qquad x = 0, 1, 2, \ldots,$$

has a unique non-negative solution. It is

$$f(x) = u(0) + u(1) + \cdots + u(x), \qquad x \geq 0,$$

$u(x)$ *being the function defined in* D18.2.

Proof: The problem is merely one of verifying the hypotheses in P3. According to the discussion of this problem preceding P3, we have

$$(1) \quad g(x,y) = u(x)v(y) + u(x-1)v(y-1) + \cdots + \begin{cases} u(0)v(y-x) & \text{if } y \geq x, \\ u(x-y)v(0) & \text{if } x \geq y. \end{cases}$$

The positivity of $g(x,y)$ on the set S (which consists of the half-line of the non-negative integers) follows from the aperiodicity of the random walk. So we shall try to prove that

$$(2) \qquad \lim_{|y|\to\infty} \frac{g(x,y)}{g(0,y)} = c[u(0) + \cdots + u(x)], \qquad x \geq 0.$$

Here c may be any positive constant, ξ has been taken to be the origin, and $|y| \to \infty$ in this problem clearly means $y \to +\infty$. In view of (1), equation (2) is equivalent to

(3) $\lim\limits_{y \to +\infty} \dfrac{1}{u(0)v(y)} [u(x)v(y) + u(x-1)v(y-1) + \cdots + u(0)v(y-x)]$
$$= c[u(0) + u(1) + \cdots + u(x)], \qquad x \geq 0,$$

and (3) will be valid if we prove

(4) $$\lim_{y \to +\infty} \frac{v(y+1)}{v(y)} = 1.$$

In fact, if (4) holds, then $f(x)$ in P3 will be a constant multiple of $u(0) + \cdots + u(x)$, and this function was shown to be regular in P19.5. Hence it remains only to prove (4).

In P18.8 a much stronger result than (4) was proved under the additional restriction that the random walk has finite variance. It was shown that $v(y)$ tends to a positive constant as $y \to +\infty$. Equation (4), being far weaker, can be proved by a simple probabilistic argument (suggested by H. Kesten) *for any recurrent random walk*. (Note, however, that the proof of (4) we are about to give runs into difficulties if the random walk is transient. See problem 11.)

For $x \geq 0$ and $y \geq 0$, we denote by $J(x,y)$ the probability that the random walk, starting at $\mathbf{x}_0 = x$, will visit y before entering the set $R - S = [x \mid x < 0]$. Also let $J(x,x) = 1$. Then

$$g(0,x) = u(0)v(x) = J(0,x)g(x,x),$$

(5) $$\frac{v(x+1)}{v(x)} = \frac{J(0,x+1)}{J(0,x)} \left[1 + \frac{u(x+1)v(x+1)}{\sum\limits_{n=0}^{x} u(n)v(n)} \right].$$

Now we need a few simple remarks.

(6) $\qquad\qquad 0 \leq u(x)v(x) \leq M$, independent of x,

(7) $$\sum_{n=0}^{\infty} u(n)v(n) = \infty,$$

(8) $J(0,x+1) \geq J(0,x)J(x,x+1), \quad J(0,x) \geq J(0,x+1)J(x+1,x),$
$$x \geq 0,$$

(9) $\qquad\qquad \lim\limits_{x \to +\infty} J(x,x+1) = \lim\limits_{x \to +\infty} J(x+1,x) = 1.$

Equation (6) comes, most directly, from the observation that $g(x,0) = u(x)v(0) = J(x,0)g(0,0)$, so that $u(x)$ is a bounded function. Similarly $v(x)$ is bounded since it plays the role of $u(x)$ in the reversed random walk. Equation (7) is a consequence of recurrence: as $x \to +\infty$, $g(x,x) \to \infty$ since the stopping time (the first time in $R - S$) tends to infinity, loosely speaking. The inequalities in (8) should also be intuitively obvious: the first one says that in going from 0 to $x + 1$, and then from $x + 1$ to x, without meeting $R - S$, the random walk goes from 0 to x without meeting $R - S$. Finally (9) is true for exactly the same reason as (7).

Combining (6), (7), (8), and (9) in the obvious manner, one observes that the right-hand side in (5) tends to one as $x \to +\infty$. But that is the statement in (4) which has been shown to guarantee the truth of E3.

In problem 12 it will be shown that the unique solution of E3 is no longer unique if one admits solutions which may oscillate in sign.

E4 Let $A(x,y) = A(0, y - x)$, $a(x) = A(x,0)$ be the potential kernel of recurrent aperiodic random walk in the plane. Then $a(x)$ *is the only non-negative solution of*

$$\sum_{y \neq 0} P(x,y)a(y) - a(x) = 0, \qquad x \neq 0.$$

To prove this we take $Q(x,y) = P(x,y)$ (P is the transition function of the random walk) for x and y in $S = R - \{0\}$. According to the results of Chapter III,

$$g_{\{0\}}(x,y) = g(x,y) > 0, \qquad x, y \in S,$$

$a(x)$ is a Q-regular function, and for arbitrary ξ in S,

$$\frac{g(x,y)}{g(\xi,y)} = \frac{a(x) + a(-y) - a(x - y)}{a(\xi) + a(-y) - a(\xi - y)}.$$

The limiting process $|y| \to \infty$ of D3 is equivalent to letting $|y| \to \infty$ in the metric (Euclidean distance) of R. And, finally,

$$\lim_{|y| \to \infty} \frac{g(x,y)}{g(\xi,y)} = \frac{a(x)}{a(\xi)}$$

follows from P12.2. Thus E4 has been obtained from P3. The analogous result for one-dimensional random walk is discussed in the next chapter. There the situation is more complicated—but in a very interesting way (see T31.1).

E5 This example[11] serves a purpose different from those in the remainder of the book. Instead of illuminating what has been done, it gives a glimpse of what lies beyond. It is intended to arouse the reader's curiosity rather than to satisfy it, and so, although we prove a few things, no attempt is made to explain what is proved. This should be so because the general potential theory of stochastic processes is a rather recent field, and much more difficult than what we do.

---

[11] This application of boundary theory is due to Hunt (Princeton lectures) and Watanabe [101]. For a generalization, see problems 9 and 10.

27. APPLICATIONS TO ANALYSIS

Let $P(x,y)$ be the transition function of the following simple unsymmetric two-dimensional random walk. If $x = (m,n)$, m and n being the coordinates of the lattice points of R, then

$$P(0,x) = \tfrac{1}{2} \text{ if } x = (0,1) \text{ and if } x = (1,0).$$

Let

(1) $\qquad f_t(x) = 2^{m+n} t^m (1-t)^n, \qquad x = (m,n), \qquad 0 < t < 1.$

Thus $f_t(x)$ is a one parameter family of functions on R. For each value of the parameter t between 0 and 1, $f_t(x)$ happens to be a regular function. This is simple to verify; it suffices to check

$$\sum_{x \in R} P(0,x) f_t(x) = \tfrac{1}{2} \cdot 2t + \tfrac{1}{2} \cdot 2(1-t) = 1 = f(0).$$

Note that the functions in (1) are also regular if $t = 0$ or 1, provided we restrict them to the first quadrant

$$R^+ = [x \mid x = (m,n), m \geq 0, n \geq 0].$$

This is very natural as R^+ is the semigroup associated with the support Σ of $P(0,x)$. From now on we shall confine our attention to those regular functions which are non-negative solutions of

(2) $\qquad f(x) = \sum_{y \in R^+} P(x,y) f(y), \qquad x \in R^+.$

Then we can assert, by linearity, that

(3) $\qquad f(x) = \int_0^1 f_t(x) \, d\mu(t) = 2^{m+n} \int_0^1 t^m (1-t)^n \, d\mu(t), \qquad x \in R^+,$

is a regular function, in the sense of equation (2), for every finite measure μ on the unit interval. (The integral in (3) is the Lebesgue-Stieltjes integral with respect to an arbitrary monotone nondecreasing function $\mu(x)$ on $0 \leq x \leq 1$. It may have mass (discontinuities) at the endpoints.)

Thus the conclusion of P3 cannot apply here, but the method of proof of P3 will apply, and we shall show that every regular function is given by (3), for some measure μ. We identify R^+ with S, $P(x,y)$ with $Q(x,y)$, and $G(x,y)$ with $g(x,y)$. Fortunately the random walk is so simple that one can write down explicitly

(4) $\qquad g(0,x) = 2^{-m-n} \binom{m+n}{m} \qquad \text{for } x = (m,n) \in R^+.$

The next step is the asymptotic formula

(5) $\qquad \lim_{|y| \to \infty} \left[\frac{g(x,y)}{g(0,y)} - 2^{m+n} \left(\frac{r}{r+s}\right)^m \left(\frac{s}{r+s}\right)^n \right] = 0.$

Here $x = (m,n)$ and $y = (r,s)$ are confined to R^+ (the first quadrant) and the proof of (5) depends on Stirling's formula

$$n! \sim n^n e^{-n}\sqrt{2\pi n} \text{ as } n \to \infty.$$

The tedious but straightforward details are omitted.

Just as in the proof of P3 we pick an arbitrary regular function (non-negative solution of (2)) with the intention of showing that it has an integral representation of the type in (3). We must now distinguish between two possibilities.

I: $h(x) = 0$ for all $x = (m,n)$ such that $m > 0$, $n > 0$.

In this case $h(x) = 2^n a$ when $m = 0$, $n \geq 1$, $h(x) = 2^m b$ when $m \geq 1$, $n = 0$, and $h(0) = a + b$; here a and b are non-negative constants. It follows that $h(x)$ is represented by (3) with a measure μ which assigns masses of a and b to the points $t = 0$ and $t = 1$. In this case, then, there is nothing to prove.

II: $h(x) > 0$ for some $x = (m,n)$ such that $m > 0$, $n > 0$.

In this case it follows from a careful scrutiny of equation (2) that $h(x) > 0$ for every $x \in R^+$. (To see this, assume the contrary. The set of zeros of h must then be of the form $[x \mid x = (m,n); m \geq m_0, n \geq n_0]$ where either $m_0 > 1$ or $n_0 > 1$ or both, since we are not in case I. Suppose therefore that $m_0 > 1$. On the vertical half-line defined by $x = (m_0 - 1, n)$, $n \geq n_0$, $h(x)$ must then be of the form $h(x) = 2^n a$, for some constant a. But now, as a simple calculation shows, it is *impossible* to extend $h(x)$ to the half-line $x = (m_0 - 2, n)$ $n \geq n_0$, in such way that h satisfies (2) and is non-negative.)

In case II then, since $h > 0$ on R^+, we can define

$$Q^h(x,y) = \frac{P(x,y)h(y)}{h(x)}, \qquad x, y \in R^+,$$

and

$$g^h(x,y) = \frac{g(x,y)h(y)}{h(x)}, \qquad x, y \in R^+,$$

just as in P3. Nor is there any difficulty in selecting $\eta \in R^+$ (η may depend on x) such that

(6) $\qquad v_n(x) = \min[e(x), ng^h(x,\eta)], \qquad x \in R^+,$

is a sequence of Q^h-potentials converging to $e(x) = 1$. Finally

(7) $\qquad v_n(x) = u^h(x) + \sum_{y \in R^+} \frac{g^h(x,y)}{g^h(0,y)} \gamma_n(y), \qquad x \in R^+,$

where

(8) $$\gamma_n(y) \geq 0, \qquad \sum_{y \in R^+} \gamma_n(y) \leq 1.$$

Of course we choose a subsequence k' such that

(9) $$\lim_{k' \to \infty} \gamma_{k'}(y) = \gamma(y), \qquad y \in R^+,$$

and a number M such that the error in (5) is less than ϵ when $|y| > M$. Then (7) implies

(10) $$\lim_{k' \to \infty} v_{k'}(x) = 1 = \sum_{|y| \leq M} \frac{g^h(x,y)}{g^h(0,y)} \gamma(y)$$
$$+ \lim_{k' \to \infty} \frac{h(0)}{h(x)} 2^{n+m} \sum_{|y| > M} \left(\frac{r}{r+s}\right)^m \left(\frac{s}{r+s}\right)^n \gamma_{k'}(y) + R(M),$$

where $x = (m,n)$, $y = (r,s)$, and $R(M)$ does not exceed $(h(0)/h(x))\epsilon$. If we let $M \to \infty$, then $R(M) \to 0$, and (10) becomes

(11) $$1 = v^h(x) + \lim_{M \to \infty} \lim_{k' \to \infty} \frac{h(0)}{h(x)} 2^{n+m} \sum_{|y| > M} \left(\frac{r}{r+s}\right)^m \left(\frac{s}{r+s}\right)^n \gamma_{k'}(y),$$

where v^h is the Q^h-potential

$$v^h(x) = \sum_{y \in R^+} \frac{g^h(x,y)}{g^h(0,y)} \gamma(y).$$

Observe now that the sums in (11) can be interpreted as integrals over the *unit interval*, with respect to measures whose masses are concentrated on the rational numbers. Equation (11) can be written

$$1 = v^h(x) + \frac{h(0)}{h(x)} 2^{m+n} \lim_{M \to \infty} \lim_{k' \to \infty} \int_0^1 t^m (1-t)^n \, d\nu_{M,k'}(t), \qquad x = (m,n),$$

where $\nu_{M,k'}$ is the sequence of measures in question. Taking again a subsequence of these measures, to assure that they *converge weakly* (see P6.8 and P9.2) to a limiting measure ν one has

(12) $$1 = v^h(x) + 2^{m+n} \frac{h(0)}{h(x)} \int_0^1 t^m (1-t)^n \, d\nu(t).$$

The last term in (12) is Q^h-regular, according to (3). Therefore the uniqueness part of the Riesz decomposition theorem shows that the potential $v^h(x)$ vanishes everywhere. Clearly then

(13) $$h(x) = h(0) 2^{m+n} \int_0^1 t^m (1-t)^n \, d\nu(t), \qquad x \in R^+,$$

so that all regular functions are of this form.

This result yields (in fact is equivalent to) the solution of a problem in classical analysis—the Hausdorff moment problem. *A sequence $c(n)$, $n \geq 0$, is called a (Hausdorff) moment sequence if*

$$c(n) = \int_0^1 t^n \, d\mu(t), \qquad n \geq 0,$$

for some measure μ on $[0,1]$.

To present Hausdorff's [41], 1921, characterization of moment sequences, we define the difference operator Δ acting on functions defined on the non-negative integers by

$$\Delta f(n) = f(n) - f(n+1), \qquad n \geq 0.$$

Its iterates are defined by $\Delta^0 f = f$, $\Delta^1 f = \Delta f$, $\Delta^2 f = \Delta(\Delta f)$, $\Delta^{n+1} f = \Delta(\Delta^n f)$.

Theorem: *The sequence $c(n)$ is a moment sequence if and only if $\Delta^k c(n) \geq 0$ whenever $k \geq 0$ and $n \geq 0$.*

Proof: Given any sequence $c(n)$, $n \geq 0$, we define a function $h(m,0)$ by

$$h(m,0) = 2^m c(m), \qquad m \geq 0.$$

Next define

$$\begin{aligned} h(m,1) &= 2^{m+1} \Delta c(m), & m &\geq 0, \\ h(m,n) &= 2^{m+n} \Delta^n c(m), & m &\geq 0, \quad n \geq 0. \end{aligned}$$

Let

$$f(x) = h(m,n), \qquad x = (m,n) \in R^+.$$

Now one verifies, using the definition of Δ and of its iterates, that $f(x)$ satisfies the equation

$$\sum_{y \in R^+} P(x,y) f(y) = f(x), \qquad x \in R^+.$$

Note that f is not necessarily regular, since it is not necessarily non-negative. But, according to (13), $f(x)$ is regular if and only if

$$f(x) = 2^{m+n} \int_0^1 t^m (1-t)^n \, d\mu(t), \qquad x = (m,n) \in R^+,$$

for some measure μ, and since $f(x)$ is determined by its values on the real axis, this is equivalent to

$$\frac{h(m,0)}{2^m} = c(m) = \int_0^1 t^m \, d\mu(t).$$

That proves the theorem, and moreover shows that the *moment problem is equivalent to the problem of characterizing the regular functions for a particular random walk.*

Problems

1. Prove P13.1 by using T24.1. (*Hint:* Show that recurrent random walk has no excessive functions, i.e., functions $f \geq 0$ such that $f \geq Pf$ on R, other than the regular functions. Then show that, given any constant $c \geq 0$, the function $u(x) = \min[f(x),c]$ is excessive, provided that $f(x)$ is regular.)

2. Prove, using D25.2, that every transient random walk has infinite recurrent sets as well as infinite transient sets.

3. For aperiodic transient random walk in one dimension which has $\mu^+ < \infty$, $\mu^- = \infty$, exhibit a transient set which is unbounded on the left (or prove that there is one!).

4. Given an aperiodic transient random walk with state space R, let $B \subseteq R$ be any additive subgroup of R. Prove that the set B is recurrent (i.e., visited infinitely often with probability one) if and only if

$$\sum_{x \in B} G(0,x) = \infty$$

Hint: The above sum is the total expected time the random walk spends in the set B.

5. Following R. Bucy [S3], prove the following strengthened form of T25.1: A set B is transient if and only if there exists a function $u \geq 0$ on B such that

$$\sum_{y \in B} G(x,y)u(y) = 1, \qquad x \in B.$$

Hint: Suppose there is such a function u and that B is recurrent. Let T_n be the time of the first visit to B after time n. Then, for $x \in B$,

$$1 = E_x \sum_{k=0}^{\infty} u(\mathbf{x}_k) = E_x \sum_{k=0}^{n} u(\mathbf{x}_k) + E_x \sum_{k=T_n}^{\infty} u(\mathbf{x}_k).$$

Now let $n \to \infty$ to arrive at the contradiction that $1 = 2$.

6. Extend P26.1, P26.2, T26.1, T26.2, and E27.2 to dimension $d \geq 3$. See Doney [S6], and also [S21].

7. Simple random walk in three dimensions. Show that the capacity $C(r)$ of the sphere $[x \mid |x| \leq r]$ varies approximately as a linear function of r for large r, and determine

$$\lim_{r \to \infty} C(r)/r.$$

8. For simple random walk of arbitrary dimension, let C denote the class of regular functions $f(x)$ with $f(0) = 1$. Following Itô and McKean [46], order the state space R in an arbitrary manner: $0 = x_0, x_1, x_2, \ldots$.

Let $C_0 = C$,
$$C_1 = \left[\varphi \mid \varphi \in C; \varphi(x_1) = \max_{f \in C} f(x_1)\right],$$
$$C_{n+1} = \left[\varphi \mid \varphi \in C_n; \varphi(x_{n+1}) = \max_{f \in C_n} f(x_{n+1})\right],$$

and show that $\bigcap_{n=0}^{\infty} C_n$ consists of a single regular function $\hat{f}(x)$. Show that this is a *minimal regular function*, i.e., that $f \in C$ and $f \leq \hat{f}$ on R implies that $f = \hat{f}$. But
$$\hat{f}(x) = \sum_{y \in R} P(x,y)\hat{f}(y) = \sum_{z \in R} P(0,z)\hat{f}(z)\frac{\hat{f}(z+x)}{\hat{f}(z)}$$

exhibits $\hat{f}$ as a convex combination of the functions
$$\varphi_z(x) = \frac{\hat{f}(z+x)}{\hat{f}(z)}$$

which are in C. Since $\hat{f}$ is minimal one can conclude that $\varphi_z = \hat{f}$ whenever $P(0,z) > 0$. Consequently $\hat{f}(z+x) = \hat{f}(z)\hat{f}(x)$ for all $x, z \in R$, which shows that $\hat{f}$ is an *exponential*. Prove that $\hat{f}$ is constant and conclude that simple random walk has only constant regular functions.

9. Problem 8 strongly suggests that every regular function for a transient random walk should be a convex combination of minimal regular functions. As in problem 8 one can show that the minimal regular functions are exponentials of the form
$$f(x) = e^{a \cdot x}$$

where a is a vector whose dimension is that of R (see [25]). In view of problem 6, random walk with zero mean is not of interest. Therefore we shall consider strongly aperiodic random walk with mean vector $\mu \neq 0$, assuming also that $P_n(0,x) > 0$ for some n which may depend on x, and that
$$\sum P(0,x)e^{a \cdot x} < \infty$$

for every vector a (these conditions can be relaxed, of course). Now define
$$A = [a \mid a \in E, \sum P(0,x)e^{a \cdot x} = 1].$$

Prove that A is the boundary of a convex body in E. (When the dimension d of E and R is one, then A of course consists of two points, one of which is the origin.) Show further that, given any non-zero vector p, there is a unique point $a = a(p) \in A$ where the outward normal is a positive multiple of p.

10. *Continuation.* If $x \to \infty$ in the direction of the mean vector μ, use the Local Central Limit Theorems P7.9 and P7.10 to show that

$$\lim_{x \to \infty} \frac{G(x,y)}{G(x,0)} = 1, \qquad y \in R.$$

If $x \to \infty$ in the direction p, choose $a = a(p) \in A$ as in problem 9, and observe that the random walk with transition function

$$Q(x,y) = P(x,y)e^{a \cdot (y-x)}$$

has its mean vector along p. Using this idea, P. Ney and the author [S21] proved

$$\lim_{|x| \to \infty} \left| \frac{G(y,x)}{G(0,x)} - e^{a(x) \cdot y} \right| = 0$$

for every $y \in R$. Now one can imitate the method of E27.5 to conclude that the regular functions of a random walk subject to the conditions in problem 9 have the representation

$$f(x) = \int_A e^{a \cdot x} \, d\mu(a), \qquad x \in R,$$

where μ is a Lebesgue-Stieltjes measure on A.

11. It is not known whether the Wiener-Hopf equation

$$\sum_{y=0}^{\infty} P(x,y) f(y) = f(x), \qquad x \geq 0,$$

has a solution for every transient random walk (excluding of course the trivial case with $P(0,x) = 0$ for $x \geq 0$ and $P(0,0) < 1$, when there is obviously no solution). Demonstrate, however, that example E27.3 does not extend to transient random walk, for the following reason. Take a random walk with negative mean, satisfying in addition the hypotheses in P19.6. Show that the unique non-negative solution of the above Wiener-Hopf equation is then of the form

$$f(x) = \sum_{k=0}^{\infty} r^k u(x - k), \qquad x \geq 0,$$

where $r > 1$ is the constant in P19.6. Consequently the function $u(0) + u(1) + \cdots + u(x)$ *cannot* satisfy the Wiener-Hopf equation for the random walks in question.

12. Show that E27.3 solves a natural potential theoretical problem, in the sense that the Wiener-Hopf equation will in general have many solutions (all but one of which must then oscillate in sign). *Hint:* Consider symmetric random walk with $P(0,x) > 0$ for $|x| \leq M$ and 0 for $|x| > M$. What is then the number of linearly independent solutions of the Wiener-Hopf equation?

13. Show that the Green function $G(x,y)$ determines the random walk (i.e., two transient random walks with the same Green function must have the same transition function). (*Hint:* Show that $G(x,y)$ determines $\Pi_A(x,y)$, and then let the set A "grow very large" so that $\Pi_A(x,y) \to P(x,y)$.)

14. One more definition of *capacity*. Given an arbitrary aperiodic transient random walk, and a finite subset $A \subset R$, consider the (random) set $\mathbf{A}_n$ which is "swept out" by the random walk in time n: formally

$$\mathbf{A}_n = [x \mid x \in \mathbf{x}_k + A \text{ for some } k, \ 0 \le k \le n].$$

Finally let $\mathbf{C}_n(A) = |\mathbf{A}_n|$ denote the cardinality of $\mathbf{A}_n$. Prove that the capacity $C(A)$ of the set A is given by

$$\lim_{n \to \infty} \frac{\mathbf{C}_n(A)}{n} = C(A)$$

with probability one.

Hint: Observe that if $A = \{0\}$, then $\mathbf{C}_n(A) = \mathbf{R}_n$, the range of the random walk, as defined in D4.1. According to E4.1 the above limit then exists and is $1 - F = G^{-1}$, the capacity of a single point. Observe further that the proof in E4.1 applies, with obvious modifications, in the present case.

15. For simple random walk in three-space or, more generally, for any random walk satisfying the hypotheses (a) through (d) in P26.1, let A be a finite subset of R and prove that

(1) $\mathbf{P}_x[\mathbf{T}_A < \infty] - \mathbf{P}_x[\mathbf{T}_A \le n] = \mathbf{P}_x[n < \mathbf{T}_A < \infty]$

$$\sim \mathbf{P}_x[\mathbf{T}_A = \infty] \frac{2C(A)}{(2\pi\sigma^2)^{3/2}} n^{-1/2}, \text{ as } n \to \infty,$$

for all $x \in R$. (This is the transient analogue of the results in section 16 concerning the rate of approach of $\mathbf{P}_x[\mathbf{T}_A \le n]$ to the stationary state solution $\mathbf{P}_x[\mathbf{T}_A < \infty]$.) As a corollary of (1) derive

(2) $\sum_{x \in R-A} \mathbf{P}_x[\mathbf{T}_A = n] - C(A) = \sum_{y \in A} \mathbf{P}_y[n < \mathbf{T}_A < \infty]$

$$\sim 2(2\pi\sigma^2)^{-3/2}[C(A)]^2 n^{-1/2}, \text{ as } n \to \infty.$$

Finally, let $\mathbf{L}_A$ be the time of the last visit to the set A, and prove that

(3) $$P_x[n < L_A < \infty] \sim \frac{2C(A)}{(2\pi\sigma^2)^{3/2}} n^{-1/2}, \text{ as } n \to \infty.$$

For help or details see S. Port [24].

Chapter VII

RECURRENT RANDOM WALK

28. THE EXISTENCE OF THE ONE-DIMENSIONAL POTENTIAL KERNEL

The first few sections of this chapter are devoted to the study of aperiodic one-dimensional recurrent random walk.[1] The results obtained in Chapter III for two-dimensional aperiodic recurrent random walk will serve admirably as a model. Indeed, every result in this section, which deals with the existence of the potential kernel $a(x) = A(x,0)$, will be *identically the same* as the corresponding facts in Chapter III. We shall show that the existence of

$$a(x) = \lim_{n \to \infty} \sum_{k=0}^{n} [P_k(0,0) - P_k(x,0)]$$

is a general theorem, valid for *every recurrent random walk*, under the very natural restriction that it be aperiodic. Only in section 29 shall we encounter differences between one and two dimensions. These differences become apparent when one investigates those aspects of the theory which depend on the asymptotic behavior of the potential kernel $a(x)$ for large $|x|$. The result of P12.2, that

$$\lim_{|x| \to \infty} [a(x + y) - a(x)] = 0, \qquad y \in R,$$

will be shown to be *false* in section 29 *for aperiodic one-dimensional recurrent random walk with finite variance*.

It will be convenient, once and for all, to classify as

[1] For a condensed but fairly complete version of sections 28 through 31, see [94]. Related studies by Hoeffding [43], and Kemeny and Snell [55] contain partial results. A potential theory for recurrent Markov chains in general has been developed by Kemeny and Snell [56], and by Orey [80]. It contains as yet no satisfactory criteria for the existence of a potential kernel corresponding to $A(x,y)$ for random walk.

D1 *Type* I: *Aperiodic recurrent random walk in one dimension, with* $\sigma^2 = \sum x^2 P(0,x) = +\infty$;
Type II: *Aperiodic recurrent random walk in one dimension, with* $\sigma^2 = \sum x^2 P(0,x) < \infty$.

It will become apparent that, although to a certain extent type I and type II exhibit the same behavior, the difference between these two cases is sufficiently sharp to require different methods of proof, even when the final results are the same! Thus it will be expedient to treat the two types separately after a certain point. Since frequent references back to Chapter III will serve to shorten the presentation considerably, the reader is asked to forgive the burden they impose on him: that of verifying the truth of remarks of the type "the proof which gave A in Chapter III is available, without any modification whatever, to give B in the present context."

In particular P11.1, P11.2, and P11.3 of Chapter III are available, and without further comment we record P11.3 as

P1 $\quad \sum_{t \in R} P_{n+1}(x,t) H_B(t,y) = H_B(x,y) - \sum_{t \in B} A_n(x,t)[\Pi_B(t,y) - \delta(t,y)],$

for $B \subset R$ *with* $1 \le |B| < \infty$, $x \in R$, $y \in B$. *Here*

$$0 \le a_n(x - t) = A_n(x,t) = \sum_{k=0}^{n} [P_k(0,0) - P_k(x,t)].$$

It will be convenient to specialize P1 to the set $C = \{0,c\}$, $c \ne 0$. A simple calculation, using the fact (see P11.2) that

$$\Pi_C(0,c) + \Pi_C(0,0) = 1,$$

gives

P2 $\quad \sum_{t \in R} P_{n+1}(x,t) H_C(t,0)$
$\qquad = H_C(x,0) + [a_n(x) - a_n(x - c)]\Pi_C(0,c), \qquad x \in R.$

Upon setting $x = c$ in P2, and noting that in view of the recurrence and aperiodicity of the random walk $\Pi_C(0,c) > 0$, one gets

P3 $\qquad 0 \le a_n(c) \le [\Pi_C(0,c)]^{-1} < \infty, \qquad c \in R.$

This bound could of course have been obtained in Chapter III, but there it was not needed, whereas we shall find it indispensable here.

Finally we shall need a lemma from Fourier analysis, which will be used both in the case of type I and type II random walk, albeit in a somewhat different fashion. It concerns the symmetrized approximations $a_n(x) + a_n(-x)$ to the potential kernel, the convergence of which is much easier to establish than that of $a_n(x)$ alone.

P4 *For every one-dimensional aperiodic recurrent random walk,*

(a) $\lim_{n \to \infty} [a_n(x) + a_n(-x)] < \infty$ *exists for each x in R. Furthermore*

(b) $\lim_{x \to +\infty} \dfrac{1}{x} \lim_{n \to \infty} [a_n(x) + a_n(-x)] = \dfrac{2}{\sigma^2} (= 0 \text{ if } \sigma^2 = +\infty),$

where

$$\sigma^2 = \sum_{x=-\infty}^{\infty} x^2 P(0,x).$$

Proof: As usual (see D6.2), letting

$$\phi(\theta) = \mathbf{E}_0[e^{i\theta x_1}] = \sum_{x \in R} P(0,x) e^{ix\theta},$$

one finds

(1) $\quad a_n(x) + a_n(-x) = \sum_{k=0}^{n} \dfrac{1}{\pi} \int_{-\pi}^{\pi} [1 - \cos x\theta] \phi^k(\theta) \, d\theta$

$\qquad\qquad\qquad\qquad = \dfrac{1}{\pi} \int_{-\pi}^{\pi} \dfrac{1 - \cos x\theta}{1 - \phi(\theta)} [1 - \phi^{n+1}(\theta)] \, d\theta.$

Before letting $n \to \infty$ we must show that for each x in R

$$[1 - \cos x\theta][1 - \phi(\theta)]^{-1}$$

is integrable on the interval $[-\pi, \pi]$. We have, for some $A < \infty$

$$\left| \dfrac{1 - \cos x\theta}{1 - \phi(\theta)} \right| \le \dfrac{|1 - \cos x\theta|}{\operatorname{Re}[1 - \phi(\theta)]} \le A \left| \dfrac{1 - \cos x\theta}{\theta^2} \right|,$$

the last inequality being the result of P7.5. Hence

$$[1 - \cos x\theta][1 - \phi(\theta)]^{-1}$$

is integrable on $[-\pi, \pi]$. Thus it is possible to let $n \to \infty$ in equation (1) to obtain

(2) $\quad \lim_{n \to \infty} [a_n(x) + a_n(-x)] = \dfrac{1}{\pi} \int_{-\pi}^{\pi} \dfrac{1 - \cos x\theta}{1 - \phi(\theta)} \, d\theta < \infty, \qquad x \in R.$

Equation (2) proved part (a) of P4. To prove part (b) decompose

$$\text{(3)} \qquad \frac{1}{x\pi} \int_{-\pi}^{\pi} \frac{1 - \cos x\theta}{1 - \phi(\theta)} d\theta = f_\delta(x,\theta) + g_\delta(x,\theta),$$

with $\delta > 0$, $x \neq 0$,

$$f_\delta(x,\theta) = \frac{1}{x\pi} \int_{-\delta}^{\delta} \frac{1 - \cos x\theta}{1 - \phi(\theta)} d\theta,$$

$$g_\delta(x,\theta) = \frac{1}{\pi x} \int_{\delta < |\theta| \leq \pi} \frac{1 - \cos x\theta}{1 - \phi(\theta)} d\theta.$$

Clearly

$$\text{(4)} \qquad \lim_{x \to \infty} g_\delta(x,\theta) = 0 \text{ for every } \delta > 0.$$

Specializing to the case when $\sigma^2 = \infty$, we choose $\delta > 0$ so small that

$$\left| \frac{\theta^2}{1 - \phi(\theta)} \right| < \epsilon \quad \text{for} \quad |\theta| < \delta.$$

This can be done for arbitrary $\epsilon > 0$, since

$$|1 - \phi(\theta)| \geq \text{Re}\,[1 - \phi(\theta)] = 2 \sum_{x \in R} P(0,x) \sin^2\left(\frac{x\theta}{2}\right)$$

$$\geq \frac{2\theta^2}{\pi^2} \sum_{[x \,|\, |x\theta| \leq \pi]} x^2 P(0,x).$$

It follows that

$$\text{(5)} \qquad \varlimsup_{x \to +\infty} |f_\delta(x,\theta)| \leq \frac{\epsilon}{\pi} \lim_{x \to +\infty} \frac{1}{x} \int_{-\delta}^{\delta} \frac{1 - \cos x\theta}{\theta^2} d\theta$$

$$= \frac{\epsilon}{\pi} \int_{-\infty}^{\infty} \frac{1 - \cos t}{t^2} dt = \epsilon.$$

Thus part (b) of P4 holds when $\sigma^2 = \infty$, as one may see by substituting (4) and (5) into (3).

When $\sigma^2 < \infty$ one chooses $\delta > 0$ in equation (3) in such a manner that

$$\text{(6)} \qquad \left| \frac{\theta^2}{1 - \phi(\theta)} - \frac{2}{\sigma^2} \right| < \epsilon \quad \text{for} \quad |\theta| \leq \delta.$$

P6.7 allows us to so choose δ, given any $\epsilon > 0$, since the random walk has mean zero (being recurrent). Using (6) to estimate $f_\delta(x,\theta)$,

$$\varlimsup_{x \to +\infty} f_\delta(x,\theta) \leq \lim_{x \to +\infty} \frac{2}{\sigma^2 \pi x} \int_{-\delta}^{\delta} \frac{1 - \cos x\theta}{\theta^2} d\theta$$

$$+ \varlimsup_{x \to +\infty} \frac{1}{\pi x} \int_{-\delta}^{\delta} [1 - \cos x\theta] \left| \frac{2}{\sigma^2 \theta^2} - \frac{1}{1 - \phi(\theta)} \right| d\theta$$

$$\leq \frac{2}{\sigma^2} \cdot \frac{1}{\pi} \int_{-\infty}^{\infty} \frac{1 - \cos t}{t^2} dt + \frac{\epsilon}{\pi} \varlimsup_{x \to +\infty} \int_{-\delta}^{\delta} \frac{1 - \cos x\theta}{\theta^2 x} d\theta$$

$$= \frac{2}{\sigma^2} + \epsilon.$$

The same argument gives the underestimate

$$\varliminf_{x \to +\infty} f_\delta(x,\theta) \geq \frac{2}{\sigma^2} - \epsilon$$

and since ϵ is arbitrary, the proof of P4 is complete.

Now we are ready to begin to prove that $a_n(x)$ converges for random walk of type I. In P2 we shall specialize to the set $C = \{0,1\}$, and introduce the notation

(1) $$\Pi_C(0,1) = \Pi, \qquad H_C(x,0) = H(x),$$

$$\sum_{t \in R} P_{n+1}(x,t) H_C(t,0) = \sum_{t \in R} P_{n+1}(x,t) H(t) = f_n(x), \qquad x \in R.$$

Then P2, with this particular set C becomes

(2) $$f_n(x) = H(x) + [a_n(x) - a_n(x-1)]\Pi.$$

We shall investigate what can be done with the aid of a subsequence n' of the positive integers such that

$$\lim_{n' \to \infty} a_{n'}(x) = \alpha'(x)$$

exists for all x in R. Such a subsequence can clearly be constructed, by the diagonal process, since we have shown in P3 above that the sequence $a_n(x)$ is bounded.

If we work with such a sequence n', then (2) becomes

(3) $$\lim_{n' \to \infty} f_{n'}(x) = H(x) + [\alpha'(x) - \alpha'(x-1)]\Pi, \qquad x \in R.$$

It is clear from the definition of $f_n(x)$ in (1) that

(4) $$f_{n'+1}(x) = \sum_{y \in R} P(x,y) f_{n'}(y), \qquad |f_n(x)| \leq 1, \qquad x \in R.$$

If we call
$$\lim_{n' \to \infty} f_{n'}(x) = f'(x),$$
it is seen from (4) that

(5) $\qquad \lim_{n' \to \infty} f_{n'+1}(x) = \sum_{y \in R} P(x,y) f'(y), \qquad x \in R.$

On the other hand, (2) gives

$$f_{n+1}(x) - f_n(x) = [a_{n+1}(x) - a_n(x)]\Pi - [a_{n+1}(x-1) - a_n(x-1)]\Pi$$
$$= [P_{n+1}(x-1,0) - P_{n+1}(x,0)]\Pi,$$

and the last term tends to zero as $n \to \infty$ (by P7.6).

Therefore
$$\lim_{n' \to \infty} f_{n'+1}(x) = \lim_{n' \to \infty} f_{n'}(x) = f'(x),$$
so that (5) becomes

(6) $\qquad f'(x) = \sum_{y \in R} P(x,y) f'(y), \qquad x \in R.$

Thus $f'(x)$ is a bounded regular function. By either P13.1 or T24.1 it is then a constant, which we shall denote f'.

The next step is to sum equation (3) over $x = 1, 2, \ldots, r$, where r is an arbitrary positive integer. The result is

(7) $\qquad rf' = \sum_{x=1}^{r} H(x) + \Pi \alpha'(r), \qquad r > 0.$

Similarly, summing (3) over $x = 0, -1, -2, \ldots, -r+1$,

(8) $\qquad rf' = \sum_{x=-r+1}^{0} H(x) - \Pi \alpha'(-r), \qquad r > 0.$

At this point let us *assume* that $\lim_{n \to \infty} a_n(x)$ *does not exist* for some type I random walk, at some point x_0 in its state space R. If this is the case, then the random walk must have the following property: there exist two subsequences n' and n'' of the positive integers such that the limits
$$\lim_{n' \to \infty} a_{n'}(x) = \alpha'(x), \qquad \lim_{n'' \to \infty} a_{n''}(x) = \alpha''(x)$$
both exist for every x in R, but such that

(9) $\qquad \alpha'(x_0) \neq \alpha''(x_0)$

at least at one point $x_0 \neq 0$ in R. (The origin is excluded since $a_n(0) = 0$ for each n.) In order to arrive at a contradiction, we shall apply the same argument which led to equations (7) and (8) to the sequence n'', and call

$$\lim_{n'' \to \infty} f_{n''}(x) = f''.$$

Now we subtract the n''-version of equation (7) from equation (7). This gives

(10) $\qquad r(f' - f'') = \Pi[\alpha'(r) - \alpha''(r)], \qquad r > 0.$

Similarly, equation (8) gives

(11) $\qquad r(f' - f'') = -\Pi[\alpha'(-r) - \alpha''(-r)], \qquad r > 0.$

Since $\Pi \neq 0$, equations (9), (10), and (11) imply that $f' \neq f''$; this may be seen by setting $r = x_0$ in (10) if $x_0 > 0$; otherwise by setting $r = -x_0$ in (11). Suppose now, as one may without loss of generality, that $f' > f''$. Then equation (10) gives

$$\alpha'(r) \geq \frac{r}{\Pi}(f' - f''), \qquad r > 0,$$

and therefore also

(12) $\qquad \dfrac{\alpha'(r) + \alpha'(-r)}{r} \geq \dfrac{1}{\Pi}(f' - f'') > 0, \qquad r > 0.$

But

$$\alpha'(r) + \alpha'(-r) = \lim_{n \to \infty} [a_n(r) + a_n(-r)]$$

which was shown to exist in P4, part (a). Therefore equation (12) implies

(13) $\qquad \lim_{r \to +\infty} \dfrac{1}{r} \lim_{n \to \infty} [a_n(r) + a_n(-r)] > 0,$

which contradicts P4. Thus we have proved

P5 *For type* I *random walk*

$$\lim_{n \to \infty} a_n(x) = a(x) < \infty \text{ exists for each } x \in R.$$

The above proof of P5 does not work for random walk of type II. Although everything runs smoothly until equation (13) above is obtained, this equation does not serve any useful purpose; it simply does not contradict P4, when the variance σ^2 is finite. Therefore we must travel by a different route. Although it will lead to the desired result, it should be noted that it uses arguments which fail for type I random walk. Specifically, we shall need to use T18.1 which asserts that the ladder random variable **Z** of a random walk of type II has finite expected value.

Our first goal is

P6 *If $H_B(x,y)$ is the hitting probability measure of a finite, nonempty subset $B \subset R$ for a random walk of type* II, *then both limits*

(a) $$\lim_{x \to -\infty} H_B(x,y) = H_B(-\infty, y)$$

(b) $$\lim_{x \to +\infty} H_B(x,y) = H_B(+\infty, y)$$

exist for every y in B.

Proof: The foundations for this theorem were carefully laid in P24.7 in the last chapter. As in D24.1, let A denote the half-line $A = [x \mid x > 0]$. Without loss of generality we may assume that the set B of P6 is a subset of A. As a simple probability argument shows, the hitting probabilities of A and of B are then related, for $x \leq 0$, by

(1) $$H_B(x,y) = \sum_{t=1}^{\infty} H_A(x,t) H_B(t,y), \qquad x \leq 0, \quad y \in B.$$

The formal proof may safely be omitted; it is based on the stopping time $\mathbf{T} = \min [n \mid n \geq 1, \mathbf{x}_n \in A]$ in the obvious way.

Now we recall from T18.1 that type II random walk has the property that

$$E_0[\mathbf{Z}] < \infty, \text{ where } \mathbf{Z} = \mathbf{x}_\mathbf{T}.$$

Thus we are in the classification (3) or equivalently (iii) in P24.7, which permits the conclusion

(2) $$\lim_{x \to -\infty} H_A(x,y) = \gamma_A(y), \qquad y \in A; \quad \sum_{y=1}^{\infty} \gamma_A(y) = 1.$$

Applying (2) to (1) one immediately has the existence of

$$\lim_{x \to -\infty} H_B(x,y) = \sum_{t=1}^{\infty} \gamma_A(t) H_B(t,y), \qquad y \in B.$$

This proves part (a) of P6, and to obtain part (b) it suffices to observe that if a random walk of type II is reversed, it remains a random walk of type II.

As a simple consequence of P6 we record

P7 *For type* II *random walk, with* $B \subset R$, $1 \le |B| < \infty$,

$$\lim_{n \to \infty} \sum_{t \in R} P_n(x,t)H_B(t,y) = \tfrac{1}{2}[H_B(-\infty,y) + H_B(+\infty,y)], \qquad y \in R.$$

Proof: We need

(1) $$\lim_{n \to \infty} P_n(x,y) = 0, \qquad x,y \in R,$$

which is P7.6, and

(2) $$\lim_{n \to \infty} \sum_{t=a}^{\infty} P_n(x,t) = 1/2$$

for every $x \in R$ and every a in R. This is a very weak form of the rarely used one-dimensional Central Limit Theorem (P6.8). Combined with P6, equations (1) and (2) evidently yield P7 by a simple truncation argument: Choosing $N > 0$ large enough so that

$$|H_B(t,y) - H_B(-\infty,y)| < \epsilon \text{ for } t \le -N,$$
$$|H_B(t,y) - H_B(+\infty,y)| < \epsilon \text{ for } t \ge N,$$

$$\overline{\lim_{n \to \infty}} \sum_{t \in R} P_n(x,t)H_B(t,y) \le \lim_{n \to \infty} \sum_{t=-N+1}^{N-1} P_n(x,t)$$
$$+ [H_B(-\infty,y) + \epsilon] \overline{\lim_{n \to \infty}} \sum_{t=-\infty}^{-N} P_n(x,t)$$
$$+ [H_B(+\infty,y) + \epsilon] \overline{\lim_{n \to \infty}} \sum_{t=N}^{\infty} P_n(x,t)$$
$$= \tfrac{1}{2}[H_B(-\infty,y) + H_B(+\infty,y)] + \epsilon.$$

An entirely similar under-estimate completes the proof of P7.

Finally we apply P7 to P2, where we make the choice $x = c$. Thus

$$\sum_{t \in R} P_{n+1}(c,t)H_C(t,0) = H_C(c,0) + a_n(c)\Pi_C(0,c).$$

Letting $n \to \infty$, the limit on the left exists by P7, so that also $a_n(c)$ must have a limit. But $C = \{0,c\}$ is an arbitrary set insofar as $c \ne 0$ is arbitrary. Since $a_n(0) = 0$ for each n, we have proved

P8 *For type* II *random walk,*

$$\lim_{n \to \infty} a_n(x) \text{ exists for every } x \text{ in } R.$$

One may summarize P5 and P8 and even P12.1 of Chapter III in the single statement that the potential kernel exists for every aperiodic recurrent random walk, regardless of dimension (remember, however, that $d = 1$ or 2 in view of T8.1!). We might as well also include in this summary the equation

$$\sum_{y \in R} P(x,y)a(y) - a(x) = \delta(x,0), \qquad x \in R,$$

which will later, in section 31, be used to characterize the potential kernel. Observe that the proof of this equation, given in P13.3, *is available, without any modification whatever*, for the one-dimensional case, now that we have P5 and P8.

Therefore we claim to have proved the following theorem.

T1 *If $P(x,y)$ is the transition function of any aperiodic recurrent random walk (either of dimension one or two), then the limit*

(a) $\qquad A(x,y) = a(x - y) = \lim_{n \to \infty} \sum_{k=0}^{n} [P_k(0,0) - P_k(x,y)] < \infty$

exists for all pairs x and y in R. Moreover $A(x,y)$ satisfies the equation

(b) $\qquad \sum_{t \in R} P(x,t)A(t,y) - A(x,y) = \delta(x,y), \qquad x,y \in R.$

29. THE ASYMPTOTIC BEHAVIOR OF THE POTENTIAL KERNEL

The results of this section are illustrated by the following examples which give an explicit formula for $a(x) = A(x,0)$ in two special cases.

E1 For *simple random walk* with $P(0,1) = P(0,-1) = 1/2$ one has, as shown in the proof of P6.6,

$$a(x) = \frac{1}{2\pi} \int_{-\pi}^{\pi} \frac{1 - \cos x\theta}{1 - \cos \theta} \, d\theta = |x|, \qquad x \in R.$$

This result will be typical of type II random walk (recurrent, aperiodic, $d = 1$, mean 0 and finite variance) in the sense that for every such random walk

$$a(x) \sim c|x| \text{ as } |x| \to \infty,$$

where c is a positive constant. The value of c will be shown to be the reciprocal of σ^2, as in fact it must be according to P28.4. We shall prove an even stronger result, in P1, concerning the asymptotic behavior of the differences $a(x + 1) - a(x)$.

29. ASYMPTOTIC BEHAVIOR OF THE POTENTIAL KERNEL

(As a harmless sort of amusement, note that the above integral representing $a(x)$ may be evaluated by a simple probabilistic argument. Applying T28.1 to P28.2 one obtains

$$\lim_{n \to \infty} \sum_{t \in R} P_{n+1}(x,t) H_C(t,0) = H_C(x,0) + [a(x) - a(x-1)] \Pi_C(0,1),$$

where C is the set consisting of 0 and 1. But clearly $H_C(x,0) = 1$ for $x \leq 0$ and 0 otherwise, while $\Pi_C(0,1) = 1/2$, so that in view of P28.7

$$a(x) - a(x-1) = 1 - 2H_C(x,0) = \begin{cases} 1 \text{ if } x > 0 \\ -1 \text{ if } x \leq 0. \end{cases}$$

Since $a(0) = 0$, it follows that $a(x) = |x|$.)

E2 The second example is a random walk we have encountered before, in E8.3 and E22.1, namely

$$P(0,0) = 1 - \frac{2}{\pi},$$

$$P(0,x) = \frac{2}{\pi} \frac{1}{4x^2 - 1} \text{ for } x \neq 0.$$

Its characteristic function is

$$\phi(\theta) = 1 - \left|\sin \frac{\theta}{2}\right|.$$

Thus it has infinite variance. Its potential kernel is

$$a(x) = \frac{1}{2\pi} \int_{-\pi}^{\pi} \frac{1 - \cos x\theta}{\left|\sin \frac{\theta}{2}\right|} d\theta = \frac{1}{\pi} \int_{-\pi}^{\pi} \frac{\sin^2 \left(\frac{x\theta}{2}\right)}{\left|\sin \frac{\theta}{2}\right|} d\theta$$

$$= \frac{4}{\pi} \left[1 + \frac{1}{3} + \frac{1}{5} + \cdots + \frac{1}{2|x| - 1}\right], \quad x \neq 0.$$

The logarithmic behavior of $a(x)$ is intimately connected with the fact that $\phi(\theta)$ varies linearly with $|\theta|$ near $\theta = 0$ and this of course is *not typical* of type I random walk ($\sigma^2 = \infty$) in general. What is typical, as we shall show, is the far weaker property that

$$\lim_{x \to \infty} [a(x) - a(x+y)] = 0 \text{ for } y \in R.$$

We start with type II random walk. That is by far the simplest case, most of the work having been done in the last section. But first we observe that T28.1, applied to P28.2, yields

P1 $\quad \lim_{n \to \infty} \sum_{t \in R} P_{n+1}(x,t) H_C(t,0) = H_C(x,0) + [a(x) - a(x-c)] \Pi_C(0,c)$

which is valid for every aperiodic recurrent random walk, regardless of dimension. Here $C = \{0,c\}$, $c \neq 0$.

P2 *For type* II *random walk*

$$\lim_{x \to +\infty} [a(x + y) - a(x)] = \frac{y}{\sigma^2},$$

$$\lim_{x \to -\infty} [a(x + y) - a(x)] = -\frac{y}{\sigma^2}, \quad y \in R.$$

Proof: P28.7 applied to P1 gives

$$\tfrac{1}{2}H_C(-\infty,0) + \tfrac{1}{2}H_C(+\infty,0) = H_C(x,0) + [a(x) - a(x-c)]\Pi_C(0,c).$$

Letting first $x \to +\infty$, and then $x \to -\infty$, one obtains the two equations

$$\tfrac{1}{2}H_C(+\infty,0) - \tfrac{1}{2}H_C(-\infty,0) = \Pi_C(0,c) \lim_{x \to +\infty} [a(x-c) - a(x)],$$

$$\tfrac{1}{2}H_C(-\infty,0) - \tfrac{1}{2}H_C(+\infty,0) = \Pi_C(0,c) \lim_{x \to -\infty} [a(x-c) - a(x)].$$

They show that the limits in P2 exist, and that one is the negative of the other. (The point c being arbitrary, we set $c = y \neq 0$.) To evaluate these limits we call

$$\lim_{x \to +\infty} [a(x + 1) - a(x)] = \alpha.$$

Then P2 is proved as soon as we show that $\alpha = (\sigma^2)^{-1}$. However, taking the Césaro mean, i.e., writing for $x > 0$

$$a(x) = \sum_{k=1}^{x} [a(k) - a(k - 1)],$$

and proceeding in a similar way for $x < 0$, we find

$$\lim_{x \to +\infty} \frac{a(x)}{x} = \lim_{x \to -\infty} \frac{a(x)}{|x|} = \alpha.$$

Thus

$$\lim_{|x| \to \infty} \frac{a(x) + a(-x)}{|x|} = 2\alpha.$$

The last limit, however, was evaluated in P28.4, where it was shown to be $2\alpha = 2(\sigma^2)^{-1}$. Thus $\alpha = (\sigma^2)^{-1}$ and P2 is proved.

The case of type I random walk requires somewhat more delicate tools. The properties of the potential kernel, described in the next three lemmas, will be crucial for the proof, in P8, that $a(x)$ increases more slowly than a linear function of $|x|$.

P3 *For type* I *random walk*

$$\lim_{|x| \to \infty} \frac{a(x)}{x} = 0.$$

Proof: P3 is an immediate consequence of T28.1 and P28.4.

P4 *For arbitrary recurrent aperiodic random walk and all* x, y *in* R

$$g_{\{0\}}(x, y) = A(x, 0) + A(0, y) - A(x, y),$$

and

$$a(x + y) \leq a(x) + a(y).$$

Proof: The formula for $g_{\{0\}}(x, y)$, defined in D10.1, was derived for two-dimensional random walk in P11.6. The same proof applies in general, as it required only information which is now in our possession even for one-dimensional random walk (all we need is contained in T28.1 and P1). The inequality in P4 follows from

$$0 \leq g_{\{0\}}(x, -y) = a(x) + a(y) - a(x + y), \qquad y \in R.$$

P5 $$\lim_{|x| \to \infty} [a(x + 1) + a(x - 1) - 2a(x)] = 0.$$

Proof: This lemma again is true in general, for every aperiodic recurrent random walk. But we shall only need it, and therefore only prove it, in the one-dimensional case. Using the definition of $a_n(x)$, straightforward calculation gives

$$a_n(x + 1) + a_n(x - 1) - 2a_n(x)$$
$$= \frac{1}{\pi} \int_{-\pi}^{\pi} \frac{1 - \cos \theta}{1 - \phi(\theta)} e^{ix\theta} [1 - \phi^{n+1}(\theta)] \, d\theta.$$

But it was observed in the proof of P28.4 that $[1 - \cos \theta][1 - \phi(\theta)]^{-1}$ is integrable on $[-\pi, \pi]$. Therefore

$$a(x + 1) + a(x - 1) - 2a(x) = \frac{1}{\pi} \int_{-\pi}^{\pi} \frac{1 - \cos \theta}{1 - \phi(\theta)} e^{ix\theta} \, d\theta.$$

The proof of P5 may now be completed by applying the Riemann Lebesgue Lemma P9.1.

It is our aim to show that for type I random walk

$$\lim_{|x| \to \infty} [a(x + 1) - a(x)] = 0.$$

Now P3 is the weak (Césaro) form of this result, and the next lemma will be recognized as a step in the direction required to strengthen P3.

P6 *For type* I *random walk*

(a) $$\varlimsup_{x \to +\infty} [a(x) - a(x-1)] \le 0,$$

(b) $$\varliminf_{x \to -\infty} [a(x) - a(x-1)] \ge 0.$$

Proof: It will suffice to prove (a). For suppose every type I random walk obeys (a). If we reverse such a random walk it will still be of type I and its potential kernel will be $a^*(x) = a(-x)$. But if $a^*(x)$ satisfies (a), then $a(x)$ is easily seen to satisfy (b). (Actually our method of proof is capable of giving both (a) and (b).)

We define

$$\Delta(x) = a(x) - a(x-1), \quad x \ge 1.$$

The inequality of P4 can then be written

(1) $$\Delta(x+1) + \Delta(x+2) + \cdots + \Delta(x+y)$$
$$\le \Delta(1) + \Delta(2) + \cdots + \Delta(y)$$

whenever $x \ge 1$ and $y \ge 1$. According to P5,

(2) $$\lim_{|x| \to \infty} [\Delta(x+1) - \Delta(x)] = 0.$$

In view of (2), letting $x \to +\infty$ in (1) gives

(3) $$y \varlimsup_{x \to +\infty} \Delta(x) \le \Delta(1) + \Delta(2) + \cdots + \Delta(y)$$

for every $y \ge 1$.

The last step consists of dividing both sides in (3) by y, and then letting $y \to +\infty$. Using P3 that leads to

$$\varlimsup_{x \to +\infty} \Delta(x) \le \lim_{y \to +\infty} \frac{1}{y} [\Delta(1) + \Delta(2) + \cdots + \Delta(y)] = \lim_{y \to +\infty} \frac{a(y)}{y} = 0,$$

which proves P6.

The result of P6 looks suspiciously unsymmetric, unless of course both the upper and lower limits in P6 have the value zero. To show that this is indeed the case, we study the probability interpretation of P6. In P1 we shall take $c = 1$, so that $C = \{0,1\}$. We also observe that

$$\lim_{n \to \infty} \sum_{t \in R} P_{n+1}(x,t) H_C(t,0) = \mu(x).$$

is a non-negative constant, which we shall call μ. (If it depended on x it would be a nonconstant bounded regular function, which is impossible. This argument should be familiar from P11.4). Now P1 can be written in the form

$$H_C(x,0) = \mu - [a(x) - a(x-1)]\Pi_C(0,c),$$

and since $\Pi_C(0,c) > 0$ we get from P6

P7 *For type* I *random walk*

$$\varliminf_{x \to +\infty} H_C(x,0) \geq \mu, \qquad \varlimsup_{x \to -\infty} H_C(x,0) \leq \mu.$$

Note that the limit of $H_C(x,0)$ exists if and only if $a(x) - a(x-1)$ tends to zero as $|x| \to \infty$. We shall now establish the existence of this limit.

P8 *For type* I *random walk*

$$\lim_{|x| \to \infty} [a(x+y) - a(x)] = 0, \qquad y \in R.$$

Proof: It clearly suffices to prove P8 for the particular value of $y = 1$. But in view of the identity preceding P7 it is enough to show that both limits in P7 exist and are equal to μ. That will be done with the aid of T18.1, which is not really essential, and of the renewal theorem in the form of P24.7, which is essential. By T18.1, since $\sigma^2 = \infty$, we know that at least one of the ladder random variables $\mathbf{Z}$ and $-\bar{\mathbf{Z}}$ (defined in D18.1) has infinite expectation. We shall assume that $\mathbf{E}[\mathbf{Z}] = +\infty$, and complete the proof only in this case. Consideration of the reversed random walk would then automatically take care of those random walks with $\mathbf{E}[\mathbf{Z}] < \infty$ but $\mathbf{E}[-\bar{\mathbf{Z}}] = \infty$.

We use the notation of D24.1, just as in the proof of P28.6, with the trivial modification that $A = [x \mid x \geq 0]$ instead of $[x \mid x > 0]$. Then we have

$$H(x) = \sum_{t=0}^{\infty} H_A(x,t)H(t), \qquad x < 0,$$

where

$$H(x) = H_C(x,0), \qquad C = \{0,1\}.$$

By P24.7

$$\sum_{t=0}^{\infty} H_A(x,t) = 1 \text{ for all } x \in R$$

and

$$\lim_{x \to -\infty} H_A(x,t) = 0 \text{ for } t \geq 0.$$

Hence
$$\varliminf_{x \to -\infty} H(x) \geq \varlimsup_{x \to +\infty} H(x).$$

When this information is combined with P7 we get
$$\lim_{x \to -\infty} H(x) = \mu = \varliminf_{x \to +\infty} H(x); \qquad \varlimsup_{x \to +\infty} H(x) \geq \mu.$$

Now there are two possibilities. Possibly $\mathbf{E}[-\overline{\mathbf{Z}}] = \infty$, in which case the present argument may be repeated for the right half-line to show that
$$\lim_{x \to -\infty} H(x) = \lim_{x \to +\infty} H(x) = \mu.$$

The other possibility is that $\mathbf{E}[-\overline{\mathbf{Z}}] < \infty$. But in this case one can also employ the renewal theorem, in fact just as we did in the proof of P28.6, to conclude that
$$\lim_{x \to +\infty} H(x) \quad \text{exists.}$$

Finally, this limit cannot exist unless it is equal to the limit inferior which we just showed to be μ. That completes the proof of P8.

P8 should be compared to P12.2 in Chapter III where, by much simpler methods, we obtained exactly the same result as P8 for arbitrary aperiodic random walk in the plane. Combining P12.2 with P8 and P2 we therefore have

T1 *For aperiodic recurrent random walk there are two possibilities.*
(1) *The dimension $d = 2$, or $d = 1$ but $\sigma^2 = \infty$. In this case*
$$\lim_{|x| \to \infty} [a(x + y) - a(x)] = 0, \qquad y \in R.$$

(2) *The dimension $d = 1$ and $\sigma^2 < \infty$, in which case*
$$\lim_{x \to \pm \infty} [a(x + y) - a(x)] = \pm \frac{y}{\sigma^2}, \qquad y \in R.$$

Thus, at least from our present vantage point, there is no difference at all between the two-dimensional random walk and the one-dimensional random walk with $\sigma^2 = \infty$. In both cases it looks, intuitively, as though the state space R is "large" enough so that the conclusion of part (1) of T1 holds. Thus the "size" of R depends on the transition probabilities—if $\sigma^2 < \infty$, the set of integers R is not so "large" as if $\sigma^2 = \infty$. This phenomenon is already familiar from the renewal theorem.

In the next section we shall encounter one single anomaly which does set one-dimensional random walk, with $\sigma^2 = \infty$, apart from two-dimensional random walk; for we shall look into the matter of finding explicit formulas for the hitting probabilities $H_A(x,y)$ and for the Green function $g_A(x,y)$ of a finite set A. For two-dimensional random walk that was done in Chapter III, using in an essential way one more property of the potential kernel, namely (P11.7) the property that

$$a(x) > 0 \text{ for } x \neq 0.$$

That, as we shall see, is simply *not true for every one-dimensional aperiodic recurrent random walk.*

30. HITTING PROBABILITIES AND THE GREEN FUNCTION

As we remarked at the end of the last section, serious difficulties are in store concerning those properties of recurrent random walk that depend on the positivity of the potential kernel $a(x)$. Therefore we shall first concentrate on some aspects of the theory that are more general. Very little beyond T29.1 of the last section is required to prove

T1 *Consider recurrent aperiodic random walk, and a subset $A \subset R$, with $1 \leq |A| < \infty$. Then either*

(1) $d = 2$, or $d = 1$ and $\sigma^2 = \infty$, in which case

$$\lim_{|x| \to \infty} H_A(x,y) \quad \text{and} \quad \lim_{|x| \to \infty} g_A(x,y)$$

both exist, the former limit for each y in A and the latter for each $y \in R$. Or

(2) $d = 1$ and $\sigma^2 < \infty$, in which case the four limits

$$\lim_{x \to +\infty} H_A(x,y), \quad \lim_{x \to -\infty} H_A(x,y),$$

$$\lim_{x \to +\infty} g_A(x,y), \quad \lim_{x \to -\infty} g_A(x,y)$$

all exist, but the limits as $x \to +\infty$ and as $x \to -\infty$ are in general not the same.

Proof: The proof of T1 is based on

P1 *For every nonempty finite set A,*

$$H_A(x,y) = \mu_A(y) + \sum_{t \in A} A(x,t)[\Pi_A(t,y) - \delta(t,y)], \qquad x \in R, \quad y \in A.$$

P1 follows from P28.1 together with T28.1. Here $\mu_A(y)$ is of course

$$\mu_A(y) = \lim_{n \to \infty} \sum_{t \in R} P_{n+1}(x,t) H_A(t,y),$$

which is independent of x, being a bounded regular function of x for each y in R. As far as the hitting probabilities $H_A(x,y)$ are concerned, the proof of T1 is almost complete. In case (1) the proof in T14.1 carries over verbatim, using of course part (1) of T29.1. In case (2) we get

$$\lim_{x \to +\infty} H_A(x,y) = \mu_A(y) + \lim_{x \to +\infty} \sum_{t \in A} a(x-t)[\Pi_A(t,y) - \delta(t,y)]$$

$$= \mu_A(y) + \lim_{x \to +\infty} \sum_{t \in A} [a(x-t) - a(x)][\Pi_A(t,y) - \delta(t,y)],$$

using P11.2, to the effect that

$$\sum_{t \in A} [\Pi_A(t,y) - \delta(t,y)] = 0, \qquad y \in A.$$

Now T29.1 implies

$$\lim_{x \to +\infty} H_A(x,y) = \mu_A(y) - \frac{1}{\sigma^2} \sum_{t \in A} t[\Pi_A(t,y) - \delta(t,y)],$$

and of course a similar argument proves the existence of the limit as $x \to -\infty$.

Finally, the simplest proof of the part of T1 which concerns the Green function comes from

$$g_A(x,y) = H_{A \cup y}(x,y) g_A(y,y), \qquad x \in R - A, \quad y \in R - A,$$

where $A \cup y$ is the set A with the point y adjoined. (When $y \in A$ there is nothing to prove as $g_A(x,y) = 0$.) In this way that part of T1 which has already been proved serves to complete the proof of T1.

T1 gains in interest by comparison with P25.3, where the hitting probabilities $H_A(x,y)$ of a finite set A were seen to possess limits even in the transient case. There the limit was zero, except in the case $d = 1$ with finite mean. But combined with our present result in T1 we have actually gained insight of rather impressive proportions. Together these theorems describe intuitively plausible, but far from obvious, regularity properties shared by *every* random walk (the aperiodicity condition being hardly a restriction, but rather an essential part of the proper formulation of such theorems). Only in the last section of this chapter, in T32.1, will we encounter another general regularity theorem of equal or even greater depth.

Now we pass at once to a curious phenomenon hinted at a little earlier, namely the classification of aperiodic recurrent random walk according to whether

$$a(x) > 0 \text{ for all } x \neq 0$$

or not. The somewhat surprising result is

P2 *All aperiodic recurrent random walks have the property that $a(x) > 0$ for every $x \neq 0$ with one single exception. This exception is left- or right-continuous random walk whose variance σ^2 is infinite.*

In the left-continuous case with $\sigma^2 = \infty$

$$a(x) = 0 \text{ for } x \geq 0 \quad \text{and} \quad a(x) > 0 \text{ for } x < 0,$$

and in the right-continuous case the situation is of course reversed.

Proof: First we suppose the random walk to be recurrent and left continuous, with $\sigma^2 = \infty$. If $A = \{0,1\}$, then by P29.1 or P1

$$H_A(x,0) = \mu_A(0) + [a(x-1) - a(x)]\Pi_A(0,1).$$

Clearly $H_A(x,0) = 0$ for $x > 0$ (the random walk can move only one step to the left at one time). If we now sum over $x = 1, 2, \ldots, r$, we obtain

$$r\mu_A(0) = a(r)\Pi_A(0,1), \qquad r \geq 0.$$

Dividing by r, and letting $r \to +\infty$, we get

$$0 \leq \mu_A(0) = \Pi_A(0,1) \lim_{r \to +\infty} \frac{a(r)}{r} = 0$$

by T29.1. Hence $a(r) = 0$ for every $r > 0$. On the other hand, clearly $H_A(x,0) > 0$ for every $x \leq 0$, so that

$$0 < a(x-1) - a(x), \qquad x \leq 0,$$

which implies that $a(x) > 0$ when $x < 0$. The right continuous random walk receives the same treatment.

To complete the proof of P2 it will now suffice to consider a recurrent aperiodic one-dimensional random walk with the property that $a(c) = 0$ for some $c > 0$, and to prove that this random walk must be left continuous with $\sigma^2 = \infty$. (Two-dimensional random walk need not be considered here since it has $a(x) > 0$ for all x in R according to P11.7.) This is done by choosing $A = \{0,c\}$. In

$$H_A(x,0) = \mu_A(0) + [a(x - c) - a(x)]\Pi_A(0,c)$$

we set $x = c$, so that $\mu_A(0) = 0$ because

$$a(c) = 0 \quad \text{and} \quad H_A(c,0) = 0.$$

But in that case

$$H_A(2c,0) = -a(2c)\Pi_A(0,c) \leq 0,$$

which is only possible if $a(2c) = 0$. Continuing this argument in the obvious manner one concludes that

$$a(nc) = 0, \qquad n = 1, 2, \ldots.$$

Hence

$$\lim_{x \to +\infty} \frac{a(x)}{x} = 0$$

which shows (by T29.1) that $\sigma^2 = \infty$.

Now it remains only to show that the present random walk must be left continuous, in addition to having $\sigma^2 = \infty$. To do so observe that, using the identity for $g_{\{0\}}(x,y)$ in P29.4,

$$g_{\{0\}}(c,-c) = a(c) + a(c) - a(2c) = 0.$$

That can happen only if there is no finite "path" of positive probability from c to $-c$, which does not pass through the origin. More precisely, it implies that every product of the form

$$P(c,x_1)P(x_1,x_2)\cdots P(x_n,-c) = 0$$

when $x_i \neq 0$ for $i = 1, 2, \ldots, n$. This is a property of left-continuous random walk, and indeed it characterizes left-continuous random walk. For suppose $P(0,a) > 0$ for some $a < -1$. Then there is a path from c to $+1$, not going through zero, another (consisting of one step) from 1 to $1 + a$, and a third from $1 + a$ to $-c$. The existence of the first and third of these paths was verified in the proof of P18.8.

Since left- and right-continuous random walk with mean zero and finite variance is perfectly well behaved as far as the positivity of $a(x)$

is concerned, one might ask if it has other interesting properties. Indeed there is a peculiar property of the potential kernel which characterizes left- and right-continuous random walk with mean 0 and $\sigma^2 < \infty$. It will be useful in the proof of T31.1.

P3 *One-dimensional aperiodic recurrent random walk has the property that*

$$a(x) = \alpha x \text{ for all } x \geq 0, \text{ where } \alpha \text{ is a positive constant,}$$

if and only if the random walk is left continuous with $\sigma^2 = \alpha^{-1}$.

Proof: Suppose first that $a(x) = \alpha x$ for $x \geq 0$, where $\alpha > 0$. Then $\sigma^2 = \alpha^{-1}$ by T29.1. Now let A be the set $A = \{-1, 0\}$. One calculates easily that

$$H_A(x, -1) = \mu_A(-1) + [a(x) - a(x+1)]\Pi_A(0, -1), \quad x \in R.$$

Setting $x = 0$ gives

$$\mu_A(-1) = a(1)\Pi_A(0, -1) = \alpha \Pi_A(0, -1).$$

Hence

$$H_A(x, -1) = [\alpha + \alpha x - \alpha(x+1)]\Pi_A(0, -1) = 0$$

for every $x \geq 1$. Thus, starting at a point $\mathbf{x}_0 = x \geq 1$, the random walk can hit the set A only at the point 0. There is no need to formalize the obvious argument by which one can now conclude that the random walk must be left continuous.

Conversely, suppose that we are given a left-continuous random walk with finite variance σ^2. By the same argument as before, one gets

$$H_A(x, -1) = [a(1) + a(x) - a(x+1)]\Pi_A(0, -1), \quad x \geq 1.$$

But now we know that $H_A(x, -1) = 0$ when $x \geq 0$, so that

$$a(x+1) - a(x) = a(1)$$

when $x \geq 0$, or $a(x) = xa(1) = x\alpha$. Thus $a(x) = \alpha x$ for $x \geq 0$, and since the variance is finite $\alpha = (\sigma^2)^{-1}$ according to T29.1.

Remark: P2 and P3 are actually manifestations of a much more general phenomenon: it can be shown (see problem 8) that *the potential kernel determines the random walk.* In other words, two aperiodic recurrent random walks with the same potential kernel $A(x,y)$ must have the same transition function $P(x,y)$. (In problem 13 of Chapter

VI, the same observation was made for transient random walk, which is determined by its Green function $G(x,y)$.)

The remainder of this section will be devoted to some interesting but completely straightforward consequences of P1 and P2. First we shall show how one can calculate $H_B(x,y)$ for left-continuous random walk with $\sigma^2 = \infty$. Given a set B with $|B| = n$, we order its elements

$$b_1 < b_2 < \cdots < b_n.$$

Then

$$H_B(x,b_k) = \mu_B(b_k) + \sum_{j=1}^{n} a(x - b_j)[\Pi_B(b_j,b_k) - \delta(j,k)]$$

for $x \in R$, $k = 1, 2, \ldots, n$. When $|B| = 1$, there is no problem. When $|B| = n > 1$, set $x = b_i$. Then

$$\delta(i,k) = H_B(b_i,b_k) = \mu_B(b_k) + \sum_{j=i+1}^{n} a(b_i - b_j)[\Pi_B(b_j,b_k) - \delta(j,k)],$$

for $1 \leq i, k \leq n$. Setting $i = k = n$ gives

$$\mu_B(b_n) = 1.$$

But from the definition of $\mu_B(y)$ as

$$\mu_B(y) = \lim_{n \to \infty} \sum_{t \in R} P_{n+1}(x,t) H_B(t,y)$$

it is clear that

$$\sum_{y \in B} \mu_B(y) = 1, \qquad \mu_B(b_k) = 0 \text{ for } k < |B| = n.$$

Now it is very easy to use

$$\delta(i,k) = \delta(n,k) + \sum_{j=i+1}^{n} a(b_i - b_j)[\Pi_B(b_j,b_k) - \delta(j,k)], \qquad 1 \leq i,k \leq n,$$

to determine the matrix Π_B in terms of the values of $a(x)$ when x is a difference of points in B. And, of course, once Π_B and μ_B are known, we have an explicit formula for $H_B(x,y)$ from P1.

Having thus dismissed left- and right-continuous random walk with $\sigma^2 = \infty$, we are again ready to copy results from Chapter III insofar as possible. In all remaining cases—this is the crucial observation— *the matrix operator $A(x,y)$, with x and y restricted to a set B with $2 \leq |B| < \infty$, has an inverse.* This was proved in P11.8 and the proof applies here too, as it depended only on the fact that $a(x) > 0$ for $x \neq 0$. As in D11.3 we call this inverse $K_B(x,y)$ and denote

$$K_B(x\cdot) = \sum_{y \in B} K_B(x,y), \qquad K_B(\cdot y) = \sum_{x \in B} K_B(x,y)$$

$$K_B(\cdot\cdot) = \sum_{x \in B} K_B(x\cdot), \qquad x, y \in B.$$

With this notation we can imitate T11.1 and T14.2 to obtain

T2 *For aperiodic recurrent one-dimensional random walk with $\sigma^2 < \infty$, or with $\sigma^2 = \infty$, but in the latter case neither left nor right continuous, and $2 \leq |B| < \infty$*

(a) $\quad \Pi_B(x,y) = \delta(x,y) + K_B(x,y) - \dfrac{K_B(x\cdot)K_B(\cdot y)}{K_B(\cdot\cdot)}, \qquad x, y \in B,$

(b) $\quad \mu_B(y) = \dfrac{K_B(\cdot y)}{K_B(\cdot\cdot)}, \quad \mu_B^*(x) = \dfrac{K_B(x\cdot)}{K_B(\cdot\cdot)}, \qquad x, y \in B,$

(c) $\quad H_B(x,y) = \mu_B(y) + \sum_{t \in B} A(x,t)[\Pi_B(t,y) - \delta(t,y)],$

$$x \in R, \quad y \in B,$$

(d) $\quad g_B(x,y) = -A(x,y) - \dfrac{1}{K_B(\cdot\cdot)}$

$$+ \sum_{s \in B} \mu_B(s) A(s,y) + \sum_{t \in B} A(x,t) \mu_B^*(t)$$

$$+ \sum_{s \in B} \sum_{t \in B} A(x,t)[\Pi_B(t,s) - \delta(t,s)]A(s,y), \qquad x, y \in R.$$

The proofs of these statements in Chapter III require no modification whatsoever. The most difficult proof was that of (d) in T14.2. There we were extremely careful[2] to refrain from using any asymptotic properties of $a(x)$ beyond the boundedness of the function $A(x,t) - A(0,t)$ for each fixed x, which in the present context follows from T29.1. The proof further made use of the equation

$$\sum_{y \in R} P(x,y) a(y) - a(x) = \delta(x,0)$$

which is also available to us in the one-dimensional case, according to T28.1.

Significant differences between the cases $\sigma^2 = \infty$ and $\sigma^2 < \infty$ manifest themselves in the asymptotic behavior of the functions in T2.

---

[2] Cf. footnote (7) of Ch. III.

For example, when $\sigma^2 = \infty$ (left- and right-continuous random walk being excluded)

(1) $$\lim_{|x|\to\infty} H_B(x,y) = \mu_B(y), \qquad y \in B,$$

(2) $$\lim_{|x|\to\infty} g_B(x,y) = \sum_{s\in B} \mu_B(s)A(s,y) - \frac{1}{K_B(\cdot\cdot)}, \qquad y \in R.$$

When $\sigma^2 < \infty$, on the other hand, the limits of H_B and g_B as $x \to +\infty$ and as $x \to -\infty$ turn out to be a little more complicated. Although they are easy to calculate on the basis of T2 and the known asymptotic behavior of $a(x)$, the remarkable thing is that (1) and (2) continue to hold in a slightly modified form. Let us call

(3) $$\operatorname*{Lim}_x f(x) = \tfrac{1}{2} \lim_{x\to+\infty} f(x) + \tfrac{1}{2} \lim_{x\to-\infty} f(x),$$

for a function $f(x)$ on the integers with the property that the two ordinary limits exist. Then it may easily be verified that (1) and (2) continue to hold when $\sigma^2 < \infty$, provided one replaces the limits by Lim as defined in (3).

Now a few words concerning the notion of *capacity*, of the *logarithmic type*, which was sketched at the end of section 14 in Chapter III. One can dismiss the left- and right-continuous random walk with $\sigma^2 = \infty$ with the statement that it does not give rise to a nontrivial potential theory. (According to any natural definition of capacity at all, three of which were discussed in Chapter III, the capacity of any finite set would be zero for left- or right-continuous random walk.)

It should be clear that for every other one-dimensional random walk with $\sigma^2 = \infty$ the associated potential theory is *exactly* of the same type as that sketched in Chapter III.

Finally, consider the case when $\sigma^2 < \infty$. If the capacity of a finite set B is defined by

$$C(B) = 0 \text{ if } |B| = 1, \qquad C(B) = \frac{1}{K_B(\cdot\cdot)} \text{ if } |B| > 1,$$

a significant difference appears in the relation between capacity and the asymptotic behavior of the Green function. While

$$C(B) = \lim_{|x|\to\infty} [a(x) - g_B(x,\infty)]$$

when $d = 2$, and when $d = 1$ and $\sigma^2 = \infty$, one gets in the case $d = 1$, $\sigma^2 < \infty$

$$C(B) = \operatorname*{Lim}_x [a(x) - \operatorname*{Lim}_y g_B(x,y)].$$

This exposition of the potential theory was kept quite brief for several reasons. First of all it is not clear why one should set any store by formal generalizations from classical potential theory under the pretext of studying probability theory. (Logarithmic capacity does not have as simple a probability interpretation as capacity for transient random walk. See problems 9 and 10 at the end of the chapter.) Secondly, according to the general plan formulated at the beginning of Chapter III, we were going to study stopping times, and in particular the hitting time $\mathbf{T}_B$ for a finite subset of the state space. With T2, and its counterpart in Chapter III, this problem is in a sense solved. But much more important than the actual explicit solution are the methods which were developed on the way. In particular the proof of part (d) of T2 was quite difficult. The reader who conscientiously traced it back to its origins in Chapter III should be somewhat reluctant to leave things as they are. The explicit formula for g_B was obtained by solving the equation

$$\sum_{t \in R} P(x,t) g_B(t,y) = H_B(x,y), \qquad x \in R - B, \quad y \in B.$$

By the substitution

$$u(t) = g_B(x,t) + A(x,t),$$

this equation was reduced to

$$\sum_{t \in R} u(t) P(t,y) - u(y) = H_A(x,y), \qquad y \in R,$$

which is a special case of Poisson's equation

$$Pf(x) - f(x) = \psi(x), \qquad x \in R, \quad \psi \geq 0 \text{ on } R.$$

The theory of Poisson's equation, however, as developed in P13.3 and T28.1 is quite crude, and was developed before we could anticipate what demands would be made on it in the future.

This theory—concerned with the existence and uniqueness of non-negative solutions to Poisson's equation, will be vastly improved in the next section. In the course of this improvement we shall use only the basic results of this chapter—namely those concerning the existence and asymptotic behavior of the potential kernel. In short we shall use only T28.1, T29.1, P2, and P3. Thus the reader may, without fear of circular reasoning, use the results of the next section to obtain a simplified proof of T2.

31. THE UNIQUENESS OF THE POTENTIAL KERNEL

In the last chapter, example E27.4, we studied the equation

$$\sum_{y \neq 0} P(x,y)f(y) = f(x), \quad x \neq 0,$$

with $P(x,y)$ the transition function of an aperiodic recurrent random walk in the plane. It was shown that all the non-negative solutions are multiples of the potential kernel $a(x) = A(x,0)$. At that time little information was available concerning the one-dimensional recurrent random walk. Now, however, it will be relatively easy to extend this result to arbitrary aperiodic recurrent random walk. In so doing we shall again discover a sharp distinction between the case of dimension one with finite variance on one hand, and all other cases on the other.

We need a preliminary lemma, which simply exhibits certain solutions, without making any claims as to uniqueness.

P1 *Recurrent aperiodic random walk. If $d = 2$ or if $d = 1$ and $\sigma^2 = \infty$, then $a(x)$ is a non-negative solution of*

(1) $$\sum_{y \neq 0} P(x,y)f(y) - f(x) = 0, \quad x \neq 0.$$

If $d = 1$ and $\sigma^2 < \infty$, then the function $f(x) = a(x) + \alpha x$ is a non-negative solution of (1) if and only if the real parameter α lies in the interval $-(\sigma^2)^{-1} \leq \alpha \leq (\sigma^2)^{-1}$.

Proof: The first part of P1 is simply a restatement of part (b) of T28.1. The second part requires proof. In view of T28.1, $a(x)$ satisfies equation (1). But also $f(x) = x$ satisfies (1)—this is clear since recurrent random walk with $\sigma^2 < \infty$ has mean zero. Thus $f(x) = a(x) + \alpha x$ satisfies (1), but it is not necessarily non-negative on $R - \{0\}$, as required. Now the condition that $|\alpha| \leq (\sigma^2)^{-1}$ is necessary, in view of T29.1, for if we want $f(x) \geq 0$ we must have

$$\lim_{x \to +\infty} \frac{a(x) + \alpha x}{x} = \frac{1}{\sigma^2} + \alpha \geq 0,$$

$$\lim_{x \to -\infty} \frac{a(x) + \alpha x}{-x} = \frac{1}{\sigma^2} - \alpha \geq 0,$$

which implies that $|\alpha| \leq (\sigma^2)^{-1}$. To prove sufficiency, suppose that $|\alpha| \leq (\sigma^2)^{-1}$. We know that

$$0 \leq g_{\{0\}}(x,y) = a(x) + a(-y) - a(x-y).$$

Using T29.1 to let $y \to +\infty$ and $y \to -\infty$, we obtain

$$0 \leq a(x) + \frac{x}{\sigma^2}, \quad 0 \leq a(x) - \frac{x}{\sigma^2}, \quad x \in R.$$

Since these functions are non-negative, so is $a(x) + \alpha x$ whenever $|\alpha| \leq (\sigma^2)^{-1}$.

We are now ready to prove uniqueness, which constitutes a considerably more delicate task.

P2 *The non-negative solutions of* (1) *exhibited in* P1 *are the only ones.*

Proof: The two-dimensional case is left out, having been settled in E27.4, but the proof for the one-dimensional case will nevertheless parallel closely that in E27.4. Consequently it will be convenient to adopt the notation in D27.4 and P27.3. At first one is tempted to let $Q(x,y) = P(x,y)$, restricted to x and y in $R - \{0\}$. Then the countable set S in P27.3 would be $S = R - \{0\}$. But this is not a good idea, since $g_{\{0\}}(x,y)$ may fail to be positive for some x or some y in $R - \{0\}$. Instead we shall choose S in a slightly more elaborate fashion, which will greatly simplify the remainder of the proof.

Given a solution $h(x) \geq 0$ (but not identically zero) on $R - \{0\}$ of

(1) $$\sum_{y \neq 0} P(x,y)h(y) = h(x), \quad x \neq 0,$$

we define

$$S = [x \mid x \in R - \{0\}, h(x) > 0].$$

The fact that the choice of S depends on $h(x)$ will cause no inconvenience. Furthermore there are really only three possibilities. Either $S = R - \{0\}$, or if this is false, then S must be either the half-line $[x \mid x > 0]$ or the half-line $[x \mid x < 0]$. That is easily verified, for suppose that $h(x_0) > 0$, for some $x_0 \in R - \{0\}$. Iteration of (1) gives

(2) $$\sum_{y \neq 0} Q_n(x,y)h(y) = h(x), \quad x \neq 0,$$

if $Q(x,y)$ is $P(x,y)$ restricted to $R - \{0\}$. Thus $h(x_0) > 0$ implies that $h(y) > 0$ whenever $g_{\{0\}}(y,x_0) > 0$, but this is certainly true of all y which have the same sign as x_0 (by a simple combinatorial argument, familiar from the proof of P18.8). This observation verifies our claim that S is either a half-line, or all of $R - \{0\}$. Let us now suppose that S is a half-line, say the positive one. Then, as we just saw, it follows that

$$g_{\{0\}}(x,y) = 0 \text{ when } x < 0, \; y > 0,$$

and in particular

(3) $$P(x,y) = 0 \text{ when } x < 0, \; y > 0,$$

as could have been observed directly from (1). This state of affairs simplifies matters considerably. We can now define, in all cases (S being either a half-line or all of $R - \{0\}$),

$$Q(x,y) = Q_1(x,y) = P(x,y), \quad Q_0(x,y) = \delta(x,y), \quad x,y \in S,$$
$$Q_{n+1}(x,y) = \sum_{t \in S} Q_n(x,t) Q(t,y), \quad x,y \in S, \quad n \geq 0,$$

and finally

$$g(x,y) = \sum_{n=0}^{\infty} Q_n(x,y), \quad x,y \in S,$$

being sure that

(4) $$g(x,y) > 0 \text{ for } x,y \in S,$$
(5) $$g(x,y) = g_{\{0\}}(x,y) \text{ for } x,y \in S.$$

Next we define the transient kernel

$$Q^h(x,y) = \frac{Q(x,y)h(y)}{h(x)}, \quad x,y \in S,$$

its iterates $Q_n^h(x,y)$, and its Green function

(6) $$g^h(x,y) = \sum_{n=0}^{\infty} Q_n^h(x,y) = \frac{g(x,y)h(y)}{h(x)},$$

and proceed along the lines of P27.3. But it is convenient at this point to separate the argument into two cases, according as $\sigma^2 = \infty$ or $\sigma^2 < \infty$.

When $\sigma^2 = \infty$ we conclude from (5) and from T29.1 that

(7) $$\lim_{|y| \to \infty} \frac{g(x,y)}{g(\xi,y)} = \lim_{|y| \to \infty} \frac{g_{\{0\}}(x,y)}{g_{\{0\}}(\xi,y)}$$
$$= \lim_{|y| \to \infty} \frac{a(x) + a(-y) - a(x-y)}{a(\xi) + a(-y) - a(\xi-y)} = \frac{a(x)}{a(\xi)},$$

where ξ is an arbitrary point in S. *Forgetting for the moment the unpleasant possibility that $a(\xi) = 0$*, we see that, in view of (5), (7), and P1, all the hypotheses in P27.3 are satisfied. Thus we are able to conclude that $h(x)$ is a multiple of $a(x)$, which proves P2 when $\sigma^2 = \infty$, provided we were able to choose ξ so that $a(\xi) > 0$. Suppose now that $\xi > 0$ and $a(\xi) = 0$ (the case $\xi < 0$ receives the same treatment). This implies, in view of P30.2, that the random walk is left continuous. Now either $S = R - \{0\}$, in which case we can modify (7) by choosing $\xi < 0$ so that $a(\xi) > 0$. Otherwise $S = [x | x > 0]$, but that is impossible, since (3) gives $P(-1, n) = 0$ for $n > 0$, whereas left continuity together with $\sigma^2 = \infty$ implies $P(-1, n) > 0$ for arbitrarily large n. Thus the proof of P2 is complete when $\sigma^2 = \infty$.

When $\sigma^2 < \infty$ we cannot hope to quote P27.3 verbatim, as it deals only with cases where there is a unique solution. But it is easy enough to retrace the steps in the proof of P27.3 until we come to the crucial argument which begins with equation (4), namely

$$(8) \qquad e(x) = 1 = \lim_{n \to \infty} \sum_{y \in S} \frac{g^h(x,y)}{g^h(\xi,y)} \gamma_n(y).$$

By the same compactness argument used there and in E27.4 and E27.5, equation (8) yields the conclusion that

$$(9) \qquad 1 = u^h(x) + \lim_{y \to -\infty} \frac{g^h(x,y)}{g^h(\xi,y)} \gamma(-\infty) + \lim_{y \to +\infty} \frac{g^h(x,y)}{g^h(\xi,y)} \gamma(+\infty), \qquad x \in S.$$

Here $u^h(x)$ is a Q^h potential, and of course only one of the two limits will be present if S is a half-line instead of all of $R - \{0\}$. (This situation can occur in the case of symmetric simple random walk.) The limits in (9) are, by a calculation based on T29.1 applied to equation (6),

$$(10) \qquad \lim_{y \to \pm \infty} \frac{g^h(x,y)}{g^h(\xi,y)} = \frac{\sigma^2 a(x) \pm x}{\sigma^2 a(\xi) \pm \xi} \cdot \frac{h(\xi)}{h(x)}, \qquad x \in S.$$

Again we first ignore the possibility that

$$\sigma^2 a(\xi) + \xi = 0 \quad \text{or} \quad \sigma^2 a(\xi) - \xi = 0.$$

As may be seen by applying P1 to (10), equation (9) represents the Q^h-regular function $e(x) \equiv 1$ as a Q^h-potential plus two Q^h-regular functions. By the Riesz decomposition theorem (T27.1) the potential must be zero. Substituting (10) into (9), and multiplying (9) through by $h(x)$ we find

$$h(x) = \gamma(-\infty) \frac{h(\xi)}{\sigma^2 a(\xi) - \xi} [\sigma^2 a(x) - x]$$
$$+ \gamma(+\infty) \frac{h(\xi)}{\sigma^2 a(\xi) + \xi} [\sigma^2 a(x) + x], \qquad x \in S.$$

This shows that every regular function $h(x)$ is a linear combination of $a(x)$ and x, which is just what we wanted to prove!

The proof of P2 will now be completed by eliminating the alarming possibility that equation (9) does not make sense because one of the denominators $\sigma^2 a(\xi) \pm \xi$ in (10) vanishes. According to P30.3 of the last section, it is only left- or right-continuous random walk that could cause trouble. For every other random walk one can choose ξ so that $\sigma^2 a(\xi) - |\xi| \neq 0$. Therefore we shall consider all the different possibilities separately.

(a) Suppose the random walk is both left and right continuous. Then $0 < P(0,1) = P(0,-1) = [1 - P(0,0)]/2 \leq 1/2$. But in this very simple case P2 can be verified directly.

(b) Suppose the random walk is left but not right continuous. Then there are several possibilities concerning the set S. If $S = R - \{0\}$ or $S = [x \mid x < 0]$ we may choose a negative value of ξ, such that, according to P30.3, both $\sigma^2 a(\xi) + \xi$ and $\sigma^2 a(\xi) - \xi$ are non-zero. Finally there is the case $S = [x \mid x > 0]$. But then an examination of equation (1) shows that this case cannot arise; equation (1) implies that if $h(x) > 0$ for $x > 0$, then $h(x) > 0$ for all x. Thus case (b) is disposed of.

(c) The case of right- but not left-continuous random walk receives the same treatment as (b) and thus the proof of P2 is complete. We summarize P1 and P2 as

T1 *If $P(x,y)$ is the transition function of aperiodic recurrent random walk, then the non-negative solutions of*

$$h(x) = \sum_{y \neq 0} P(x,y)h(y), \qquad x \neq 0,$$

are multiples of

(1) $\qquad a(x)$ if $d = 2$ or $d = 1$ and $\sigma^2 = \infty$,

(2) $\qquad a(x) + \alpha x$, where $|\sigma^2 \alpha| \leq 1$, if $d = 1$ and $\sigma^2 < \infty$.

For many applications it is important to modify T1 to give information about potentials, i.e., about non-negative solutions of

$$\sum_{y \in R} P(x,y)f(y) - f(x) = \delta(x,0), \qquad x \in R,$$

or, more generally, about functions $f(x) \geq 0$ on R such that

$$Pf(x) - f(x) = \psi(x), \qquad x \in R,$$

with

$$\psi(x) = 0 \quad \text{when} \quad x \in R - A.$$

This reformulation is accomplished by means of a *minimum principle* which is the recurrent analogue of the maximum principle for transient random walk in part (a) of P25.9 in Chapter VI.

T2 *Suppose that $P(x,y)$ is the transition function of aperiodic recurrent random walk, that $f(x) \geq 0$ on R, and that*

$$\sum_{y \in R} P(x,y)f(y) - f(x) \leq 0 \text{ on } R - A,$$

where A is a given subset of R. Then

$$f(x) \geq \inf_{t \in A} f(t), \qquad x \in R.$$

Proof: Let $Q(x,y) = P(x,y)$ when x,y are in $R - A$, and let $Q_n(x,y)$ be the iterates of the transient kernel Q defined on the set $S = R - A$. (Q is a transient kernel if, as we may assume, the set A is nonempty, since the original random walk is recurrent.) Observe that $f(x)$ is excessive, relative to Q on S, because

(1) $\quad f(x) - \sum_{y \in S} Q(x,y)f(y) \geq \sum_{t \in A} P(x,t)f(t) = w(x) \geq 0, \qquad x \in S.$

Proceeding as in the proof of the Riesz decomposition theorem, as one may since $Pf - f \leq 0$ on S, one iterates the application of Q to (1) n times to obtain

(2) $\quad f(x) - \sum_{y \in S} Q_{n+1}(x,y)f(y) \geq \sum_{y \in S} [Q_0(x,y) + \cdots + Q_n(x,y)]w(y).$

If we call

$$g(x,y) = \sum_{n=0}^{\infty} Q_n(x,y), \qquad x,y \in S,$$

the right-hand side in (2) has the limit, as $n \to \infty$,

(3) $\quad \sum_{y \in S} g(x,y)w(y) = \sum_{y \in S} \sum_{t \in A} g(x,y)P(y,t)f(t) = \sum_{t \in A} H_A(x,t)f(t).$

Here $H_A(x,t)$ is of course the hitting probability measure of the set A, for the random walk given by $P(x,y)$ on R. Now it follows from (2) that

$$f(x) \geq \sum_{t \in A} H_A(x,t) f(t)$$

$$\geq \left[\inf_{t \in A} f(t) \right] \sum_{t \in A} H_A(x,t) = \inf_{t \in A} f(t), \qquad x \in S.$$

That proves T2. We shall actually use only the very special case when the set A is a single point. Nevertheless the full strength of T2 is indispensable in developing further the logarithmic type potential theory which was sketched at the end of the last section.

Letting $A = \{0\}$ in T2 the following extension of T1 is immediate

P3 *The non-negative solutions of*

(1) $$\sum_{y \in R} P(x,y) f(y) - f(x) = \delta(x,0), \qquad x \in R$$

are exactly the following functions: if $d = 2$ or if $d = 1$ and $\sigma^2 = \infty$, then

$$f(x) = a(x) + c, \text{ where } c \geq 0;$$

if $d = 1$ and $\sigma^2 < \infty$, then

$$f(x) = a(x) + \alpha x + c$$

where $-(\sigma^2)^{-1} \leq \alpha \leq (\sigma^2)^{-1}$ *and* $c \geq 0$.

Proof: In view of T2 every non-negative solution $f(x)$ of (1) assumes its minimum at the origin. Letting $h(x) = f(x) - f(0)$, it is clear that $h(x) \geq 0$ and

$$\sum_{y \neq 0} P(x,y) h(y) - h(x) = 0, \qquad x \neq 0.$$

Hence $h(x)$ must be a constant multiple of the solutions given in T1. Finally, this multiplicative constant is determined by use of part (b) of T28.1.

The result of P3 gives an analytic characterization of the potential kernel $a(x)$—with the exception of the case of finite variance. There an additional requirement is needed; for instance, a symmetry condition such as $f(-x) \sim f(x)$ as $x \to +\infty$ will serve to pick out $a(x)$ from the one-parameter family of solutions in P3. It is easy to extend P3 to a larger class of potentials.

T3 When $P(x,y)$ is aperiodic and recurrent, let $A \subset R$, $1 \le |A| < \infty$, $\psi = 0$ on $R - A$, $f \ge 0$ on R, and suppose that

(1) $$Pf(x) - f(x) = \psi(x) \ge 0, \qquad x \in R.$$

Then
$$f(x) = \sum_{y \in A} A(x,y)\psi(y) + c_1 \quad \text{when } d = 2 \text{ or } d = 1 \text{ with } \sigma^2 = \infty,$$

and
$$f(x) = \sum_{y \in A} A(x,y)\psi(y) + c_2 x + c_3 \quad \text{when } d = 1 \text{ with } \sigma^2 < \infty.$$

Here c_1, c_2, c_3 are constants satisfying
$$c_1 + \min_{x \in A} \sum_{y \in A} A(x,y)\psi(y) \ge 0,$$

$$\sigma^2 |c_2| \le \sum_{x \in A} \psi(x), \qquad c_3 + \min_{x \in A} \left[\sum_{y \in A} A(x,y)\psi(y) + c_2 x \right] \ge 0.$$

Conversely every $f(x)$ of this form is non-negative and satisfies (1).

Proof: If
$$g(x) = \sum_{t \in R} A(x,t)\psi(t), \qquad \sum_{x \in A} \psi(x) = c,$$

then, according to T29.1,
$$\lim_{|x| \to \infty} [g(x) - ca(x)] = 0 \quad \text{when } d = 2 \text{ or } d = 1 \text{ with } \sigma^2 = \infty,$$

$$\lim_{x \to \pm\infty} [g(x) - ca(x)] = \mp \frac{1}{\sigma^2} \sum_{t \in A} t\psi(t) \quad \text{when } d = 1 \text{ with } \sigma^2 < \infty.$$

In either case $g(x) - ca(x)$ is bounded. By T28.1,
$$h(x) = f(x) + ca(x) - g(x) + \gamma$$

is then a non-negative solution of
$$Ph(x) - h(x) = c\delta(x,0),$$

provided $Pf(x) - f(x) = \psi(x)$ and γ is a sufficiently large constant. Consequently the problem is reduced to that of finding $h(x)$, which may be done using P3. When $d = 2$ or $d = 1$ with $\sigma^2 = \infty$, $h(x) = ca(x) + \text{constant}$, so that
$$f(x) = h(x) - ca(x) + g(x) - \gamma = \sum_{t \in A} A(x,t)\psi(t) + \text{constant}.$$

When $d = 1$ with $\sigma^2 < \infty$, one obtains

$$f(x) = \sum_{t \in A} A(x,t)\psi(t) + \alpha x + \text{constant}, \qquad |\alpha| \leq (\sigma^2)^{-1}.$$

It is easily verified that the only possible values of the constants c_1, c_2, c_3 are those given in T3.

One can extend T3 without difficulty to hold for charges $\psi(x)$ whose support is an infinite subset A of R or even all of R, but only if ψ is a-integrable, i.e., if $\sum_{t \in R} a(x - t)|\psi(t)| < \infty$ for $x \in R$. To show that one cannot hope for much more, consider the example

E1 *Two-dimensional simple random walk*, where we write

$$P(0,z) = \begin{cases} \frac{1}{4} & \text{for } z = 1, -1, i, -i, \\ 0 & \text{for all other } z \in R. \end{cases}$$

Let
$$f(z) = |\text{Re}(z)|, \qquad A = [z \mid \text{Re}(z) = 0].$$

Then
$$\sum_{\zeta \in R} P(z,\zeta)f(\zeta) - f(z) = \psi(z),$$

where
$$\psi(z) = \begin{cases} \frac{1}{2} & \text{for } z \in A \\ 0 & \text{for } z \in R - A. \end{cases}$$

These facts are easily checked. Now P3 or T3, if valid in this case, would give

$$f(z) = \tfrac{1}{2} \sum_{\zeta \in A} A(z,\zeta)$$

which is false, the sum on the right being divergent since

$$A(0,z) \sim \frac{2}{\pi} \ln |z| \text{ as } |z| \to \infty.$$

T3 shows that, when $d = 2$ or $d = 1$ with $\sigma^2 = \infty$, the potential theory sketched in sections 30 and 14 is the "correct" and only analogue of classical logarithmic potential theory. However, when $d = 1$ with $\sigma^2 < \infty$, P3 and T3 show that there is a whole *one parameter family of different potential theories*: for each α, every finite subset B of R will have a different equilibrium charge, capacity, and so forth. Their properties will be much like those in logarithmic potential theory, the minimum principle (T2) making possible the entire development once the equilibrium charge $\mu_B^*(x)$ of a set B is found. It is easy to verify that every finite set B has the one-parameter family of equilibrium charges

$$\mu_\alpha^*(t) = \frac{1+\alpha}{2} H_B^*(+\infty,t) + \frac{1-\alpha}{2} H_B^*(-\infty,t), \qquad t \in B, \quad |\alpha| \leq 1,$$

in other words that

$$\sum_{t \in B} \left[A(x,t) + \frac{\alpha}{\sigma^2}(x - t) \right] \mu_\alpha^*(t) = \text{constant}$$

for $x \in B$. Here H_B^* denotes the hitting probability measure for the reversed random walk. When we set $\alpha = 0$ we get the simplest (symmetric) equilibrium charge

$$\mu_0^*(t) = \mu_B^*(t) = \tfrac{1}{2} H_B^*(+\infty,t) + \tfrac{1}{2} H_B^*(-\infty,t) = \frac{K_B(t\,\cdot)}{K_B(\cdot\,\cdot)},$$

satisfying

$$\sum_{t \in B} A(x,t) \mu_B^*(t) = \text{constant} = \frac{1}{K_B(\cdot\,\cdot)} \quad \text{for } x \in B.$$

32. THE HITTING TIME OF A SINGLE POINT

In T16.1 of Chapter III it was shown that

$$\lim_{n \to \infty} \frac{\mathbf{P}_x[\mathbf{T} > n]}{\mathbf{P}_0[\mathbf{T} > n]} = a(x), \qquad x \neq 0,$$

for aperiodic two-dimensional random walk. Here $a(x)$ is the usual potential kernel, and

$$\mathbf{T} = \min[n \mid 1 \leq n \leq \infty; \mathbf{x}_n = 0]$$

is the time of the first visit, after time 0, of the random walk to the origin.

The theory of transient random walk in Chapter VI and of one-dimensional recurrent random walk in this chapter now offers us the opportunity to assess precisely under which conditions this theorem is valid. It will turn out, fortunately, that we are dealing here with a *completely general property* of random walk, *recurrent or transient*, subject only to the entirely natural restriction of aperiodicity.

The remainder of this section will be devoted to the proof of this assertion, which we state formally as

T1 *For arbitrary aperiodic random walk, recurrent or transient, of dimension $d \geq 1$*

$$\lim_{n\to\infty} \frac{\mathbf{P}_x[T > n]}{\mathbf{P}_0[T > n]} = \lim_{n\to\infty} \sum_{k=0}^{n} [P_k(0,0) - P_k(x,0)]$$

for every $x \neq 0$. Both limits are finite and non-negative.

First (in P1) the proof will be given for transient random walk. The next item on the agenda will be the proof of T1 for one-dimensional recurrent random walk with finite variance σ^2. In this case we shall in fact obtain a result which is stronger than T1, as P4 will show that

$$\mathbf{P}_x[T > n] \sim a(x)\mathbf{P}_0[T > n] \sim a(x)\sqrt{\frac{2}{\pi n}}\,\sigma, \qquad \text{as } n \to \infty, \text{ for } x \neq 0.$$

Since T1 was proved in Chapter III for two-dimensional recurrent random walk, the only case then remaining after P4 will be the one-dimensional recurrent case with $\sigma^2 = \infty$. Here we shall encounter the greatest difficulty. It will be resolved with the aid of the recent ([60], 1963) theorem of Kesten, concerning $F_n = \mathbf{P}_0[T = n]$:

T2 *For strongly aperiodic recurrent random walk in one dimension*

$$\lim_{n\to\infty} \frac{F_{n+1}}{F_n} = 1.$$

T2 is also correct in two dimensions (problem 4) but we will not need it in this case. The proof of T2 will be given in propositions P5 through P8. We defer it for a while in order to discuss the very easy, and not particularly interesting transient case—never to return to it again.

P1 *For every aperiodic transient random walk*

$$\lim_{n\to\infty} \frac{\mathbf{P}_x[T > n]}{\mathbf{P}_0[T > n]} = a(x), \qquad x \neq 0,$$

if we define (as is natural)

$$a(x) = G(0,0) - G(x,0) = \sum_{n=0}^{\infty} [P_n(0,0) - P_n(x,0)].$$

Proof: This simple proof could easily have been given in Chapter I, as we shall use only the methods and notation of section 1. Keeping in mind that

$$\mathbf{P}_x[\mathbf{T} > n] = \mathbf{P}_x[n < \mathbf{T} \leq \infty],$$

we have

$$\lim_{n \to \infty} \frac{\mathbf{P}_x[\mathbf{T} > n]}{\mathbf{P}_0[\mathbf{T} > n]} = \frac{\mathbf{P}_x[\mathbf{T} = \infty]}{\mathbf{P}_0[\mathbf{T} = \infty]} = \frac{1 - \sum_{n=1}^{\infty} F_n(x,0)}{1 - F(0,0)}, \qquad x \neq 0.$$

From P1.2 one obtains

$$\sum_{n=1}^{\infty} t^n F_n(x,0) = \frac{\sum_{n=0}^{\infty} t^n P_n(x,0)}{\sum_{n=0}^{\infty} t^n P_n(0,0)}, \qquad x \neq 0, \; 0 \leq t < 1.$$

Letting $t \to 1$, and using P1.5, we find that

$$F(x,0) = \sum_{n=1}^{\infty} F_n(x,0) = \frac{G(x,0)}{G(0,0)}, \qquad x \neq 0.$$

Hence

$$\frac{1 - \sum_{n=1}^{\infty} F_n(x,0)}{1 - F(0,0)} = G(0,0) - G(x,0) = a(x), \qquad x \neq 0.$$

That proves T1 *in the transient case*, and in the remainder of this section we shall always work with aperiodic recurrent random walk in one dimension. We shall adhere to the notation of D16.1, so that

$$\mathbf{T}_x = \min[n \mid n \geq 1, \mathbf{x}_n = x], \qquad \mathbf{T}_0 = \mathbf{T},$$
$$F_n = \mathbf{P}_0[\mathbf{T} = n], \qquad R_n = \mathbf{P}_0[\mathbf{T} > n], \qquad U_n = \mathbf{P}_0[\mathbf{x}_n = 0],$$
$$n \geq 0.$$

Our next lemma shows that T1 holds for every recurrent random walk in a "weak" sense, namely in the sense of Abel summability.

P2 *For recurrent random walk*

$$\lim_{t \nearrow 1} \frac{\sum_{n=0}^{\infty} t^n \mathbf{P}_x[\mathbf{T} > n]}{\sum_{n=0}^{\infty} t^n \mathbf{P}_0[\mathbf{T} > n]} = a(x) \; \text{for} \; x \neq 0.$$

Proof: Observe first that

$$(1 - t) \sum_{n=0}^{\infty} t^n \mathbf{P}_0[T > n] = 1 - \mathbf{E}_0[t^{T_0}],$$

$$(1 - t) \sum_{n=0}^{\infty} t^n \mathbf{P}_x[T > n] = 1 - \mathbf{E}_x[t^{T_0}] = 1 - \mathbf{E}_0[t^{T_{-x}}],$$

for $x \in R$, $0 \le t \le 1$. Therefore P2 will be proved if

(1) $$\lim_{t \nearrow 1} \frac{1 - \mathbf{E}_0[t^{T_x}]}{1 - \mathbf{E}_0[t^{T_0}]} = a(-x) \text{ for all } x \ne 0.$$

Now we define

$$R_t(x,y) = \sum_{n=0}^{\infty} t^n P_n(x,y),$$

and

$$\phi_0(x) = 1 \text{ if } x = 0, \ 0 \text{ otherwise.}$$

Then, for $t < 1$,

$$\mathbf{E}_0\left[\sum_{n=0}^{T_x - 1} t^n \phi_0(\mathbf{x}_n)\right] = \mathbf{E}_0\left[\sum_{n=0}^{\infty} t^n \phi_0(\mathbf{x}_n)\right] - \mathbf{E}_0\left[\sum_{n=T_x}^{\infty} t^n \phi_0(\mathbf{x}_n)\right]$$

$$= R_t(0,0) - \mathbf{E}_0\left[\sum_{k=0}^{\infty} t^{T_x + k} \phi_0(\mathbf{x}_{T_x + k})\right].$$

Using the fact that T_x is a stopping time, the last expectation may be decomposed and one has

(2) $$\mathbf{E}_0\left[\sum_{n=0}^{T_x - 1} t^n \phi_0(\mathbf{x}_n)\right] = R_t(0,0) - \mathbf{E}_0[t^{T_x}] R_t(x,0),$$

for $0 \le t < 1$, $x \ne 0$. Similarly (we omit the proof) one finds

(3) $$\mathbf{E}_x\left[\sum_{n=0}^{T_x - 1} t^n \phi_0(\mathbf{x}_n)\right] = R_t(x,0) - \mathbf{E}_0[t^{T_0}] R_t(x,0).$$

With the aid of (2) and (3) the ratio in (1) is

(4) $$\frac{1 - \mathbf{E}_0[t^{T_x}]}{1 - \mathbf{E}_0[t^{T_0}]}$$

$$= \left\{R_t(x,0) - R_t(0,0) + \mathbf{E}_0\left[\sum_{n=0}^{T_x - 1} t^n \phi_0(\mathbf{x}_n)\right]\right\} \left\{\mathbf{E}_x\left[\sum_{n=0}^{T_x - 1} t^n \phi_0(\mathbf{x}_n)\right]\right\}^{-1},$$

for $0 \le t < 1$, $x \ne 0$. Now, by Abel's theorem

(5) $\quad \lim\limits_{t \nearrow 1} [R_t(x,0) - R_t(0,0)] = \sum\limits_{n=0}^{\infty} [P_n(x,0) - P_n(0,0)] = -a(x),$

where the last step used part (a) of T28.1. Next we observe that

(6) $\quad \lim\limits_{t \nearrow 1} \mathbf{E}_0 \left[\sum\limits_{n=0}^{T_x - 1} t^n \phi_0(\mathbf{x}_n) \right] = \mathbf{E}_x \left[\sum\limits_{n=0}^{T_x - 1} \phi_0(\mathbf{x}_n) \right] = g_{\{x\}}(0,0),$

according to the definition of the Green function in D10.1. But by P29.4

(7) $\quad g_{\{x\}}(0,0) = a(x) + a(-x).$

Substitution of (5), (6), and (7) into (4) shows that P2 is true, provided

(8) $\quad \lim\limits_{t \nearrow 1} \mathbf{E}_x \left[\sum\limits_{n=0}^{T_x - 1} t^n \phi_0(\mathbf{x}_n) \right] = 1.$

This fact, however, is known from P10.3. (The frequent reference to rather difficult results in this proof is misleading. The proof could in fact be made quite elementary, at the cost of adding a little to its length.)

P3 *For one-dimensional aperiodic recurrent random walk with finite variance $\sigma^2 < \infty$,*

$$\lim\limits_{n \to \infty} \sqrt{n}\, \mathbf{P}_0[\mathbf{T} > n] = \sqrt{\frac{2}{\pi}}\, \sigma.$$

Proof: We write, for $0 \le t < 1$

(1) $\quad U(t) = \sum\limits_{n=0}^{\infty} U_n t^n = \sum\limits_{n=0}^{\infty} t^n P_n(0,0) = \frac{1}{2\pi} \int_{-\pi}^{\pi} \frac{d\theta}{1 - t\phi(\theta)},$

where $\phi(\theta)$ is the characteristic function of the random walk. Our immediate goal is to show that

(2) $\quad \lim\limits_{t \nearrow 1} \sqrt{1-t}\, U(t) = \frac{1}{\sqrt{2}\,\sigma}.$

Given any $\epsilon > 0$ we can choose $0 < \delta < \pi$ so that

$$\left[\frac{\theta^2 \sigma^2}{2} + 1 - t\right](1 - \epsilon) \le 1 - t\phi(\theta) = 1 - \phi(\theta) + (1-t)\phi(\theta)$$

$$\le \left[\frac{\theta^2 \sigma^2}{2} + 1 - t\right](1 + \epsilon)$$

when $|\theta| \leq \delta$. By T7.1 the integral in (1), taken from δ to π and from $-\pi$ to $-\delta$, remains bounded as $t \to 1$ since the random walk is aperiodic. Hence those parts of the integral in (1) are error terms which go to zero when multiplied by $\sqrt{1-t}$, and (2) will be true if

$$(3) \qquad \lim_{t \nearrow 1} \sqrt{1-t} \, \frac{1}{2\pi} \int_{-\delta}^{\delta} \frac{d\theta}{\frac{\theta^2 \sigma^2}{2} + (1-t)} = \frac{1}{\sqrt{2}\,\sigma}$$

for every $\delta > 0$, and that is easily verified by making the change of variable from θ to $x = \theta\sigma[2(1-t)]^{-1/2}$.

The next step uses the identity

$$\sum_{n=0}^{\infty} R_n t^n = [(1-t)U(t)]^{-1}$$

which follows from P1.2. Combined with (2) it gives

$$(4) \qquad \lim_{t \nearrow 1} \sqrt{1-t} \sum_{n=0}^{\infty} R_n t^n = \sigma\sqrt{2}.$$

Observe that the sequence $R_n = \mathbf{P}_0[T > n]$ is monotone nonincreasing, so that the stage is set for the application of Karamata's theorem. The strong form of this theorem, given in P20.2 implies that

$$\lim_{n \to \infty} \sqrt{n}\, R_n = \sigma\sqrt{2/\pi},$$

which proves P3.

If we combine equation (4) in the proof of P3 with P2, we have

$$(5) \qquad \lim_{t \nearrow 1} \sqrt{1-t} \sum_{n=0}^{\infty} t^n \mathbf{P}_x[T > n] = \sigma\sqrt{2}\, a(x)$$

for every $x \neq 0$. Karamata's theorem may again be applied to (5), and the conclusion is

P4 $\qquad \lim_{n \to \infty} \sqrt{n}\, \mathbf{P}_x[T > n] = \sigma\sqrt{2/\pi}\, a(x)$ *for* $x \neq 0$.

Of course P2 and P4 together yield

$$\lim_{n \to \infty} \frac{\mathbf{P}_x[T > n]}{\mathbf{P}_0[T > n]} = a(x)$$

for $x \neq 0$, so that *we have proved* T1 *for aperiodic recurrent random walk in the case* $d = 1$, $\sigma^2 < \infty$.

32. THE HITTING TIME OF A SINGLE POINT

Now it remains only to prove T1 in one dimension, when $\sigma^2 = \infty$. It is natural to look back at the proof of T16.1, inquiring whether the present case can be handled by the same methods. In the proof of T16.1 we arrived at the equation

(1) $$\frac{\mathbf{P}_x[\mathbf{T} > n]}{\mathbf{P}_0[\mathbf{T} > n]} = \frac{R_{n+1}}{R_n} \sum_{t \neq 0} g(x,t) v_n(t), \qquad x \neq 0.$$

Here $v_n(t)$ was a sequence of probability measures on $R - \{0\}$, and we were able to show that

(2) $$\lim_{n \to \infty} v_n(t) = 0, \qquad t \in R - \{0\}$$

using the result of P16.1, namely

(3) $$\lim_{n \to \infty} \frac{R_n}{R_{n+1}} = 1.$$

The proof of T16.1 was then completed by combining (2) and (3) with the asymptotic result that

(4) $$\lim_{|t| \to \infty} g(x,t) = a(x), \qquad x \in R - \{0\}.$$

In view of P29.4 and T29.1 equation (4) is also valid for aperiodic recurrent one-dimensional random walk with $\sigma^2 = \infty$. (It is false when $\sigma^2 < \infty$ and that is why this case required a different proof, given in P2, P3, and P4.) Consequently the proof of T1 would be complete if we had a proof of (3). That difficulty will be resolved by proving the far stronger statement in T2, which implies (3). (If a random walk is strongly aperiodic, then obviously $F_n/F_{n+1} \to 1$ entails

$$\frac{R_n}{R_{n+1}} = 1 + \frac{F_{n+1}}{R_{n+1}} \to 1.$$

And if it is not strongly aperiodic, then by P5.1, $F_k = 0$ unless $k = ns$ for some integer $s > 1$. But in that case the random walk with transition function $P_s(x,y)$ is strongly aperiodic, with some subgroup of R as its state space, and one gets $F_{ns}/F_{(n+1)s} \to 1$, which suffices to conclude that (3) holds).

The proof of T2 begins with

P5 *For every ϵ, $0 < \epsilon < 1$, there is some $N = N(\epsilon)$ such that*

$$F_n \geq (1 - \epsilon)^n, \qquad n \geq N.$$

Proof: We start with an arbitrary ϵ. Given any integer a we have

$$\mathbf{P}_0[\mathbf{x}_n = a] = \mathbf{P}[S_n = a] \geq (1 - \epsilon)^n$$

for all large enough n, according to P5.2. The random walk being recurrent, we can choose $a_1 > 0$ and $a_2 > 0$ so that

$$P(0,a_1) > 0 \quad \text{and} \quad P(0,-a_2) > 0.$$

Then

(1) $\quad \mathbf{P}[\mathbf{S}_1 = a_1, \ \mathbf{S}_{n-1} - \mathbf{S}_1 = a_2 - a_1, \ \mathbf{S}_n = 0] \geq (1 - \epsilon)^n$

for sufficiently large n. This is obvious since the above probability is nothing but $P(0,a_1)\mathbf{P}[\mathbf{S}_{n-2} = a_2 - a_1]P(0,-a_2)$. The notation of partial sums $\mathbf{S}_k = \mathbf{X}_1 + \cdots + \mathbf{X}_k$ of Chapter IV was chosen because it facilitates the following combinatorial argument. Consider the sequence $\mathbf{X}_2, \mathbf{X}_3, \ldots, \mathbf{X}_{n-1}$, and subject it to all possible cyclic permutations (of which there are $n - 2$, including the identity). If p is one of these cyclic permutations (acting on the integers $2, 3, \ldots, n - 1$) we get $\mathbf{X}_{p(2)}, \ldots, \mathbf{X}_{p(n-1)}$. For each cyclic permutation p consider the polygonal line $\mathbf{L}(p)$ in the plane formed by connecting the points

$$(1,a_1), (2,a_1 + \mathbf{X}_{p(2)}), (3,a_1 + \mathbf{X}_{p(2)} + \mathbf{X}_{p(3)}), \ldots, (n - 1,a_2).$$

For at least one of the $n - 2$ permutations the polygonal line will lie *entirely above or on the chord* connecting $(1,a_1)$ to $(n - 1,a_2)$. This may easily be inferred from the illustration below, where $n = 6$ and p is the cyclic permutation $(3, 4, 5, 2)$. Which one of the permutations

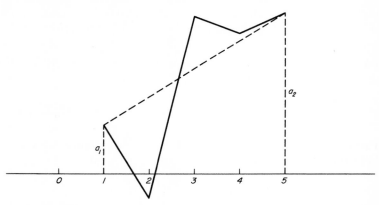

has this property depends of course on the values of the random variables $\mathbf{X}_2, \ldots, \mathbf{X}_{n-1}$. But that is irrelevant. What does matter is that the probability in (1) is invariant under these permutations, as the random variables $\mathbf{X}_k$ are independent and identically distributed, i.e., their joint probability measure is invariant under permutation of coordinates. Thus each cyclic permutation of the integers $2, \ldots, n-1$ is equally likely to yield such a polygon.

Now we are nearly finished. Call $E_n(p)$ the event that $\mathbf{X}_1 = a_1$, $\mathbf{X}_n = -a_2$ and that $\mathbf{L}(p)$ lies above its chord. Then, for each p, the probability of $E_n(p)$ is at least as large as the probability in (1) divided by $n-2$. If e denotes the identity permutation, then

$$\mathbf{P}_0[E_n(e)] \geq \frac{(1-\epsilon)^n}{n-2},$$

for large enough n. On the other hand, the event $E_n(e)$ implies that the first return of the random walk $\mathbf{x}_n$ with $\mathbf{x}_0 = 0$ to the origin occurs at time n. Hence

$$F_n \geq \mathbf{P}_0[E_n(e)] \geq \frac{(1-\epsilon)^n}{n-2}$$

for sufficiently large n, and that proves P5.

Remark: At this point, using P5, it is possible to complete the proof of T2 in a few lines, under the additional hypothesis that $P(0,0) > 0$. This may be done just as in the proof of the ratio ergodic theorem in P5.4. (Wherever we used P5.2, giving the exponential lower bound $P_n(0,0) \geq (1-\epsilon)^n$ in proving P5.4, one may now use P5 in exactly the same way.) However, it does not seem possible to eliminate the hypothesis that $P(0,0) > 0$ when working with the sequence F_n instead of $U_n = P_n(0,0)$. Therefore a different approach is necessary, which begins with

P6 *For every integer m and every $\epsilon > 0$ there is some $M = M(m,\epsilon)$ such that for each $A > 0$*

$$\varlimsup_{n \to \infty} \frac{1}{n} \sum_{k=1}^{n-M} \mathbf{P}[|\mathbf{S}_k| \leq A \text{ or } \mathbf{S}_{k+j} \neq \mathbf{S}_k \text{ when } m \leq j \leq M | \mathbf{T} = n] \leq \epsilon.$$

Proof: In the statement of P6, $\mathbf{T}$ is the first return time to zero. In addition to conditional probabilities of the form

$$\mathbf{P}[A \mid B] = \frac{\mathbf{P}[A \cap B]}{\mathbf{P}[B]}$$

when $\mathbf{P}[B] > 0$ (as will always be the case), we shall use conditional expectations, defined by

$$\mathbf{E}[\mathbf{f} \mid B] = \frac{1}{\mathbf{P}[B]} \mathbf{E}[\mathbf{f}; B].$$

We define the random variables (characteristic functions of events) $\mathbf{J}_k(A,n) = 1$ if $|\mathbf{S}_k| \leq A$ and $\mathbf{T} = n$, 0 otherwise; $\mathbf{L}_k(m,M,n) = 1$ if $\mathbf{S}_{k+j} \neq \mathbf{S}_k$ for $m \leq j \leq M$, 0 otherwise. For any pair of events A_1, A_2, $\mathbf{P}[A_1 \cup A_2 \mid B] \leq \mathbf{P}[A_1 \mid B] + \mathbf{P}[A_2 \mid B]$, so that the sum in P6 is bounded above by

$$(1) \quad \mathbf{E}\left[\sum_{k=1}^{n-M} \mathbf{J}_k(A,n) \mid \mathbf{T} = n\right] + \mathbf{E}\left[\sum_{k=1}^{n-M} \mathbf{L}_k(m,M,n) \mid \mathbf{T} = n\right].$$

Given $\eta > 0$ (to be given the value $\epsilon/2$ later) we pick M so large that

$$\mathbf{P}[\mathbf{S}_{k+j} \neq \mathbf{S}_k \text{ for } m \leq j \leq M] = \mathbf{P}[\mathbf{S}_j \neq 0 \text{ for } m \leq j \leq M] \leq \eta.$$

This may be done, since m is fixed and the random walk is recurrent. Then

$$(2) \quad \mathbf{E}\left[\sum_{k=1}^{n-M} \mathbf{L}_k(m,M,n) \mid \mathbf{T} = n\right]$$

$$= \mathbf{E}\left[\sum_{k=1}^{n-M} \mathbf{L}_k; \sum_{k=1}^{n-M} \mathbf{L}_k \leq 2n\eta \mid \mathbf{T} = n\right]$$

$$+ \mathbf{E}\left[\sum_{k=1}^{n-M} \mathbf{L}_k; \sum_{k=1}^{n-M} \mathbf{L}_k > 2n\eta \mid \mathbf{T} = n\right]$$

$$\leq 2n\eta + \frac{n}{\mathbf{P}[\mathbf{T} = n]} \mathbf{P}\left[\sum_{k=1}^{n-M} \mathbf{L}_k(m,M,n) \geq 2n\eta\right].$$

To continue the estimation observe that $\mathbf{L}_k = \mathbf{L}_k(m,M,n)$ depends on $\mathbf{X}_{k+1}, \mathbf{X}_{k+2}, \ldots, \mathbf{X}_{k+M}$, but not on any of the other increments $\mathbf{X}_n = \mathbf{S}_n - \mathbf{S}_{n-1}$ of the random walk. Thus we can write

$$\sum_{k=0}^{n-M} \mathbf{L}_k = (\mathbf{L}_0 + \mathbf{L}_M + \mathbf{L}_{2M} + \cdots) + (\mathbf{L}_1 + \mathbf{L}_{M+1} + \cdots) + \cdots$$
$$+ (\mathbf{L}_{M-1} + \mathbf{L}_{2M-1} + \cdots) = \mathbf{I}_1 + \mathbf{I}_2 + \cdots + \mathbf{I}_M.$$

Now each sum $\mathbf{I}_r$, $r = 1, \ldots, M$, contains at most $[n/M]$ terms. Further, each $\mathbf{I}_r$ is a sum of independent random variables, each of which assumes the value 1 with probability at most η. If $\sum_{k=1}^{n-M} \mathbf{L}_k > 2n\eta$, then one of the $\mathbf{I}_r$ must exceed $2n\eta/M$, so that

$$(3) \quad \mathbf{P}\left[\sum_{k=1}^{n-M} \mathbf{L}_k(m,M,n) \geq 2n\eta\right]$$

$$\leq \sum_{r=1}^{M} \mathbf{P}\left[\mathbf{I}_r \geq 2\eta \frac{n}{M}\right] \leq M\mathbf{P}\left[\mathbf{I}_1 \geq 2\eta \frac{n}{M}\right].$$

According to P5.3, if $\mathbf{I}$ is a Bernoulli sum of j independent random variables, each of them being one with probability η and zero otherwise, then the probability that $\mathbf{I}$ exceeds $2j\eta$ is less than a multiple of e^{-cj}. Hence

(4) $$\mathbf{P}\left[\mathbf{I}_1 \geq 2\eta \frac{n}{M}\right] \leq c_1 e^{-c_2(n/M)},$$

where c_1, c_2 are positive constants, depending only on η. Now one can combine (2), (3), and (4) to get

(5) $$\mathbf{E}\left[\sum_{k=1}^{n-M} \mathbf{L}_k(m,M,n) \mid \mathbf{T} = n\right] \leq 2n\eta + \frac{Mn}{F_n} c_1 e^{-c_2(n/M)}.$$

Turning to the estimation of the sum involving $\mathbf{J}_k(A,n)$, let $\boldsymbol{\phi}_k$ be the random variable which is 1 if $|\mathbf{S}_k| \leq A$ and 0 otherwise. Truncating just as was done in (2), one gets

(6) $$\mathbf{E}\left[\sum_{k=1}^{n-M} \mathbf{J}_k(A,n) \mid \mathbf{T} = n\right] \leq n\eta + \frac{n}{F_n} \mathbf{P}\left[\sum_{k=1}^{n} \boldsymbol{\phi}_k \geq n\eta \text{ and } \mathbf{T} = n\right]$$
$$\leq n\eta + \frac{n}{F_n} \mathbf{P}\left[\mathbf{T} \geq n \mid \sum_{k=1}^{n} \boldsymbol{\phi}_k \geq n\eta\right].$$

Now we choose some real $\chi > 0$ and an integer $r > 0$ such that

(7) $$\min_{|a| \leq A} \mathbf{P}[\mathbf{S}_k = a \text{ for some } k < r] \geq \chi > 0.$$

Then

$$\mathbf{P}[\mathbf{T} \geq j + r \mid |\mathbf{S}_j| \leq A] \leq 1 - \chi,$$

and since the event that $\boldsymbol{\phi}_1 + \cdots + \boldsymbol{\phi}_n \geq n\eta$ implies that one of the subsequences $\boldsymbol{\phi}_j + \boldsymbol{\phi}_{j+r} + \cdots$ of $\boldsymbol{\phi}_1 + \cdots + \boldsymbol{\phi}_n$ must exceed $n\eta/r$, we conclude that

(8) $$\mathbf{P}\left[\mathbf{T} \geq n \mid \sum_{1}^{n} \boldsymbol{\phi}_k \geq n\eta\right] \leq (1 - \chi)^{(n\eta)/r}.$$

Combining (5), (6), and (8),

(9) $$\frac{1}{n}\sum_{k=1}^{n-M} \mathbf{P}[|\mathbf{S}_k| \leq A \text{ or } \mathbf{S}_{k+j} \neq \mathbf{S}_k \text{ for all } n \leq j \leq M | \mathbf{T} = n]$$
$$\leq 2\eta + \frac{M}{F_n} c_1 e^{-c_2(n/M)} + \frac{1}{F_n}(1-\chi)^{(n\eta)/r}.$$

Now we set $2\eta = \epsilon$ and then, taking the limit superior as $n \to \infty$, P6 will be proved if the two exponential terms in (9) go to zero faster than F_n. But that follows from P5, no matter what are the positive constants c_2 and χ, and therefore P6 is true.

P7 *Let $\epsilon_1 > 0$, m and A be given. Then there exist positive integers k and M such that*

$$\left| \frac{q_n}{F_n} - 1 \right| \leq \epsilon_1 \quad \text{for all sufficiently large } n,$$

where

$$q_n = \mathbf{P}[\mathbf{T} = n, |\mathbf{S}_k| > A \text{ and } \mathbf{S}_{k+j} = \mathbf{S}_k \text{ for some } m \leq j \leq M].$$

Proof: Take ϵ in P6 to be $\epsilon = \epsilon_1^2$ and choose M so that P6 holds. P6 implies that for every $\delta > 0$, and sufficiently large n,

$$\sum_{k=1}^{n-M} \mathbf{P}[\{|\mathbf{S}_k| \leq A \text{ or } \mathbf{S}_{k+j} \neq \mathbf{S}_k \text{ for } m \leq j \leq M\} \cap \{\mathbf{T} = n\}]$$
$$\leq (\epsilon_1^2 + \delta)nF_n,$$

which shows that there are at least $\epsilon_1 n$ values of k for which

$$\mathbf{P}[\{|\mathbf{S}_k| \leq A \text{ or } \mathbf{S}_{k+j} \neq \mathbf{S}_k \text{ for } m \leq j \leq M\} \cap \{\mathbf{T} = n\}] \leq \epsilon_1 F_n.$$

If we pick such a value of k in the definition of q_n (which depends on k), then

$$(1 - \epsilon_1)F_n \leq q_n \leq F_n,$$

which proves P7.

P8 *For sufficiently large m and A there exist positive integers k and M such that*

$$|q_n - \tilde{q}_{n+1}| \leq \epsilon_1 q_n \quad \text{for sufficiently large } n,$$

with ϵ_1 and q_n defined in P7 and

$\tilde{q}_{n+1} = \mathbf{P}[\mathbf{T} = n + 1, |\mathbf{S}_k| > A, \mathbf{S}_{k+j} = \mathbf{S}_k$ *for some* $m + 1 \leq j \leq M + 1]$.

Proof: Remember that $U_n = \mathbf{P}[\mathbf{S}_n = 0]$, and

$$\lim_{n \to \infty} \frac{U_{n+1}}{U_n} = 1,$$

according to T5.1, which is crucial for the proof. We choose m so that

(1) $\qquad |U_{k+1} - U_k| \leq \frac{\epsilon_1}{4} U_k$ for $k \geq m$,

and A so large that

(2) $\qquad \mathbf{P}[|\mathbf{S}_j| \geq A \text{ for some } j \leq M + 1] \leq \frac{\epsilon_1}{4} \min_{m \leq k \leq M+1} U_k.$

To simplify the notation let

$\mathbf{P}[\mathbf{S}_{k+j} = a, \mathbf{S}_{k+r} \neq 0 \text{ for } r = 1, \ldots, j - 1 \mid \mathbf{S}_k = a] = c_{k,j}(a) = c_{k,j}$

$\mathbf{P}[\mathbf{T} = n - k - j; \mathbf{S}_p \neq a \text{ for } p = 1, 2, \ldots, M - j \mid \mathbf{S}_0 = a]$
$\qquad\qquad\qquad\qquad\qquad = d_{n-k,j}(a, M) = d_{n-k,j}{}^3$

Then, decomposing q_n according to the possible values of $\mathbf{S}_k$ and according to the last time $k + j$ when $\mathbf{S}_{k+j} = \mathbf{S}_k$,

$$q_n = \sum_{|a|>A} \mathbf{P}[\mathbf{S}_k = a; \mathbf{T} > k] \sum_{j=m}^{M} c_{k,j} d_{n-k,j}.$$

A similar decomposition for $\tilde{q}_{n+1}$ yields

$$\tilde{q}_{n+1} = \sum_{|a|>A} \mathbf{P}[\mathbf{S}_k = a; \mathbf{T} > k] \sum_{j=m}^{M} c_{k,j+1} d_{n-k,j},$$

so that

(3) $\qquad q_n - \tilde{q}_{n+1} = \sum_{|a|>A} \mathbf{P}[\mathbf{S}_k = a; \mathbf{T} > k] \sum_{j=m}^{M} (c_{k,j} - c_{k,j+1}) d_{n-k,j}.$

To estimate this difference observe that

$c_{k,j} = \mathbf{P}[\mathbf{S}_j = 0; \mathbf{S}_r \neq -a \text{ for } r = 1, 2, \ldots, j - 1]$
$\qquad = U_j - P[\mathbf{S}_j = 0, \mathbf{S}_r = -a \text{ for some } r = 1, 2, \ldots, j - 1].$

Since $|a| > A$,

$\qquad |c_{k,j} - U_j| \leq \mathbf{P}[|\mathbf{S}_r| \geq A \text{ for some } r \leq M + 1],$

and from (2) one gets

(4) $\qquad |c_{k,j} - U_j| \leq \frac{\epsilon_1}{4} U_j, \quad m \leq j \leq M.$

---

[3] According to D3.1, $\mathbf{S}_0 = 0$ by definition. Hence $\mathbf{P}[\ldots \mid \mathbf{S}_0 = a]$ is meaningless—it should be interpreted as $\mathbf{P}_a[\ldots]$.

In the same way one proves

(5) $\qquad |c_{k,j+1} - U_{j+1}| \le \frac{\epsilon_1}{4} U_j, \qquad m \le j \le M.$

Combining (4) and (5), we have

$$|c_{k,j+1} - c_{k,j}| \le \frac{\epsilon_1}{2} U_j + |U_{j+1} - U_j|,$$

and now one concludes from (1) that

$$|c_{k,j+1} - c_{k,j}| \le \frac{3\epsilon_1}{4} U_j.$$

Using (4) again, we obtain

(6) $\qquad |c_{k,j+1} - c_{k,j}| \le \frac{3\epsilon_1}{4} \left(1 - \frac{\epsilon_1}{4}\right)^{-1} c_{k,j} \le \epsilon_1 c_{k,j},$

at least if, as we may assume, ϵ_1 is sufficiently small.

Applying (6) to equation (3) we have

$$|q_n - \tilde{q}_{n+1}| \le \epsilon_1 \sum_{|a|>A} \mathbf{P}[\mathbf{S}_k = a; \mathbf{T} > k] \sum_{j=m}^{M} c_{k,j} d_{n-k,j} = \epsilon_1 q_n,$$

which proves P8.

We are now ready to complete the proof of T2. By definition $\tilde{q}_{n+1} \le F_{n+1}$, and so by P8 and P7,

$$F_{n+1} \ge \tilde{q}_{n+1} \ge (1 - \epsilon_1) q_n \ge (1 - \epsilon_1)^2 F_n$$

for large enough n, which amounts to saying that

$$\lim_{n \to \infty} \frac{F_{n+1}}{F_n} \ge 1.$$

But an obvious modification of P7 gives

$$\left| \frac{\tilde{q}_{n+1}}{F_{n+1}} - 1 \right| \le \epsilon_1$$

so that

$$F_n \ge q_n \ge \frac{1}{1 + \epsilon_1} \tilde{q}_{n+1} \ge \frac{1 - \epsilon_1}{1 + \epsilon_1} F_{n+1}.$$

That implies
$$\varlimsup_{n\to\infty} \frac{F_{n+1}}{F_n} \le 1,$$
which completes the proof of T2. In view of the remarks preceding P5, we have also *completed the proof of* T1.

Remark: One of the possible extensions of T1 was mentioned in Chapter III. It concerns the limit of the ratios
$$\frac{\mathbf{P}_x[\mathbf{T}_B > n]}{\mathbf{P}_0[\mathbf{T} > n]}, \text{ as } n \to \infty.$$
The step from T1 to the theorem (proved in [60]) that these ratios have a limit, as $n \to \infty$, for every finite set B, requires considerable effort. As the discussion at the end of section 16 might lead one to expect, the difficulty consists in showing that the ratios R_n/R_{2n} are bounded for every recurrent random walk. This is done in [60] by using very delicate estimates concerning the *dispersion function* of P. Lévy ([71], section 16).

Other aspects of the asymptotic behavior of the sequences R_n, U_n, and F_n are still shrouded in mystery, giving rise to challenging conjectures. Kesten has shown ([61]) that every strongly aperiodic recurrent one-dimensional random walk in the domain of attraction of a symmetric stable law (i.e., satisfying
$$0 < \lim_{\theta \to 0} |\theta|^{-\alpha}[1 - \phi(\theta)] = Q < \infty$$
for some α, $1 \le \alpha \le 2$) has the property that

(1) $$\lim_{n\to\infty} \frac{1}{F_n} \sum_{k=0}^{n} F_k F_{n-k} = 2.$$

He conjectures that (1) holds for every strongly aperiodic recurrent random walk.[4] This conjecture is related, in an obvious way, to the random variables

$$\mathbf{T}_k = \text{time of the } k^{\text{th}} \text{ return to } 0.$$

Equation (1) simply states that $\mathbf{P}_0[\mathbf{T}_2 = n]/\mathbf{P}_0[\mathbf{T}_1 = n]$ tends to 2 as $n \to \infty$, but it also follows from (1) that

(2) $$\lim_{n\to\infty} \frac{\mathbf{P}_0[\mathbf{T}_r = n]}{\mathbf{P}_0[\mathbf{T}_s = n]} = \frac{r}{s},$$

for every pair r,s of positive integers.

---

[4] This is now proved under the additional assumption that the random walk is symmetric [S18].

Problems

1. For one-dimensional aperiodic recurrent random walk with absolute moment of order $3 + \delta$ for some $\delta > 0$, show that

$$\lim_{|x| \to \infty} \left[a(x) - \frac{|x|}{\sigma^2} \right]$$

exists and is finite. Is this true if only σ^2 is finite? (To obtain a sharp result, use problem 2 in Chapter II.)

2. Here and in problem 3 the random walk is one dimensional and aperiodic, with a characteristic function $\phi(\theta)$ such that

$$0 < \lim_{\theta \to 0} \frac{1 - \phi(\theta)}{|\theta|^\alpha} = Q < \infty$$

for some α, $1 \leq \alpha \leq 2$. Show that all such random walks are recurrent. Prove that

$$\lim_{|x| \to \infty} \mathbf{P}_0[\mathbf{T}_{-x} < \mathbf{T}_x] = 1/2,$$

where $\mathbf{T}_x$ is the time of the first visit to x.

3. Continuation. Show that

$$\lim_{|x| \to \infty} \mathbf{P}_0[\mathbf{T}_x < \mathbf{T}_{2x}] = 2^{\alpha - 2}.$$

4. Extend T32.2 by showing that

$$\lim_{n \to \infty} \frac{F_{n+1}}{F_n} = 1$$

for every strongly aperiodic recurrent random walk in the plane.

5. Here and in the next problem the random walk is one dimensional and aperiodic, with mean 0 and $\sigma^2 < \infty$. Use P32.3 to conclude that the ratios R_n/R_{2n} are bounded in n. Then use the discussion at the end of Chapter III to prove that

$$\lim_{n \to \infty} \frac{\mathbf{P}_x[\mathbf{T}_B > n]}{\mathbf{P}_0[\mathbf{T} > n]} = \operatorname*{Lim}_y g_B(x,y) = \tfrac{1}{2} \left[\lim_{y \to +\infty} g_B(x,y) + \lim_{y \to -\infty} g_B(x,y) \right]$$

for all $x \in R - B$, for every finite set B.

6. Use the result of problem 5 together with P32.3 to evaluate the limit

$$\lim_{n \to \infty} \sqrt{n}\, \mathbf{P}_0[\mathbf{T}_{\{x, -x\}} > n] = f(x),$$

where $x \neq 0$ is an arbitrary point in R (but not the origin). Show that either $f(x) \equiv 0$ for $x \neq 0$ or $f(x) > 0$ for all $x \neq 0$, and characterize those random walks which give $f(x) \equiv 0$.

7. The mean square distance of a random walk $\mathbf{x}_n$ from the origin, subject to the condition of no return to the origin in time n, is

$$D_n = \mathbf{E}_0[|\mathbf{x}_n|^2 \mid \mathbf{T} > n] = \frac{1}{R_n} \mathbf{E}_0[|\mathbf{x}_n|^2; \mathbf{T} > n].$$

Suppose that the random walk is genuinely d-dimensional, with mean vector zero and second moment $m_2 = \mathbf{E}_0[|\mathbf{x}_1|^2] < \infty$. When the dimension $d = 1$, prove that

$$D_n \sim 2nm_2 = 2\mathbf{E}_0[|\mathbf{x}_n|^2] \quad \text{as } n \to \infty,$$

whereas for dimension $d \geq 2$

$$D_n \sim nm_2 = \mathbf{E}_0[|\mathbf{x}_n|^2] \quad \text{as } n \to \infty.$$

Hint: To obtain these asymptotic relations, first derive the identity

$$\mathbf{E}_0[|\mathbf{x}_n|^2; \mathbf{T} > n] = (R_0 + R_1 + \cdots + R_{n-1})m_2, \qquad n \geq 1.$$

8. Prove that the potential kernel $A(x,y)$ of a recurrent random walk determines the transition function uniquely.

Hint: In the spirit of problem 13, Chapter VI, show that for every subset $B \subset R$, $\Pi_B(x,y)$ is determined by $A(x,y)$. This follows from T11.1 and T30.2 in all cases except for left-, or right-continuous random walk. (But even then $A(x,y)$ does determine $\Pi_B(x,y)$, according to the discussion preceding T30.2.) Finally, let $B_r = [x \mid |x| \leq r]$ and show that

$$P(x,y) = \lim_{r \to \infty} \Pi_{B_r}(x,y).$$

9. The logarithmic capacity $C(A)$ of a finite set $A \subset R$ was defined for arbitrary recurrent aperiodic random walk in the discussion following T30.2. Using part (d) of T30.2 one can interpret $C(A)$ as the *difference in size between the set A and the single point $\{0\}$, as seen from infinity*. Define the difference in size as seen from the point x, as

$$C(A;x) = g_{\{0\}}(x,x) - g_A(x,x).$$

Thus we measure the size of a set A, seen from x, as the expected time spent at x between visits to A. Prove that

$$C(A) = \lim_{|x| \to \infty} C(A;x),$$

when $d = 2$ or when $d = 1$ and $\sigma^2 = \infty$, while the situation is much more complicated when $d = 1$ and $\sigma^2 < \infty$.

10. For still another characterization of capacity define the *energy dissipated by a finite set A* (or the *expected area swept out by the set A*) as

$$E_n(A) = \sum_{x \in R} \mathbf{P}_x[\mathbf{T}_A \le n].$$

Investigate whether $E_n(A) - E_n(B)$ may be normalized appropriately so as to approximate $C(A) - C(B)$ for arbitrary recurrent random walk.

Hint: Consider first the asymptotic Abel sums, showing that

$$\sum_{n=0}^{\infty} t^n [E_n(A) - E_n(\{0\})] \sim \left[\sum_{n=0}^{\infty} R_n t^n \right]^2 C(A) \text{ as } t \nearrow 1.$$

Here, as usual, $R_n = \mathbf{P}_0[\mathbf{T}_0 > n]$. Conclude that

$$\lim_{n \to \infty} \frac{E_n(A) - E_n(B)}{\sum_{k=0}^{n} R_k R_{n-k}} = C(A) - C(B)$$

whenever this limit exists. (S. Port has shown [S23] that it exists for arbitrary recurrent random walk.)

11. Prove, using the properties of $a(x)$, that for an arbitrary recurrent random walk the expected number of distinct points visited between successive visits to zero is always infinite.

12. Prove that the potential kernel of arbitrary two-dimensional recurrent random walk satisfies

(i) $\lim_{|x| \to \infty} a(x) = \infty.$

13. *Continuation.* Prove that (i) also holds in one dimension when $\sigma^2 < \infty$. However, when $\sigma^2 = \infty$, show that either (i) holds, or

(ii) $\lim_{x \to +\infty} a(x) = M < \infty, \quad \lim_{x \to -\infty} a(x) = +\infty,$

or

(iii) $\lim_{x \to +\infty} a(x) = +\infty, \quad \lim_{x \to -\infty} a(x) = M < \infty.$

Note: The exact criteria for this classification are still unknown.

BIBLIOGRAPHY

1. Ahlfors, L. V., Complex analysis, McGraw-Hill Book Co., Inc., New York, 1953.
2. Baxter, G., An analytic approach to finite fluctuation problems in probability, *J. d'Analyse Math.* **9**, 31–70, 1961.
3. Baxter, G., A norm inequality for a "finite-section" Wiener-Hopf equation, *Ill. J. Math.* **7**, 97–103, 1963.
4. Blackwell, D., Extension of a renewal theorem. *Pacific J. Math.* **3**, 315–320, 1953.
5. Bochner, S., Lectures on Fourier integrals, Princeton University Press, Princeton, N.J., 1959.
6. Borel, E., Les probabilités dénombrables et leurs applications arithmétiques, *Rend. Circ. Mat. Palermo* **27**, 247–271, 1909.
7. Brelot, M., Éléments de la théorie classique du potentiel, Les cours de Sorbonne, Paris, 1959.
8. Choquet, G., and J. Deny, Sur l'equation $\mu = \mu * \sigma$, *Comptes Rendus Ac. Sc. Paris* **250**, 799–801, 1960.
9. Chung, K. L., Markov chains with stationary transition probabilities, J. Springer, Berlin, 1960.
10. Chung, K. L., On the renewal theorem in higher dimensions, *Skand. Aktuarietidskrift* **35**, 188–194, 1952.
11. Chung, K. L., and P. Erdös, Probability limit theorems assuming only the first moment, *Mem. Am. Math. Soc.* **6**, 1951.
12. Chung, K. L., and W. H. J. Fuchs, On the distribution of values of sums of random variables, *Mem. Am. Math. Soc.* **6**, 1951.
13. Chung, K. L., and D. Ornstein, On the recurrence of sums of random variables, *Bull. Am. Math. Soc.* **68**, 30–32, 1962.
14. Ciesielski, Z., and J. Taylor, First passage times and sojourn times for Brownian motion in space and exact Hausdorff measure of the sample path, *Trans. Am. Math. Soc.* **103**, 434–450, 1962.
15. Courant, R., K. Friedrichs, and H. Lewy, Über die partiellen Differenzengleichungen der mathematischen Physik, *Math. Ann.* **100**, 32–74, 1928.
16. Courant, R., and D. Hilbert, Methods of mathematical physics, Vols. I and II, Interscience Publishers, Inc., New York, 1953 and 1962.

17. Cox, D. R., and W. L. Smith, Queues, Methuen & Co., London, 1961.
18. Cramér, H., Mathematical methods of statistics, Princeton University Press, Princeton, N.J., 1946.
19. Darling, D. A., The maximum of sums of stable random variables, *Trans. Am. Math. Soc.* **83**, 164–169, 1956.
20. Darling, D. A., and M. Kac, On occupation times for Markoff processes, *Trans. Am. Math. Soc.* **84**, 444–458, 1957.
21. Derman, C., A solution to a set of fundamental equations in Markov chains, *Proc. Am. Math. Soc.* **5**, 332–334, 1954.
22. Donsker, M., An invariance principle for certain probability limit theorems, *Mem. Am. Math. Soc.* **6**, 1951.
23. Doob, J. L., Stochastic processes, John Wiley and Sons., Inc., New York, 1953.
24. Doob, J. L., Discrete potential theory and boundaries, *J. Math. and Mech.* **8**, 433–458, 1959.
25. Doob, J. L., J. L. Snell and R. E. Williamson, Application of boundary theory to sums of independent random variables, Contr. to Probability and Statistics (Hotelling Anniversary Volume) Stanford University Press, Stanford, Calif., pp. 182–197, 1960.
26. Dudley, R. M., Random walk on Abelian groups, *Proc. Am. Math. Soc.* **13**, 447–450, 1962.
27. Dvoretzky, A., and P. Erdös, Some problems on random walk in space, Second Berkeley Symp. on Statistics and Probability, pp. 353–368, 1951.
28. Erdös, P., A problem about prime numbers and the random walk II, *Ill. J. Math.* **5**, 352–353, 1961.
29. Erdös, P., and M. Kac, On certain limit theorems of the theory of probability, *Bull. Am. Math. Soc.* **52**, 292–302, 1946.
30. Fejér, L., Über trigonometrische Polynome, *J. f. d. reine u. angew. Math.* **146**, 53–82, 1915.
31. Feller, W., An introduction to probability theory and its applications, Vol. I, 2nd ed. John Wiley and Sons, Inc., New York, 1957, and Vol. II (forthcoming).
32. Feller, W., and S. Orey, A renewal theorem, *J. Math. and Mech.* **10**, 619–624, 1961.
33. Fisher, M. E., and M. F. Sykes, Excluded volume problem and the Ising model of ferromagnetism, *Phys. Rev.* **114**, 45–58, 1959.
34. Gnedenko, B. V., The theory of probability, Chelsea Publishing Co., New York, 1962.
35. Gnedenko, B. V., and A. N. Kolmogorov, Limit distributions for sums of independent random variables, Addison–Wesley Publishing Co., Inc., Reading, Mass., 1954.
36. Grenander, U., and G. Szegö, Toeplitz forms and their applications, University of California Press, Berkeley, 1958.

37. Halmos, P. R., Measure theory, D. Van Nostrand Co., Inc., Princeton, N.J., 1950.
38. Halmos, P. R., Finite-dimensional vector spaces, D. Van Nostrand Co. Inc., Princeton, N.J., 1958.
39. Hardy, G. H., Divergent Series, Clarendon Press, Oxford, England, 1949.
40. Harris, T. E., Branching processes, J. Springer, Berlin, 1963.
41. Hausdorff, F. Summationsmethoden und Momentfolgen, I, II, *Math. Zeitschrift* **9**, 74–109 and 280–299, 1921.
42. Hewitt, E., and L. J. Savage, Symmetric measures on Cartesian products, *Trans. Am. Math. Soc.* **80**, 470–501, 1955.
43. Hoeffding, W., On sequences of sums of independent random vectors, Fourth Berkeley Symp. on Statistics and Probability Vol. II, pp. 213–226, 1961.
44. Hunt, G. A., Some theorems on Brownian motion, *Trans. Am. Math. Soc.* **81**, 294–319, 1956.
45. Hunt, G. A., Markoff chains and Martin boundaries, *Ill. J. Math.* **4**, 313–340, 1960.
46. Itô, K., and H. P. McKean, Jr., Potentials and the random walk, *Ill. J. Math.* **4**, 119–132, 1960.
47. Itô, K., and H. P. McKean, Jr., Diffusion processes and their sample paths, J. Springer, Berlin, 1965.
48. Kac, M., Random walk and the theory of Brownian motion, *Am. Math. Monthly* **54**, 369–391, 1947.
49. Kac, M., Toeplitz matrices, translation kernels, and a related problem in probability theory, *Duke Math. J.* **21**, 501–510, 1954.
50. Kac, M., A class of limit theorems, *Trans. Am. Math. Soc.* **84**, 459–471, 1957.
51. Kac, M., Probability and related topics in physical sciences, Interscience Publishers, Inc., New York, 1959.
52. Karlin, S., On the renewal equation, *Pacific J. Math.* **5**, 229–257, 1955.
53. Katz, M., The probability in the tail of a distribution, *Ann. Math. Stat.* **34**, 312–318, 1963.
54. Kellogg, O. D., Foundations of potential theory, Dover Publications, Inc., New York, 1953.
55. Kemeny, J. G., and J. L. Snell, A note on Markov chain potentials, *Ann. Math. Stat.* **32**, 709–715, 1961.
56. Kemeny J. G., and J. L. Snell, Boundary theory for recurrent Markov chains. *Trans. Am. Math. Soc.* **106**, 495–520, 1963.
57. Kemperman, J. H. B., The passage problem for a stationary Markov chain, University of Chicago Press, Chicago, 1961.
58. Kesten, H., On a theorem of Spitzer and Stone and random walks with absorbing barriers, *Ill. J. Math.* **5**, 246–266, 1961.

59. Kesten, H., Random walks with absorbing barriers and Toeplitz forms, *Ill. J. Math.* **5**, 267–290, 1961.
60. Kesten, H., and F. Spitzer, Ratio theorems for random walk, *Journal d'Analyse Math.* **11**, 285–322, 1963.
61. Kesten, H., Ratio theorems for random walk II, *Journal d'Analyse Math.* **11**, 323–379, 1963.
62. Khinchin, A. I., Sur la loi des grands nombres, *Comptes Rendus Ac. Sci. Paris* **188**, 477–479, 1929.
63. Kiefer, J., and J. Wolfowitz, On the characteristics of the general queueing process with applications to random walk, *Ann. Math. Stat.* **27**, 147–161, 1956.
64. Kingman, J. F. C., and S. Orey, Ratio limit theorems for Markov chains, *Proc. Am. Math. Soc.*, **15**, 907–910, 1964.
65. Kochen, S. B., and C. J. Stone, A note on the Borel–Cantelli problem *Ill. J. Math*, to appear in 1964.
66. Kolmogorov, A. N., Grundbegriffe der Wahrscheinlichkeitsrechnung, J. Springer, Berlin, 1933.
67. Kolmogorov, A. N., Anfangsgründe der Theorie der Markoffschen Ketten mit unendlich vielen möglichen Zuständen, *Mat. Sbornik N.S.* **1**, 607–610, 1936.
68. König, H., Neuer Beweis eines klassischen Tauber-Satzes, *Archiv der Math.* **9**, 278–281, 1960.
69. Kreĭn, M. G., Integral equations on the half-line with a difference kernel, *Uspehi Mat. Nauk* **13**, 3–120, 1958; also *Am. Math. Soc. Translations* Series 2, **22**, 1962.
70. Lamperti, J., Wiener's test and Markov chains, *J. Math. An. and Appl.* **6**, 58–66, 1963.
71. Lévy, P., Théorie de l'addition des variables aléatoires, Gautier-Villars, Paris, 1937.
72. Lévy, P., Processus stochastiques et mouvement Brownien, Gautier-Villars, Paris, 1948.
73. Loève, M., Probability theory, 3rd ed., D. Van Nostrand Co., Inc., Princeton, N.J., 1963.
74. Loomis, L. H., An introduction to abstract harmonic analysis, D. Van Nostrand Co., Inc., Princeton, N.J., 1953.
75. Martin, R. S., Minimal positive harmonic functions, *Trans. Am. Math. Soc.* **49**, 137–172, 1941.
76. McCrea, W. H., and F. J. W. Whipple, Random paths in two and three dimensions, *Proc. Royal Soc. Edinburgh* **60**, 281–298, 1940.
77. McKean, H. P., Jr., A problem about prime numbers and the random walk I, *Ill. J. Math.* **5**, 131, 1961.
78. Munroe, M. E., Introduction to measure and integration, Addison–Wesley Publishing Co., Inc., Reading, Mass., 1953.

79. Nevanlinna, R. H., Eindeutige analytische Funktionen, 2nd ed., J. Springer, Berlin, 1953.
80. Orey, S., Potential kernels for recurrent Markov chains, *J. Math. Anal. and Appl.*, to appear in 1964.
81. Paley E. A. C., and N. Wiener, Fourier transforms in the complex domain, *Am. Math. Soc. Coll. Publ.*, **XIX**, 1934.
82. Pollaczek, F., Über eine Aufgabe der Wahrscheinlichkeitstheorie I–II, *Math. Zeitschrift* **32**, 64–100 and 729–750, 1930.
83. Polya, G., Über eine Aufgabe der Wahrscheinlichkeitsrechnung betreffend die Irrfahrt im Strassennetz, *Math. Ann.* **84**, 149–160, 1921.
84. Rényi, A., Wahrscheinlichkeitsrechnung, D.V.d. Wissenschaften, Berlin, 1962.
85. Rényi, A., Legendre polynomials and probability theory, *Ann. Univ. Sc. Budapestiniensis* **III–IV**, 247–251, 1960–61.
86. Riesz, F., Sur les fonctions surharmoniques et leur rapport à la theorie du potentiel 2, *Acta Math.* **54**, 1930.
87. Riesz, F., and B. Sz.-Nagy, Lecons d'Analyse Fonctionelle, Kiadó, Budapest, 1952.
88. Rosén, B., On the asymptotic distribution of sums of independent identically distributed random variables, *Arkiv för Mat.* **4**, 323–332, 1962.
89. Smith, W. L., A frequency function form of the central limit theorem. *Proc. Cambridge Phil. Soc.*, **49**, 462–472, 1953.
90. Smith, W. L., Renewal theory and its ramifications, *J. Royal Stat. Soc. Series B* **20**, 243–302, 1958.
91. Sparre Andersen, E., On the fluctuations of sums of random variables, *Math Scand.* **1**, 263–285, 1953, and **2**, 195–223, 1954.
92. Spitzer, F. L., A combinatorial lemma and its application to probability theory, *Trans. Am. Math. Soc.* **82**, 323–339, 1956.
93. Spitzer, F. L., Recurrent random walk and logarithmic potential, Fourth Berkeley Symp. on Statistics and Probability, Vol. II, pp. 515–534, 1961.
94. Spitzer, F. L., Hitting probabilities, *J. Math. and Mech.* **11**, 593–614, 1962.
95. Spitzer, F. L., and C. J. Stone, A class of Toeplitz forms and their application to probability theory, *Ill. J. Math.* **4**, 253–277, 1960.
96. Stöhr, A., Über einige lineare partielle Differenzengleichungen mit konstanten Koeffizienten, I, II, and III, *Math. Nachr.* **3**, 208–242, 295–315, and 330–357, 1949–50.
97. Szegö, G., Beiträge zur Theorie der Toeplitzschen Formen, I, II, *Math. Zeitschrift* **6**, 167–202, 1920, and **9**, 167–190, 1921.
98. Täcklind, S., Sur le risque de ruine dans des jeux inéquitables, *Skand. Aktuarietidskrift* **25**, 1–42, 1942.

99. Waerden, B. L. van der, Algebra II, 4th Ed., J. Springer, Berlin, 1959.
100. Watanabe, T., On the theory of Martin boundaries induced by countable Markov processes, *Mem. Coll. Science, University Kyoto, Series A* **33**, 39–108, 1960.
101. Watanabe, T., Probabilistic methods in Hausdorff moment problem and Laplace-Stieltjes transform, *J. Math. Soc. of Japan*, **12**, 192–206, 1960.
102. Watanabe, T., On a theorem of F. L. Spitzer and C. J. Stone, *Proc. Japan Academy* **37**, 346–351, 1961.
103. Watson, G. N., A treatise on the theory of Bessel functions, 2nd Ed., Cambridge University Press, 1944.
104. Watson, G. N., Three triple integrals, *Oxford Qu. J. of Math.* **10**, 266–276, 1939.
105. Widom, H., Toeplitz matrices, *Math. Assoc. of America Studies in Mathematics*, Vol. 3, 179–209, 1965.
106. Zygmund, A., Trigonometric Series, I, II, Cambridge University Press, 1959.

SUPPLEMENTARY BIBLIOGRAPHY

1. Breiman, L., Probability, Addison-Wesley, 1968.
2. Brunel, A. and D. Revuz, Marches Recurrentes au sense de Harris sur les groupes localement compactes I, *Ann. Sc. Ecole Norm. Sup.* **7**, 273–310, 1974.
3. Bucy, R. S., Recurrent sets, *Ann. Math. Stat.* **36**, 535–545, 1965.
4. Chung, Kai Lai, A course in probability theory, 2nd ed. Academic Press, 1974.
5. Darling, D. A. and P. Erdös, On the recurrence of a certain chain, *Proc. Am. Math. Soc.* **19**, 336–338, 1968.
6. Doney, An analogue of the renewal theorem in higher dimensions, *Proc. London Math. Soc.* **16**, 669–684, 1966.
7. Dym, H. and H. P. McKean, Fourier series and integrals, Academic Press, New York, 1972.
8. Dynkin E. B. and A. A. Yushkevich, Markov processes—theorems and problems, Plenum Press, New York, 1969.
9. Feller, W., An introduction to probability theory and its applications, Vol. II, 2nd ed. John Wiley and Sons, New York, 1971.
10. Flatto, L. and J. Pitt, Recurrence criteria for random walks on countable Abelian groups, III, *J. Math.* **18**, 1–20, 1974.
11. Heyde, C. C., Revisits for transient random walk, *Stoch. Proc. Appl.* **1**, 33–52, 1973.
12. Jain, N. C. and S. Orey, On the range of a random walk, *Israel J. Math.* **6**, 373–380, 1968.
13. Jain, N. C. and W. E. Pruitt, The range of recurrent random walk in the plane, *Z.f. Wahrscheirlichkeitstheorie* **16**, 279–292, 1970.
14. Jain, N. C. and W. E. Pruitt, The range of transient random walk, *J. d'Analyse Math.* **24**, 369–393, 1971.
15. Jain, N. C. and W. E. Pruitt, The range of random walk, *Proc. Sixth Berkeley Symp. Probability Statistics* **3**, 31–50, 1972.
16. Kesten, H., On the number of self avoiding walks, I and II, *J. Math. Phys.* **4**, 960–969, 1963; and **5**, 1128–1137, 1964.
17. Kesten, H., The Martin boundary of recurrent random walks on countable groups, *Proc. Fifth Berkeley Symp. Statistics Probability* **2**(2), 51–74, 1967.

18. Kesten, H., A ratio limit theorem for symmetric random walk, *J. d'Analyse Math.* **23**, 199–213, 1970.
19. Kesten, H., Sums of independent random variables—without moment conditions, *Ann. Math. Stat.* **43**, 701–732, 1972.
20. Kesten, H. and F. Spitzer, Random walk on countably infinite Abelian groups, *Acta Math.* **114**, 237–265, 1965.
21. Ney, P. and F. Spitzer, The Martin boundary for random walk, *Trans. Am. Math. Soc.* **121**, 116–132, 1966.
22. Ornstein, D. S., Random walks I and II, *Trans. Am. Math. Soc.* **138**, 1–43; and 45–60, 1969.
23. Port, S. C., Limit theorems involving capacities for recurrent Markov chains, *J. Math. An. Appl.* **12**, 555–569, 1965.
24. Port, S. C., Limit theorems involving capacities, *J. Math. Mech.* **15**, 805–832, 1966.
25. Port, S. C. and C. J. Stone, Potential theory of random walks on Abelian groups, *Acta Math.* **122**, 19–114, 1969.
26. Stone, C., On the potential operator for one dimensional recurrent random walks, *Trans. Am. Math. Soc.* **136**, 413–426, 1969.

INDEX

Abel's theorem, 199
Abelian groups, 92, 281
Absorption problem for
  simple random walk, 237
  random walk with mean 0,
    $\sigma^2 < \infty$, 244
Actuarial science, 174
A. e. convergence, 24
Aperiodic random walk, 20, 67
Arc-sine law, 227, 233, 258
Area swept out by a set, 394

Bachelier, L., 270
Baxter, G., 174, 220, 271
Bernoulli random walk
  definition, 2
  exponential bound, 45
  Green function, 8, 12
  regular functions, 131
Bernstein's estimate, 45
Bessel functions, 104, 235, 236
Bessel's inequality, 57
Beta function, 256
Biorthogonal system, 271
Birkhoff's ergodic theorem, 38, 53
Blackwell, D., 195, 282, 284
Borel, É., 32
Borel-Cantelli lemma, 317
Boundary, 327
Branching process, 234
Brownian motion, 40, 173, 219, 270, 312, 320

Cantor's diagonal process, 277, 347
Capacity
  definitions (transient random walk), 300, 303, 341
  electrostatic, 40
  inequalities, 304
  logarithmic (recurrent random walk), 366, 393, 394
  of finite sets, 40, 301
  of infinite sets, 305
Capacitory potential, 300

Catalan's constant, 124
Cauchy distribution, 156
Central limit theorem, 64, 79, 196, 342, 351
Césaro mean, 354
Characters
  of Abelian groups, 91
  orthogonality of, 92
Characteristic function, 57
Charge distribution, 137, 302
  potential due to, 137, 301
Chebychev's inequality, 37
Choquet, G., 276
Chung, K. L., 23, 41, 99, 288, 307, 322
Ciesielski, Z., 312
Coin tossing at random times, 235
Combinatorial method, 174
Complete orthonormal set, 315
Completely continuous operator, 241, 315
Conformal mapping, 258
Conjecture, of H. Kesten, 391
Convergence
  weak, 63
  a.e. (with probability one), 24
Convex body, 340
Convolution, 22
Coset, 44
Courant, R., 257
Courant-Hilbert, 242
Covariance matrix, 80
Cylinder set, 25

Darling, D. A., 231, 232
Deny, J., 276
Diagonal
  hitting probability of, 89
Diagonalization, 237
Difference equations, 9, 239, 257
Diffusion problem, 158
Dirichlet problem
  exterior, 134, 157, 175, 257
  interior, 173
Dispersion function, 391

INDEX

Distribution function, 63
Domain of attraction, 391
Dominated convergence theorem, 25, 58
Doob, J. L., 192, 298, 327
Dudley, R. M., 95
Dvoretzky, A., 38

Eigenfunction expansion, 242
Eigenvalues
  of quadratic forms, 70
  for simple random walk, 239
  distribution of, 272
Elliptic integral, 103
Entrance probability, 290
Equilibrium charge for
  recurrent random walk, 146, 377
  transient random walk, 300
Equilibrium potential for
  recurrent random walk, 146
  transient random walk, 300
Erdös, P., 38, 41, 99, 232, 270
Ergodic theory, 32
Escape probability, 290
Euler's constant, 124, 127
Excessive function, 323
Expectation, 26
Exponential bound
  for Bernoulli sums, 45, 385, 387
Exponential regular function, 340
Extension theorem, 26
Exterior Fourier series, 179, 220

Factorization, 180
Fatou's lemma, 84
Favorable game, 217, 227
Fejér representation, 182
Feller, W., 18, 99, 195, 224, 225, 237, 288
Finite set
  capacity of, 301
  energy dissipated by, 394
  volume swept out by, 40, 341
Fluctuations, 218
Fourier series, 54
Fourier transform, 63
Fraction of points visited, 271
Friedrichs, K., 258
Fubini's theorem, 25

Function
  excessive, 323
  integrable, 25
  measurable, 24
  regular, 130, 323
  simple, 24
Fundamental parallelogram
  volume of, 69

Gambler's ruin, 217
Gauss' mean value theorem, 314
Gaussian density, 79
Genuinely d-dimensional random walk, 72
Geometric
  distribution, 110
  random walk, 232, 270
Gnedenko, B. V., 76
Gram-Schmidt process, 248
Green (or red) points, 52, 272
Green function, 274
  asymptotic behavior of, 288, 307, 359
  determines the random walk, 341
  of exterior of interval, 238, 240, 251, 259
  finite set, 140, 143, 365
  half-line, 209, 327, 332
  half-plane, 153
  quadrant, 156
Green's theorem, 79

Halmos, P. R., 84, 298
Hardy and Littlewood's theorem, 228
Hardy, G. H., 225
Harmonic analysis, 54, 91
Harmonic functions, 129
  mean value theorem for, 129
Hausdorff, F., 338
Hausdorff moment problem, 328, 338
Hewitt, E., 298
Hilbert space, 249
Hitting probability measure, 107, 113, 135
  of diagonal, 89
  exterior of interval, 238
  finite set, 365
  half-line, 174, 195
  half-plane, 155

quadrant, 157, 173
triangle, 172
asymptotic behavior, 293, 312, 359
Hitting time, 29, 106
of exterior of interval, 242, 244, 262
half-line, 189, 191, 195, 196
single point, 377
Hoeffding, W., 343
Hunt, G. A., 159, 327, 334

Ideal, 65
Images, method of, 154
Imbedded random walk, 257
Independent
functions, 25
random variables, 27
Infinite transient set, 339
Inner
function, 179, 220
product, 248
Integrable function, 25
Interior Fourier series, 179, 220
Interval
absorption outside, 237
Invariance principle, 270
Inventory, 174
Isomorphism, 65
Isotropic random walk, 308
Itô, K., 173, 316, 321, 339

Kac, M., 159, 171, 231, 232, 237, 270, 272
Karamata's theorem, 225, 382
Katz, M., 232
Kemeny, J. G., 120, 343
Kemperman, J. H. B., 174, 185, 233
Kesten, H., 40, 255, 256, 333, 378, 391
Khinchin, A. I., 22
Kiefer, J., 209
Kolmogorov, A. N., 24, 32, 99
König, H., 225
Kreĭn, M. G., 212
Kronecker's lemma, 254

Ladder random variables, 195, 284, 350, 357
Lagrange, J. L., 234
Lamperti, J., 317

Laplace
equation, 128, 257
operator, 80, 159
transform, 273
Lattice point, 1
Law of large numbers
strong, 32
weak, 22
Lebesgue, H.,
integral, 25
measure, 55
Lebesgue-Stieltjes integral, 335
Left (right)—continuous random walk, 21, 185, 194, 208, 227, 234, 289, 361, 363, 364
Legendre polynomials, 235
Lévy, P., 173, 219, 391
Lévy continuity theorem, 63, 156
Lewy, H., 258
Lifetime, 100
Light bulbs, 93, 103
Lightning rod, 301
Linearly independent set, 66, 71
Liouville's theorem, 129, 180
Local central limit theorem, 79, 308, 340
Loève, M., 263
Logarithmic capacity, 147
Logarithmic potential theory, 129, 374
Long thin set
capacity of, 301
Loomis, L. H., 92

McCrea, W. H., 149, 150, 154
McKean, H. P. Jr., 173, 316, 321, 339
Markov chain, 20, 99, 322
Martin, R. S., 327
Martin boundary, 327
Martingale, 32
system theorem, 191
Maximum modulus principle, 179
Maximum partial sum, 205, 206, 207, 269
moments of, 209
Maximum principle for
transient kernel, 325
transient random walk, 302, 321

Mean of random walk, 21
Mean recurrence time, 99
Mean square distance, 393
Measurable function, 24
Measure theory, 24
Mercer's theorem, 242, 268, 315
Method of moments, 233, 263
Minimum principle, 373
Minimal regular functions, 340
Minimal solution
  of Poisson's equation, 296
Minkowski's lemma, 70
Mittag-Leffler distribution, 231
Module, 65
Moments, method of, 233, 263
Moment problem, 263, 338
Morera's theorem, 180

Nagy, B. Sz., 242
Newtonian potential kernel, 307, 314
Ney, P., 341
Normal distribution, 53

Occupation time problems, 231, 256
Orey, S., 46, 288
Ornstein, D., 23
Orthogonal polynomials, 247
Orthogonality of
  characters, 92
  eigenfunctions, 241
  exponentials, 55
  polynomials, 247
Outer function, 179, 220

Parseval's identity, 56
Permutation
  cyclic, 384
  of coordinates, 298
Persistent random walk, 7
Plus operator, 220
Poisson's equation, 136, 294, 367
  minimal solution, 296
Poisson kernel, 185
Poisson process, compound, 235
Pollaczek, F., 174
Pollard, H., 99
Polya's theorem, 52

Population
  in branching process, 234
Positive partial sums,
  number of, 219
Potential kernel, 123
  asymptotic behavior, 123, 124, 358, 392
  determines the random walk, 363, 393
  inverse of, 120, 364
  positivity of, 118, 361
  satisfies Poisson's equation, 137, 374
  uniqueness of, 368, 372, 374
Potential
  with respect to transient kernel, 323
  for recurrent random walk, 146
  for transient random walk, 301
  maximum principles, 302, 321, 325
  minimum principle, 146, 373
Potential theories
  Newtonian and logarithmic, 40, 128, 274, 374
  one-parameter family of, 376
Principal ideal ring, 65
Probability measure, 24
Probability space, 24, 26
Projection, 220
Putnam competition, 103

Quadratic form
  positive definite, 70
  of second moments, 61
Queuing theory, 174, 234

Random flight, 104
Random walk
  aperiodic, 20
  characteristic function of, 54
  definition, 2, 29
  first moment of, 21
  general property of, 377
  genuinely d-dimensional, 72
  left (right)—continuous, 21
  range of, 35
  reversed, 111
  strongly aperiodic, 42
Random subset of R, 53

occupation time of, 53
Random variable, 26
  expectation of, 26
Range of random walk, 35, 52, 342
  strong law, 38
  weak law, 35
Ratio ergodic theorem
  strong, 40, 49
  weak, 10
Rational numbers
  random walk on, 95
Rayleigh, Lord, 104
Rectangle,
  random walk restricted to, 270
Recurrence criteria, 23, 82, 83, 85
Recurrent random walk
  definition, 7
  existence of potential kernel, 343, 352
Recurrent set, 275, 296, 300
  characterization of, 316, 321
Red (or green) points, 52, 272
Reflection principle, 154
Regular function, 130, 276, 294
  for transient kernel, 323
  integral representation for, 337, 341
  nonexistence of, 329
Renewal theorem, 99, 100, 103, 201, 203, 282, 284, 288
Rényi, A., 224, 235, 237
Reproducing kernel, 250
Reversed random walk, 111
Riemann approximating sum, 88, 227, 261, 268
Riemann Lebesgue lemma, 85, 96, 127, 355
Riesz, F., 242, 324
Riesz decomposition
  of excessive functions, 324, 326, 331, 337
Right continuous random walk, 21, 33, 289
  see also left continuous random walk
Robin's constant, 147
Rosén, B., 199
Rouché's theorem, 186

Savage, L. J., 298
Schwarz inequality, 62, 319
Second difference operator, 128
Self-avoiding random walk, 105, 168
Semigroup, 15
Set function
  countably additive, 24
Set, swept out by random walk, 40, 341
Sigma-field, 24
Simple function, 24
Simple random walk, 2, 12
  absorption problem for, 237
  asymptotic behavior of transition probability, 78
  Fejér factorization, 183
  harmonic polynomials, 131
  in the plane, 89, 102, 148, 167, 168, 170, 172, 173
  in three-space, 103, 308
  positive partial sums, 224
  potential kernel, 148, 352
  regular functions, 276, 339
Singular points, 320
Size of set
  seen from infinity, 394
Smith, W. L., 79
Snell, J. L., 120, 343
Sparre Andersen, E., 218, 219
Spatial homogeneity, 1
Spectral theorem, 240
Sphere
  capacity of, 339
  first time outside, 312
  total time spent in, 312
Spherical shell
  potential inside, 314
Spherically symmetric functions, 104, 314
Stable law, 391
State space, 1
Stationary state, 159, 342
Stirling's formula, 13, 336
Stöhr, A., 131
Stopping time, 29, 105
Storage problems, 174
Strictly stationary process, 53

Strongly aperiodic random walk, 42, 75
Subgroup, 65, 339
Superharmonic function, 323
Symmetric random walk, 73, 87, 169, 182, 192, 204, 223, 224
Szegö, G., 247

Täcklind, S., 217
Tauber's theorem, 199
Taylor, J., 312
Temperature, 158
Theta function, 273
Three dimensional random walk
  Green function, 103, 308
  regular functions, 332
  transient sets, 316, 321
Time dependent behavior, 159
Time of first visit to
  a finite set, 391, 393
  a point, 377, 392
Trace of transition matrix, 272
Traffic, 174
Transient kernel, 322
Transient random walk, 7
  Green function of, 274
Transient set, 275, 296, 300
  characterization of, 316, 321
Transition function, 1

asymptotic behavior of, 72
Triangular lattice, 66
Two-dimensional recurrent random walk, 113
Type I and II random walk, 344

Van der Pol, B., 128
Vector module, 65
Vibrating string, 241

Wald's lemma, 191
Watanabe, T., 271, 327, 334
Watson, G. N., 103
Weak convergence, 63
Weierstrass' approximation theorem, 272
Weight function, 249
Whipple, F. J. W., 149, 150, 154
Whitman, W., 40, 53
Widom, H., 212
Wiener, N., 320
Wiener's test, 320, 321
Wiener-Hopf equation, 212, 332, 341
Wolfowitz, J., 133, 209

Zero-one law of
  Kolmogorov, 39
  Hewitt and Savage, 298
Zygmund, A., 54